THE ETHANOL PAPERS

THE ETHANOL PAPERS

The Whole Story on Ethanol Fuel

Marc J. Rauch

Dialog Press

Washington, D.C.

Dedicated to all the American military personnel who lost their lives or good health fighting in wars for oil.

Production and Design by Wordsmithy, LLC

ISBN 978-0-914153-50-4
This book is printed on acid-free paper.
Printed in the United States of America
First edition. First Printing

Table of Contents

Foreword....................i
Acknowledgments....................iii
Introduction....................v

THE BIG PICTURE....................1
Life As We Might Have Known It: What If Ethanol Was Our Primary Engine Fuel....................2
The Rise & Fall of General Motors and the Subjugation of the Industrialized World....................8
Our 'Opium War' with Gasoline....................19
Gasoline Companies Should Fund the Big Three Bailout........27
2008 LA Auto Show: Is This the Start of the Big Three Farewell Tour?....................32
Why It's Essential for Congress to Provide Financial Stimulus to the U.S. Auto Industry....................35
The Ethanol Dividend....................39
Why Is Getting US Off Gasoline So Important to The Auto Channel?....................45
No New Gasoline-Powered Vehicles in the U.S. By 2014... Can It Be Done?....................47
No New Gasoline-Powered Vehicles In The U.s. By 2014...How Can It Be Done?....................52
The Hypocrisy of Big Oil and API....................58
Breaking News: No American Servicemen Die Defending Ethanol... Again....................69

MYTHS & LEGENDS....................71
Ethanol Does NOT Suck Water Out of the Air....................72
The Irrelevance of BTU Rating - Big Oil's Gimmick to Hoodwink the Public....................76

Detroit News Story Goes Overboard in Slamming Ethanol.....84
Why Do Small Engines Suffer From Ethanol Problems?........87
Rolling Oil and Gasoline Bombs..100
The Figurative Ethanol Blend Wall is a Fictional Blend Wall..103
The Oil Industry Has Been Poisoning Us For Decades, and They've Always Known It..106
Latest CARB Rules Screw Californians Yet Again..................110
Motorcycles and Ethanol (E10) - The Shocking Truth............118
California Government Seeks New Ways To Screw Public Using Energy Concerns..120
Taking On EPA, CAL Air Resources Board, Obama, and Big Oil..123
Today We're Spotlighting Big Oil's Favorite Alternative Fuel - Electric..128
The Irrelevance Of BTU Rating - Revisited............................131
Ethanol Chronicles - An Almost Daily Blog - July through December, 2015..134
Ethanol Chronicles - SPECIAL EDITION: David Are You There?..217
Gasoline vs. Ethanol - Ten Pictures That Tell The Story.........231

MUGS, PUGS, THUGS...CON MEN AND OIL MEN 237
The Auto Channel Fights for the Truth About Ethanol Versus Gasoline..236
Thank Limbaugh, Beck, Hannity, Prager and Savage for Rising Gasoline Prices..254
An Open Letter to Tim Searchinger and All the other Gasoline Whores..258
Big Oil's Anti-E15 Campaign Filled with Gross Exaggerations and Misinformation..261
Biofuels Are Worth Every Effort and Cost..............................268
AAA Blunder on Ethanol Sets Off Firestorm of Criticism......273
Oil Industry Stages Another Sham Anti-Ethanol Press Conference..290
Why Does AAA and Big Oil Feel They Must Lie to America?..293
Ethanol Honesty is the Best Energy Policy..............................301

Why Is Jay Leno Misrepresenting Ethanol?...............................308
Lauren Fix Takes Ethanol Opposition to a New Level
of Stupidity..314
Open Letter To Mark Levin About Ethanol, Donald Trump
and Ted Cruz..324
Open Letter to Jillian Kay Melchor and the
National Review About the Iowa Primary.............................326
Open Letter to Barry Ritholtz and *Bloomberg* About
Ethanol..330
Meet The New 'Tool' In Big Oil's Efforts to Scuttle Ethanol..356
Are Ethanol Opponents Sniffing Glue?....................................363
Biofuel Blunder? Renewable Fuel Mandate Is Having
A Negative Effect on Home Runs...370
The Case Against Ethanol Opponents: They Are
Simply Incorrect 374
It's Time To Rethink The Value of the PhD..............................378
Uncovering The Gas Roots of Contemporary Ethanol
Opposition...386
Unmasking The Gas Roots of Contemporary Ethanol
Opposition - Round 1..397
Unmasking The Gas Roots of Contemporary Ethanol
Opposition - Round 2..407
SHOCKING NEWS: Existence of Parallel Universes
Explain Support of Petroleum Oil...415
Three Degrees Still Can't Make Ethanol Slam Stick................421
Fox News' Revolting Take on Ethanol.....................................425
The Immorality of Arguing That There's a Moral Case
for Fossil Fuels..433
American Petroleum Institute Goes Bonkers..........................445
Big Oil Uses New Dopes To Spread Their Old Dope.............449
Ethanol Criticism Polluted By Oiliness....................................451
Open Letter To *Washington Examiner* & John Siciliano
About Ethanol..458
We Need A Lasting Solution to the Lies Told By
Big Oil and API...461
UK's Dept. for Transport Uses Boogeyman Allusions
to Sidetrack E10 Adoption...466

TRUTH ABOUT ETHANOL: A BOOK WITHIN A BOOK...475
Truth About Ethanol – Part 1...476
Truth About Ethanol – Part 2...487
Truth About Ethanol – Part 3...504
Truth About Ethanol – Part 4...513
Truth About Ethanol – Part 5...528
Truth About Ethanol – Part 6...543

CONCLUSION...559
MAJOR SOURCES...562

Foreword

The way history follows success stories, you would think that coal, oil, and leaded gasoline have been among the crowning achievements of an advanced civilization. And you would have to think that renewable energy, particularly ethanol, has been a colossal failure.

Search any big library or online bookseller. Thousands of histories have been written about fossil fuels. Only a handful have been written about ethanol, solar, wind or any renewable energy alternatives.

And yet, despite the fractured historical record, ethanol, solar and wind have been wildly successful. They have become cheaper, cleaner, healthier and better for national security than the fossil fuels they are swiftly replacing.

Why is this success so little known? Part of the answer is that discussions about the history and policy issues surrounding renewable energy have been fogged by the oil industry and its public relations campaigns.

Despite the historical gaps this has created, it seems we are approaching a transitional moment. Our energy priorities are shifting, and our views of them are swiftly changing as well.

At moments like these, advocates like Marc J. Rauch have a special role to play in skirmishing ahead of the main currents of discourse and puncturing the inflated self-images that once-dominant industries are still trying to project.

Although ethanol is not perfect, we are far too accustomed to thinking about these issues as either black or white. Ethanol, from any public interest perspective, is closer to beige and silver, and rather distant from the crude oil-colored end of the spectrum.

Mr. Rauch's collection of personable, spirited essays helps us see a broader picture while taking a deep look at the technological and political arguments surrounding a complex set of fuel-related issues.

William Kovarik, PhD

co-author of *The Forbidden Fuel: A History of Power Alcohol*, with Hal Bernton and Scott Sklar (University of Nebraska Press) and author of "Henry Ford, Charles Kettering and the Fuel of the Future," *Automotive History Review*.

Acknowledgments & Thanks

I owe a great deal to lots of people who have done fantastic work documenting, producing, experimenting with, and teaching about alternative fuels. First and foremost are William Kovarik and David Blume. And when I thank Dave Blume and Bill Kovarik, it means that I also thank and acknowledge the teams they have worked with. In Dave's case, there are at the very least, Tom Harvey and Ian "Scotty" Crawford; and on Bill's side, there are co-authors Hal Bernton and Scott Sklar of their seminal book The Forbidden Fuel - A History of Power Alcohol. Throughout this book, as well as in the bibliography at the conclusion, I provide links to various materials produced by Bill and Dave.

Equal thanks and acknowledgements to Edwin Black, Bobby Likis, Tom Quinn, Adam Khan and Classy Evans; Mark Fulmer, Robert Rapier, Ron Lamberty, Brad Brunner, Robert Falco, Anne Korin, Carol Black, Josh and Rebecca Tickell; Steve Vander Griend, Ted Chipner, Jeff Broin and Poet Ethanol; Robert White, Andy Wicks, Andrew Randolph, Michael Carolan, Alan Anderson, Jamie Kitman, Jacob Kaul, Mickey Shay, Jay Berry, Chuck and Cindy Zimmerman; Dana Fletcher, Ron Bengtson; everyone at Growth Energy; the team at Dialog Press; everyone at Renewable Fuels Association; everyone at American Coalition of Ethanol; everyone at South Dakota Farm Bureau; everyone at Fuel Freedom Foundation; everyone at Urban Air Initiative, all the other associations, and, of course, my very great friend and business partner Bob Gordon. A special thanks to Randall Doyal at Al-Corn Clean Fuel. Links to all these people are provided in various places in the book and in the bibliography at the end of the book.

I also thank all my adversaries and the ethanol-haters. They played a pivotal role in so many of my essays and in this book. Without antagonists and their unknowledgeable, sometimes deceitful, often lame-brained, occasionally corrupt opinions and comments, the protagonists would not be able to shine so brightly. You'll come to know the names of the oil-adoring individuals and the organizations that place profit above all else as you read this book, and you'll be able to judge for yourself.

All of the people mentioned by name above, comprise what I consider to be my peer-review committee. I submit all my papers to these people and willfully accept, change, delete, add, and explain any information and issues that require alterations. I dutifully submit all my published criticisms of anti-ethanol commentaries to those authors, allowing them ample opportunity to respond and provide explanations or substantiation of their claims. I don't often have to go back and make material changes in the information I've presented over the years because very, very, very few challenges have ever been mounted by my adversaries; they may bark a little at what I write, but they never bite.

Marc Rauch
January 10, 2020

Introduction

In 1996, a book titled *The Cigarette Papers* rocked the world. It exposed what many people knew, what many people talked about for years...that tobacco smoking is deadly, and how the tobacco industry manipulated politicians, the media, and the medical community to spread their lies.

The Ethanol Papers is meant to also rock the world and counter the continuous lies and disinformation spread by the oil industry. It's a rough-and-tumble, no-holds-barred examination of the ethanol vs. gasoline issue, written in plain jargon that can be understood by scientists, academics, and politicians alike. This is the most in-depth, complete explanation of the issue, and it's targeted for smart people who really want to know the truth.

This book is a compilation of most of the papers I've written about ethanol and alternative fuels since 2008, along with some new commentary and a chronological reshuffling to improve the flow of information. *The Ethanol Papers* is divided into three sections:

THE BIG PICTURE - Historical based essays, compositions, and monographs that set forth how we arrived where we are.

MYTHS & LEGENDS - Examines and explains the un-science that has been used to discredit ethanol and other alternative energy solutions for a century and more.

MUGS, PUGS, THUGS, NITWITS, HALFWITS, DIMWITS, SNIPERS, VIPERS, CON MEN AND OIL MEN - No further explanation needed.

I first learned about the potential of alternative engine fuels in the late 1970's while doing some unrelated marketing research in the New York Public Library on Fifth Avenue. I came across a book that described turning corn and other farm crops into ethanol and methanol.

For a guy who grew up primarily in Brooklyn and Queens, farming was as alien to me as the dark side of the moon. Nevertheless, the story was riveting because it presented economic possibilities that made my head spin: replacing foreign petroleum oil fuels with domestically produced fuels from crops...WOW!

With the 1970's oil crisis still very much on my mind (and on the minds of most Americans), it was a thrilling discovery. But I was busy trying to build an advertising agency, so the last thing I could focus on was where to build silos in New York City to house all the harvested crops needed to produce biofuels.

Alas, it would take two decades before I could re-examine the potential for alternative fuels. Along the way, I co-founded The Auto Channel and TheAutoChannel.com with Bob Gordon, and added greatly to my personal automotive experience from owning a variety of new and classic vehicles.

After the September 11, 2001 attacks and the dramatic rise in the price of petroleum oil fuels, The Auto Channel began examining alternative fuels in greater detail—Bob and I, and our long-time business associate and friend Mark Fulmer and some of our correspondents who also caught the alt-fuel bug.

I have often written in previous articles, that although we were all well aware of smog and airborne health problems, our interest in alternative fuels centered on energy independence and the wondrous economic potential of alternative fuels. We read more, we attended and covered as many alternative energy events as we could, and we discussed the subject with many people...lots of people.

We were feverishly fascinated. We delved into all forms of alternative energy and fuels, including solar, wind, and even compressed air—that's right, compressed air—to power a passenger vehicle. (See: "Zero-Emissions are Up in The Air at MDI" (*https://www.theautochannel.com/news/2009/03/04/452456.html*)).

Our position on "energy independence" was and is based on two points:

1. Freeing the country of dependence and extortion from foreign sources for engine fuels. This is the normal stated goal of energy independence.

2. Freeing individuals from the philosophical and societal anguish they experience because they think their automobile desires or requirements are existentially irresponsible, if not downright globally catastrophic. This concern grew out of the 1970's alarm that big cars were rapidly draining the world's petroleum supply while choking us all with poisonous fumes.

Of course, most of us now know that the global supply issue was a canard, while the health and economic issues were and are serious problems that must be overcome.

We believe that people should own and drive what they want. If they want a large vehicle that burns lots of fuel and they can afford it, then they should have the freedom to do so. The only legitimate concern is the amount of harmful emissions the vehicle spews into the air. If a fuel could be easily substituted for gasoline or diesel that reduces harmful emissions, regardless of the size of the vehicle and its engine, then vehicle size is irrelevant.

By the mid-2000's, we were reporting and opining on alternative fuel and energy issues almost as often as we did on conventional automotive issues; however, we hit an intellectual roadblock. Although we gained great book knowledge, we had little personal practical knowledge and experience to back up the book learning.

We looked for additional opportunities to test drive realistic alternative-fuel vehicles powered by compressed natural gas, hydrogen, propane, electric, and ethanol. Obviously, our ownership of The Auto Channel opened the door for us to do quite a lot of this.

In addition, even when not test-driving press fleet flex-fuel (ethanol) vehicles, I experimented with using higher blends of ethanol-gasoline in hundreds of non-flex fuel vehicles. When traveling and driving rental cars, I did this a lot. What I discovered was that every non-flex fuel gasoline-powered vehicle I drove could use high levels of ethanol-gasoline blends with little or no change in vehicle performance. And, of course, no damage was ever done to the vehicles. I splash-blended – that is, manually blended various level of ethanol such as E20, E25, E30, E35, E40, and so on by first pumping in E85 and then topping off with E10.

Naturally, we couldn't do the same level of experimentation with Compressed Natural Gas (CNG) and hydrogen since it wasn't possible to test dedicated CNG, propane, and hydrogen press fleet vehicles with the same regularity (and no car rental company offered these types of cars for rent).

While this experimentation increased our knowledge and gave us some practical experience, we were still pretty much on the outside looking in. We could answer lots of questions, provide somewhat intelligent arguments to the haters of alt-fuels and vehicles, but we were at another roadblock—our knowledge and experience at that time could only take us so far.

Ten years ago, when the petroleum oil industry still treated natural gas and propane as ugly stepchildren, they circulated many of the same lies about CNG and propane as they did about ethanol. They claimed that CNG/propane damaged engines, provided less power and mileage, caused greater pollution, and so on. We were constantly fending off emails from the haters. And many of the negative claims were complete nonsense. One of my favorite stupid claims was that it took thirty minutes or more to fill the CNG tanks of a passenger car.

In 2008, I wrote my first long paper on alternative fuels. It was a two-parter, titled respectively, "No New Gasoline-Powered Vehicles in the U.S. by 2014...Can It Be Done?" and "No New Gasoline-Powered Vehicles in the U.S. by 2014...How It Can Be Done?"

This story was published prior to the 2008 elections, and about two months prior to Barack Obama's erstwhile energy speech when he was competing for the Democratic nomination to run for president. As we all know, none of the items presented by Mr. Obama in his energy speech ever came to pass during his two terms as president, but that's something to discuss at another time.

Two alternative fuels rose to the top of our list: CNG and ethanol.

I started shopping for a CNG-powered vehicle. I wanted to put the pedal to the metal, and really get a taste of the pudding, so to speak, for what CNG is all about. Unfortunately, by 2008 the only automobile company still making CNG cars for America was Honda (a CNG version of their Civic compact car), and Honda's extremely limited production capacity of the CNG Civic meant that I couldn't buy a new CNG car. I had to resort to looking for a used vehicle.

I finally located a used factory converted CNG Dodge Ram Van at a dealership in the Los Angeles area. I flew into L.A., rented a car and drove to the dealership. As it turned out, I needed a van for business reasons (to use as a video production vehicle), so I felt that getting a full-size van would come in very handy, in addition to allowing me to experiment with CNG.

After buying the van, the long drive home was very uneventful. Outside of having to chart where I could buy CNG, everything was very routine. The vehicle performed perfectly (as perfect as any full-size van could be expected).

I had no sensation of loss of power, and had no trouble accelerating to the legal speed limit (plus the extra 5-15 mph that everyone drives at). Climbing the "grapevine" north of Los Angeles proved no problem; and each time I stopped to fill up, it took no longer than it would take to fill a gasoline tank. The only aspect that was different was that most CNG filling stations are just basic filling facilities. If I wanted a coffee or use a restroom, I had to also stop at a conventional convenience store.

This experience with CNG—this mundane experience—was the same experience I had for the next eight years every time I drove the van. The van's engine wasn't destroyed by CNG use, my vehicle always passed smog tests with flying colors (and with lower harmful emissions), I saved money on fuel, and because the van was approved for HOV lane use, I could drive in carpool lanes even when I was alone. In many cases, I didn't have to pay bridge or tunnel tolls.

I definitely proved that CNG is a great fuel and that CNG-powered vehicles are as good or better than any comparable gasoline-powered vehicle.

However, there are three problems with CNG. The first is that no automobile manufacturers are mass producing CNG vehicles for the United States. Even Honda ceased their very limited production of CNG Civics. The second problem is that federal and most state regulations make it almost economically impossible to convert an existing gasoline passenger vehicle to run on CNG. In some states, it is not legal to do the conversion regardless of how much you spend. Therefore, CNG as an engine fuel, while abundant and domestically produced, has been rendered irrelevant.

The third problem is that CNG is brought to us by the same people (the petroleum oil industry) who have lied to us, poisoned us, betrayed our nation, and ripped us off with their phony supply and demand tactics. The last thing we need to do is to place our trust in the jaws of these wolves.

ETHANOL - The Single Solution Fuel

Shortly after buying the van, I started looking for a used non-flex-fuel gasoline-powered car to buy so I could conduct extended long-term ethanol tests.

One of the best aspects of using ethanol is that all gasoline-powered vehicles on the road, and all brand-new gasoline vehicles, can run on ethanol or ethanol-gasoline blends without any engine modifications.

I had the opportunity to buy a friend's used 2002 non-flex-fuel Ford Taurus. He was asking $3,500. I immediately bought it. I figured that one way or another, the purchase price would be worth its weight in editorial gold by providing The Auto Channel with some great stories and practical ethanol experience.

I also purchased a small still to see if it is possible for average people to produce their own ethanol, if need be. As I learned, it is possible, and rather simple.

Over the past few years, my Ford Taurus has been the subject of several articles I've written about ethanol, and it plays an important part in the speeches I've given at industry and government events.

On multiple occasions I've related how I fueled up on every ethanol-gasoline blend from E10 to E85, with the only problem being that the "check engine" light might illuminate. I could save up to $1.00 per gallon of fuel by mixing E85 with standard E10 fuel. I never experienced any engine or performance problems related to the use of ethanol.

In fact, the single most noteworthy experience I had with my 2002 non-flex-fuel Ford Taurus was the first time I took the vehicle for a smog test after using high ethanol-gasoline blends for about two years.

The measured PPM results were an astounding "0" for CO (carbon monoxide), "0" for NO (nitrogen oxide), and "1" for HC (hydrocarbon). In other words, the high use of ethanol fuel made my car super clean. If the general public followed my example all future emissions' goals would be met today, not in 2030 or 2040...right now. And consumers would save money, too.

All that has to happen is for the president of the U.S. to publicly acknowledge that high ethanol-gasoline blends are approved for all modern gasoline-powered passenger vehicles (mid-1990s and newer), and to direct the EPA to remove all cautionary notices to the contrary. A move like this would allow America to stop importing all foreign petroleum oil for fuel, make us energy independent, and be a great financial boon to the U.S. economy.

In conducting my own personal tests, I did what few, if any, other automotive reporters have done: put my money where my mouth is. More importantly, in all the online skirmishes I've engaged in with alternative-fuel haters. I've never come across one person who spoke from personal experience. They only quote the lies told to them by the oil industry. On many occasions I've asked them if they've ever tried using a high ethanol-gasoline blend in their personal vehicles or rental cars, but they never respond. It would be the simplest thing to do to prove their claims that ethanol is bad by just renting a car for a few days and pumping some E85 into the car. But they don't do it.

I've issued challenges to publicly debate these impassioned nay-sayers or at least prove my findings wrong, but there have been no takers. I've written about these challenges and tried to shame them into responses. In some instances, there are brief responses where they ask me for some additional information; but after I provide the information, they disappear and, in almost all instances, never address the subject again in the media. (My most enjoyable literary skirmishes are included in this book.)

In the egregious case of Robert Bryce, author of *Gusher of Lies*, he touts the same erroneous information he spouted in 2008, when he wrote his shoddy book, even though everything negative he claimed then has since been disproved or made redundant by evolving technologies.

In the case of Robert L. Bradley, co-author of *Energy - The Master Resource*, for the first time, I met an alt-fuel hater who was truly looking for, and willing to accept, other information. Rob Bradley and I corresponded several times. In the end, he acknowledged that he had not considered the information I presented when he wrote his treatise *Oil, Gas, and Government: The U.S. Experience* (1996), and said he would include the information he learned from me in his treatise going forward. He also asked me to write a short essay for his website to explain why and how gasoline wrongfully became our primary engine fuel. He published the piece in early 2016; and since that time, not one of his readers or followers has attempted to challenge or contradict the conclusions I presented.

Moving On

The nation is at an interesting point in time, and I am at a new crossroad in my experimentation with ethanol.

Most new gasoline-powered passenger vehicles are openly approved for E15, which means there is no threat of voiding the warranty for these new cars if the owner uses E15, assuming one can find it. A government mandate to use E15, E20 or E25 would be an appropriate and safe action for the Trump Administration to take.

Next-generation internal combustion engines will require higher ethanol-gasoline blends to meet harmful emission mitigation goals while allowing vehicles to meet performance specifications. This will include E20 to E50 or higher. The oil industry will have no option but to acquiesce and will probably wind up investing in ethanol plants to stay in control. It will be fun to see how they change their tune and start singing the praises of ethanol, just as Standard Oil and Cities Service used to do in the United Kingdom starting in the 1930s (a fact little known to American consumers).

Smog Check Vehicle Inspection Report (VIR)

Vehicle Information

'ime: 04/20/2013 @ 05:04 PM

2002	Make: FORD	Model:	TAURUS
[redacted]	State: California	VIN:	[redacted]
3.00 L	Type: Passenger	Transmission:	Automatic
0	Test Weight: 3625	Cylinders:	6
128831	Certification: California	VLT Record #:	30759
Gasoline	Exhaust: Single	Inspection Reason:	Biennial
		Engine Year:	0000

Overall Test Results

atulations! Your vehicle passed the enhanced Smog Check inspection, which helps California reach its c
ing an extra 100 tons of smog-forming emissions from the air. Thank you for keeping your vehicle well
ive Visual Inspection: PASS Functional Check: PASS Emissions Test: P
:ck Certificate Number: OU947374C
Number: [redacted]

Your Smog Check certificate will be electronically transmitted to DMV
Please keep this copy for your records

Emission Control Systems Visual Inspection/Functional Check Results

Functional tests are used to assist in the identification of oxides of nitrogen, crankcase and cold start emission which are not measured during

ECS	RESULT	ECS	RESULT	ECS
PCV	N/A-	Thermostatic Air Cleaner	PASS-	Fuel Ev:
Catalytic Converter	N/A-	Air Injection System	PASS-	Fuel Tar
Visual EGR	PASS-	Ignition Spark Controls	N/A-	Carbure
Fuel Injection	PASS-	O2 Sensors and Connectors	PASS-	Wiring (
Vacuum Line Connections	PASS-	Other Emission Related		Senso
	N/A-	Functional EGR	N/A-	Ignition
MIL	N/A-	Fill Pipe Restrictor	PASS-	Liquid F
Fuel Evap Test				

ASM Emission Test Results

	CO2 (%)	O2 (%)	HC (PPM)			CO(%)			NO (PPM)		
RPM	MEAS	MEAS	MAX	AVE	MEAS	MAX	AVE	MEAS	MAX	AVE	MEAS
1640	15.26	0.03	52	4	1	0.49	0.01	0.00	424	27	0
1677	15.30	0.03	36	4	1	0.46	0.01	0.00	711	26	0

imum Allowable Emissions AVE = Average Emissions For Passing Vehicles MEAS

2013 Smog Test Ford Taurus - APEC 2016

As for me, I'm moving on, too. I traded in my non-flex-fuel Ford Taurus for a late model flex-fuel Dodge Grand Caravan and retired my CNG Dodge Ram van. Needless to say, I'll continue to write and speak about the benefits of ethanol, and my testing of ethanol blends will focus on which level of ethanol-gasoline blends give my new van the best and most economical performance. (Many people say that E30 to E50 will be best. We'll see, and I'll let you know.)

And for fun, my wife's non-flex-fuel Nissan Altima will serve as our test vehicle for various ethanol-gasoline blends in a non-flex-fuel car.

Notes on Text

The Ethanol Papers is a technological breakthrough in publishing that would not have been possible even a few years before its publication. This work began as a collection of online investigations, exchanges, debates, and commentaries at The Auto Channel website, known for its hard-hitting expertise on alternative fuel and automotive technology. These materials were accumulated into an online flipbook called *The Ethanol Papers,* brimming with active links to videos, online reports, and documents. The flipbook was then adapted for this international paperback, simultaneously released by booksellers in 190 countries, paired with a first-ever online companion flipbook, which provides active links to readers of the physical book. In addition, an e-book version is available for virtually any device.

Some online exchanges and articles were lightly edited or abridged for clarity and length.

As of this printing, the author's original companion flipbook can be found at: *https://www.theautochannel.com/news/2018/10/12/632678-ethanol-papers-massive-book-provides-whole-story-ethanol-fuel-free.html,* hosted by the Auto Channel. As the internet is constantly changing, if that link does not work, any search engine should return a current location for "Ethanol Papers."

Modern book text is produced using a variety of collaborative software, including automated spellcheckers and text revision tools. It is possible for such software and machine-to-machine transfers to create typographical errors or other textual changes beyond the control of the author, editors, or publisher. Any changes, corrections, or additions to this book should be reported to, and/or can be found at *dialogpress.com.*

Web site addresses (URLs), titles, and statistics were accurate as of original press time. Neither the author nor the publisher is responsible for URLs that have expired or moved since the manuscript was prepared.

The opinions, assertions, commentaries, critiques, conclusions, and editorializing expressed in *The Ethanol Papers* are solely those of the author and do not represent Dialog Press, its employees, associates, or global partners.

The Editors

THE BIG PICTURE

**Historically based essays, compositions,
and monographs
that set forth how we
arrived where we are.**

Life as We Might Have Known It: What If Ethanol Was Our Primary Engine Fuel

Originally published March 7, 2016

A couple of weeks ago, I had the opportunity to exchange emails regarding ethanol and energy with Robert L. Bradley, Jr. Rob is the author of several books, including *Energy—The Master Resource; Capitalism at Work: Business, Government and Energy,* and *Edison to Enron: Energy Markets and Political Strategies.* Rob is also CEO and founder of the Institute for Energy Research and an adjunct scholar of the Cato Institute.

Our exchange began when he wrote that government should not decide which fuel is the winner; that fossil fuel's dominance over alternative fuels was due to consumer choice, and that society owes a great thanks to fossil fuels for giving us such a high quality of living.

I responded by agreeing that the government should not dictate which fuels win and added that our government had very definitely determined who the winner was. "The government did this," I wrote, "by placing burdensome taxes on alcohol right after the founding of the United States. The burdensome taxes became onerous in 1862. Alcohol was priced out of the market by the government. When the taxes were finally lifted for a short period after 1907, alcohol began to compete with gasoline, and virtually all objective scientists agreed it was a better fuel than gasoline." I added that government interference culminated with the national prohibition of alcohol, which ended the competition.

I also wrote, "Consumers never had the opportunity to vote on which fuel they wanted; there was never a national referendum. And in the years since the repeal of prohibition, the public has never been told the truth about petroleum oil fuels, nor has the public been given the opportunity to vote to discontinue petroleum oil subsidies."

On his point about thanking fossil fuels for our high quality of living, I wrote that any thanks owed was owed to machines and their inventors; alternative fuels were used and could continue to be used. The internal combustion engine wasn't invented because excess gasoline and diesel fuel were lying around, but gasoline and diesel fuel were developed because there was a glut of virtually tax-free cheap petroleum oil.

This response seemed to have caught Rob off guard. He replied, "I am very interested in the gasohol (ethanol) argument and taxation that you refer to... I have not studied it as much as you have. I need to."

Frankly, Rob's forthright reply caught me off guard; but it was a welcome change to the stonewalling or silence that I usually get in response to these points.

We continued our exchanges; and after providing Rob with answers to his questions, he wrote, "Okay, I am finally getting educated. This is something I did not sufficiently understand when I wrote my treatise, *Oil, Gas, and Government: The U.S. Experience* (1996) ... I have written notes in my treatise and will remember this going forward."

Rob then kindly asked if I would present my thesis in less than 1,000 words, in a more formal manner, and that he would post it on his MasterResource.org website. The following is my position:

Life as We Might Have Known It: What If Ethanol Was Our Primary Engine Fuel

Events of the American Civil War set in stone social and economic conditions that have been with us ever since, and they are certain to be with us long into the foreseeable future.

Perhaps the least considered events were the Revenue Acts of 1861 and 1862, passed by the U.S. Congress to fund the Union's military efforts. Ramifications of these bills included the first Federal income tax statute and the levying of excise taxes (often referred to as "sin taxes") on tobacco and ethyl alcohol (ethanol). Alcohol's tax rose to more than $2 per gallon, regardless of whether the alcohol was intended for personal consumption or industrial purposes. In today's money, that's purchasing power equivalent to about $120. In 1860, rent for four rooms was $4.45/month; land was selling for $3 to $5 an acre; and a laborer's wage was 90 cents a day.

In 1872, some taxes were repealed—income tax, for example. The alcohol tax remained in place for decades until removed by the 1906 Free Alcohol Act.

The high tax on alcohol priced alcohol out of consideration for use as lamp and kitchen stove fuel, colloquially referred to as "burning fluid" or sometimes "camphene." The alcohol-turpentine fuel had come into widespread use earlier in the 1800's when whale oil became too expensive and hard to come by.

The beneficiary of alcohol's forced decline in popularity was the nascent petroleum oil industry and their principal product, kerosene. Up until the imposition of the tax, alcohol burning fluid and kerosene were similarly priced. After the tax was levied, alcohol burning fluid cost as much as 10 times more than kerosene. Kerosene had also been taxed to help the war effort: a paltry 10 cents per gallon.

The earliest efforts at building stationary and mobile internal combustion engines used alcohol fuels: Samuel Morey in 1826; Nicholas Otto in 1860; Henry Ford's Quadricycle in 1896. Other builders of horseless carriages, such as Charles Duryea, used the new fuel "gasoline" because of its dramatically lower price. However, early automobile racers all used alcohol or alcohol blends for more power.

In countries without domestic petroleum resources and onerous alcohol taxes, alcohol fuels became the preferred engine fuel. In 1899 Germany, the price of gasoline and alcohol fuel were both equivalent to approximately 27 cents. Tests conducted in Germany and France showed that alcohol-optimized engines were at least as efficient as gasoline-optimized engines.

The first decade of the 20th century saw significant improvements in internal combustion engines for the new-fangled automobile, as well as powered aviation. Higher compression engines were required for better performance and heavier vehicles. But higher compression engines revealed a gasoline problem: Gasoline in high compression engines caused engine knock—abnormal detonation that was so severe it could shake the engine apart. There were two solutions to this problem. The first was 100% ethanol. The second was to add ethanol to gasoline. Both solutions came with their own problems. Cost: The tax-bloated price of ethanol made it unattractive and threatened efforts to portray the automobile as a device that could be afforded by the masses.

There was another problem—a problem for the petroleum industry. They couldn't control ethanol production and supply. Ethanol can be produced by anyone, anywhere, from a variety of raw materials. But the oil industry was in the cat-bird seat; they had price on their side until President Teddy Roosevelt came along.

The history of Teddy Roosevelt's campaign against corporate monopolies is well known. Less well known were the Congressional Ways and Means Committee Hearings held in 1906, during his administration. These hearings resulted in the Free Alcohol Act that removed the exorbitant tax on alcohol production. By removing the tax, for the first time, ethanol fuel became less expensive than gasoline. This meant that top scientists and engine builders of the day could work toward creating better performing vehicles that relied on low-cost ethanol and ethanol blends. Henry Ford's Model T was the most famous of the new vehicles that could now be powered by gasoline or ethanol. The Model T had two simple adjustments located on the dashboard and steering wheel column.

Dave Blume with Ford Model T.

The ethanol fuel industry would now get its time in the sun, except for three obstacles.

The first was the petroleum industry's head start. Three decades of kerosene dominance gave the oil industry an overwhelming financial war chest to be used for marketing and political tutelage.

https://www.youtube.com/watch?v=5qDYoEupI28

The second obstacle was the First World War. Although the war ignited the need for mechanical innovation, the innovations were primarily used for military purposes. Had the war not occurred, higher compression engines in consumer automobiles would have been introduced, and these engines would have required greater quantities of anti-knock ethanol.

The third obstacle was National Prohibition. With the war out of the way, and the public ready to roar into the Roaring 20s, ethanol was eliminated from the scene; and nothing can kill competition like a law banning one of the two competitors.

In the early 1920s, ethanol still had its supporters in the top industrial scientists of the day, including those at General Motors, the world's largest automobile manufacturer. Ethanol was the only solution to engine knock. However, after the GM scientists invented leaded gasoline (using tetraethyl-lead) and combined their patents with Standard Oil and Dupont Chemical, the world's two largest companies in their respective industries, the prospect of billions of dollars in profits from leaded gasoline put aside any consideration of ethanol-gasoline blends in America.

In countries that did not have Prohibition, such as Great Britain, ethanol did compete successfully for many years. In fact, Standard Oil (and its derivative companies) marketed ethanol and ethanol-gasoline blends in those countries, advertising them as superior to leaded or unleaded gasoline. *The Forbidden Fuel*, written by Bernton, Kovarik & Sklar, points out how ethanol-gasoline blends were labeled as "the most perfect motor fuel the world has ever known," and provided "extra power, extra economy and extra efficiency." A Standard Oil promotional pamphlet advises that with 10 gallons of their ethanol-gasoline blend in a fuel tank, it is possible to pour almost a pint of water into the tank "without the slightest trouble—in fact, in some circumstances with better running."

The end of Prohibition in America did mean that ethanol could once again compete with petroleum oil fuels, but what had been a significant head start for the petroleum industry prior to the Free Alcohol Act was now an insurmountable advantage. It would require industry acknowledgement that tetraethyl-lead, and then MTBE (methyl tert-butyl ether), was deadly; that economic devastation could be caused by foreign oil dominance; and that there would be considerable environmental damage before the government would be compelled to cautiously encourage ethanol use.

I want to reiterate that the U.S. government did choose the winner in the energy fuel war, and that its continued interference has made it impossible for a free engine fuel market to exist.

What if the government had not taxed alcohol in 1862? What if alcohol could have competed freely during the second half of the 19th century and all through the 20th century without Prohibition (or without Prohibition affecting ethanol production for fuel)? What might have happened?

We might not have engaged in so many wars; we might not have lost tens of thousands of servicemen in those wars; we might not have had to spend trillions of dollars to engage in those wars. We might not have needed to pour money into the Middle East oil countries to secure their oil because we would not have needed it. Instead of making terrorist regimes rich—allowing them to fund terrorism—they might all have remained sleepy backwater countries.

If ethanol fuel was allowed to prosper, our farmers might have required less aid. Our skies might never have become smog filled. We might have had fewer catastrophic oil spills and disasters. If we didn't have National Prohibition, perhaps organized crime wouldn't have become so organized.

The Rise & Fall of General Motors and the Subjugation of the Industrialized World

How the perfect storm of commercial forces created the world's largest corporation, enslaved us to gasoline, and ultimately brought the car maker to its knees

It was once said "If it's good for General Motors, it's good for the U.S.A."
• Have you ever wondered why and how General Motors, the world's largest corporation, could fail and go bankrupt?
• Have you ever wondered why gasoline is the primary fuel for passenger cars and trucks for most of the world?
• Have you ever wondered exactly how and why the United States could owe 50% or more of its national debt to oil producing countries?

Originally published April 6, 2012)

The need for speed and power

In the early years of the automobile industry there was uncertainty as to which fuel, and what type of engine would power the new vehicles. The choices were steam, electric and internal combustion engines. Although steam power had proven its viability in the first stationary engines and in the first round of land and sea vehicles or vessels, by the late 1800s, diesel was already replacing steam because of its ability to produce higher torque (needed to move heavy vehicles/vessels) and because it simplified the process of locomotion by not requiring a team of men to start the coal or wood fires and keep the water boiling. By comparison, diesel allowed faster starting and required less continuous attention to keep the engine running.

Electric motors could also provide high torque, but the motors required a constant flow of electricity, which could only be guaranteed if the vehicle was tethered to an electrical source.

This was good for light and heavy trains that operated along a fixed roadway, but battery technology and affordability were impractical for personal transportation devices with the ability to travel in all directions and to non-electrified locations.

The internal combustion engine (ICE) was the solution. There were two types available: spark-ignited and compressed hot air-ignited. Compressed hot air ignited ICE uses diesel fuel. Spark ignited internal combustion engines today are most commonly associated with gasoline, but ethanol, methanol, natural gas and propane can also work.

Diesel engines were heavier, bulkier and cost more to produce. While they could move heavier loads (torque), they couldn't quickly provide higher speeds. Additionally, diesels required fuel-injector technology which at the time was costlier and less reliable than fuel-air mix carburetors. Consequently, spark-ignited internal combustion engines became the dominant engine for passenger cars and light trucks.

At this point, the two most available fuels for spark-ignited ICE were liquid: either alcohol (ethanol or methanol) or gasoline. Alcohol enjoyed wide support from automobile pioneers, such as Henry Ford and General Motors' top scientists, because it could be produced almost anywhere by almost anyone (alcohol distillation technology has been in the public domain for hundreds of years). Alcohol fuels also produced superior performance compared to gasoline. Alcohol-powered engines allowed for higher piston compression, which delivered more speed and power. Gasoline caused a knock in high-compression engines that would literally "knock" the engine into self-destruction. Only low-compression, lower speed engines could safely use gasoline. Early on, as speeds were measured against human walking or horse riding, this was acceptable. But as roads were leveled and paved, and consumers wanted bigger, faster vehicles, this became a huge limitation. One of the solutions to gasoline knock was to blend ethanol into every gallon of gasoline. The alcohol quieted the knock, allowing the gasoline-ethanol blend to be used in better performing engines.

Unfortunately, alcohol suffered from two major impediments to universal acceptance as a pure engine fuel. The first was cost. Alcohol production has been heavily taxed since the founding of the United States, initially to raise revenues to pay for the War of Independence. As usually happens with taxes imposed to pay for one specific thing, it is nearly impossible to retire the tax even after the initial purpose has been fulfilled. The extra dollars are just too attractive to politicians who need to fund pork-barrel projects to ensure re-election, and there was always another war that had to be paid for. (Prior to Congress passing a national individual income tax 100 years ago, tax on alcohol routinely accounted for as much as 40% of Federal revenue.)

The Federal tax on alcohol reached new heights during the American Civil War, rising to over $2 per gallon. This was a devastatingly high tax. Average weekly wages in the 1860s was about $12. In respect to today's average wages, that would be like a tax of $129 per gallon. (The current Federal tax on spirits—drinkable alcohol—is $13.50 per gallon).

Kerosene was also taxed to help pay war costs: A paltry 10 cents per gallon!

The alcohol tax was not retired until Theodore Roosevelt's presidency about 40 years later, when the Free Alcohol Act brought the price of corn ethanol down to 14 cents per gallon and molasses ethanol to 9.5 cents per gallon (versus gasoline at 22 cents per gallon).

I mentioned that there were two impediments to universal acceptance of alcohol as a universal engine fuel. The second problem was, ironically, alcohol's easy production and ubiquitous availability; it is too simple and easy to make.

By the middle of the 1800s, as the world's whale population drastically declined due to over-hunting, and the price of whale oil sky-rocketed, an alcohol-turpentine blend was used as a less expensive replacement for indoor lighting and heating. But with the development of kerosene in the 1850s, followed by the imposition of the much higher alcohol tax in 1862, a new use was found for the sticky, smelly black goo that freely seeped up from the ground in Pennsylvania (areas of Poland and Azerbaijan experienced the same phenomena). This was the start of the petroleum oil boom. Although kerosene has an obnoxious smell and black smoke, at least it was cheap.

As the world's interest in the new-fangled automobiles soared throughout the world, people discovered kerosene could be used as the basis for engine fuel; and after additional refining, the oil pioneers developed gasoline. By this time, one man and his company rose to worldwide prominence—John D. Rockefeller and Standard Oil. His enormous wealth was contingent upon oil and its by-products; and in the time-honored tradition of the wealthy, he used his money to secure political and commercial support of his products.

The political bribes paid by Rockefeller catapulted support for the Volstead Act, the 18th Amendment to the Constitution, over the top, insuring its passage. This amendment, commonly referred to as Prohibition, outlawed the production of alcohol with very few exceptions. Consequently, in the crucial period just after World War I, as Americans were adopting a whole new mobile lifestyle, the one fuel that could challenge Rockefeller's gasoline on both a cost and performance basis—alcohol—was declared illegal. Ethanol was rendered dead as a competitor on the commercial battlefield.

Before and during Prohibition, Henry Ford expressed his belief that alcohol (ethanol) was the fuel of the future. His Model T, the product that is said to have given birth to moving assembly line production, was designed and built to use ethanol or gasoline by giving the driver an adjustable carburetor and spark plug advance controls to optimize the performance of the fuel used.

Even after Prohibition commenced, General Motors' top scientists, Charles F. Kettering, Thomas A. Midgley and T.A. Boyd, continued to believe that ethanol was the fuel of the future. Considering that it was illegal to even produce the small amounts to conduct tests, alcohol was still being experimented with as the best alternative solution to gasoline's knock problem.

However, in 1921, GM's scientists discovered that by adding tetra-ethyl lead to gasoline, the knock was subdued, and the new lead-gasoline fuel could be used in advanced higher compression engines. This was in the early days of the Roaring

Twenties; and in order to really roar, the public needed a fuel to set them free.

The leaded gasoline gave GM the ability to build vehicles with higher performing engines, and it gave them a unique process to patent. GM combined their process with similar processes being tested by DuPont Chemicals and Standard Oil, which gave GM three cents on every gallon of leaded gasoline sold anywhere in the world. They quickly determined that their share of profits in the sale of leaded gasoline would be worth many billions of dollars over the next decades.

The significance to understanding what this meant to General Motors is not just calculated by profits earned, more importantly it explains why General Motors, the world's largest company and automaker, would come to have no interest in developing vehicles that could get better mileage per gallon of gasoline. It is essentially the same reason why tobacco companies had no desire to lessen the addictive qualities of cigarette smoking. The more gallons of fuel a vehicle used translated to more profits for GM. In essence, the vehicles could be used as a loss-leader to stimulate gasoline sales.

This new stream of nearly incalculable profits meant that they didn't need to compete on an even playing field with other auto manufacturers:

- They could undercut the competitions' products because they didn't need to make profits from their vehicles and could thereby drive the competition into bankruptcy.
- They could accept worker demands they knew the other car companies would not be able to live with on a long-term basis.
- Moreover, even if they didn't undercut the price of their competition or permit unrealistic employee compensation packages, the added revenue gave them a marketing war-chest that could not be overcome—ultimately helping to drive a long list of car makers into obscurity.

Any of the above could have given GM a huge edge over the competition, but taken together, most competitors found the circumstances impossible to withstand.

The Worm Turns

Ironically, however, I believe that this exalted position led to their inconceivable plummet.

The gasoline profits allowed General Motors to become a bloated, inflexible entity, incapable of—and disinterested in—meeting the challenges of new design, technologies, demographic shifts, societal attitudes about the environment, and the up-start band of oil-dictators called OPEC. As the years passed, and as new faces came onboard, the knowledge of this gasoline-profit advantage was lost. They thought they were invincible. They thought they were successful solely because they built a better car.

Normally a strong market competitor will strive to set a pace via product improvement that catches the competition off guard. Typical marketing thrusts would be to announce new and improved formulations that deliver more for the consumer's money, a longer-lasting ingredient, or something that makes it better for your health. Up until the mid-1970s, when it was too late, General Motors only promoted "bigger" or "more powerful" vehicles. And, of course, "bigger" and "more powerful" translated to more gasoline used. American brand competitors followed the lead, tripping over the low bar set by GM. Where GM survived and prospered because of the billions in gasoline profits, the other car makers had to try and survive only on product features and benefits, but it wasn't enough for even the most creative auto makers and most innovative efforts.

(The GM team of Kettering and Midgley also invented Freon for refrigeration. Kettering patented a refrigerating apparatus to use the gas; this was issued to Frigidaire, a wholly owned subsidiary of General Motors. In 1930, General Motors and DuPont formed Kinetic Chemicals to produce Freon. This invention and accompanying profits added to GM's financial stability.)

Now, I will agree that most Americans probably wanted bigger and more powerful vehicles in the 30s, 40s, 50s and 60s, much as we still do today during the gasoline-price crisis. However, GM did little or nothing to deliver the bigger, more powerful vehicles that would consume less fuel. Again, up until the mid-1970s, GM mostly scoffed at the smaller more fuel-efficient vehicles coming from Europe and Japan. If they designed a small sporty car to compete with a foreign import, it was more powerful and it used more gasoline.

At the same time, GM filled its executive offices with inept or redundant managers, who allowed the best and brightest new technology or design people to go to foreign manufacturers. For decades, they retained the services of advertising and marketing consultants whose only talent was in knowing how to waste more money each new model year, and they built walls around themselves until they couldn't see the figurative and literal tsunami that came from Japan.

Perhaps, if the 70s gasoline crisis didn't happen, and if the environmental movement failed to take hold, or if the gasoline companies could have continued to lie about the negative effects of leaded-gasoline and get away with it, then General Motors might have been able to exist without their share of profits from the patents. But the combination of these factors created the perfect storm that set the stage for GM's tumble.

I readily acknowledge that others before me have cited GM's lack of effective action as the cause of their near demise and current shaky existence, but I believe that no one has ever put together and articulated the reason GM permitted itself to fall into this position. Many analysts opined that GM would be okay if it wasn't for the heavy pension packages they committed to. Some analysts suggest that as foreign carmakers open factories in the U.S., they will succumb to the same pension problems and end up struggling like GM.

I think GM might never have survived the competition after WWII if they didn't have those billions of extra dollars. I also believe that Mercedes, BMW, Honda, Toyota, Kia, Hyundai, Nissan and every other foreign brand that establishes a factory here will not go through the same debilitating pension mistakes because they just don't have the extra financial wherewithal to get themselves into the same hot water. They may fail, but it's more likely that they would fail from their own original, stupid mistakes.

When General Motors rushed to Washington in 2008 to beg for financial assistance, Presidents Bush and Obama, and all of Congress, should have said no. They should have told GM to get the money from their buddies in the oil industry, which had just experienced a record-breaking profit year. In case anyone out there is uncertain of how the oil industry makes most of their money, it is from the sale of gasoline for automobiles. Instead of the American public being threatened with "GM is too big for us to lose," the oil industry should have rightly been assigned that responsibility, particularly since we were already paying for the second and third Middle-East oil wars and providing security for the entire world's ocean-going oil tankers.

Then, as if Washington hadn't already mishandled the GM financial collapse by giving away our money, the carmaker was allowed to escape any responsibility to common stock shareholders.

Steven Rattner was appointed "Car Czar" by President Obama to oversee the GM bailout. He was a member of Exxon's Board of Directors and was once a New York Times reporter on economics who became an investor in communications companies when he was tapped by President Obama to lead the GM and Chrysler bailouts. How much could he possibly have known about the auto industry? I think not much.[1]

GM continues to be deluded into thinking that its tired brand names have any real market value and has bet its future on becoming a significant player in a giant communist country. It is chasing an electric technology fantasy that is still decades away from being meaningful and has pretty much abandoned the flex-fuel position that it spent several years and untold dollars trying to establish. It finally built more powerful cars that used less gasoline. I was present at numerous GM presentations when they pushed and pushed and pushed E85 and flex-fuel vehicles. If it was a wrong direction—and perhaps it was, since GM has made so many wrong decisions—then why are many of the executives who were involved in those decisions still with the company? Why bring Robert Lutz, who believes in electric vehicles, back into the fold?

Just Good Capitalism

As a devoted capitalist and marketing guy, I have to say the GM action in seeking an inexpensive, exclusive fuel to power the growth of the automobile age was, at first blush, a great example of what capitalism is about. The problem is that capitalism allowed to run amuck is nearly as bad as unchecked socialism. It may be a convenient inexpensive capitalist idea to dump waste materials from a chemical plant into the river that flows behind the factory; but if the water is also used for drinking by the community down-river—who you hope will buy your products, the free-market idea becomes a dead-market reality. In effect, that is the result of General Motors' patented process as the leaded gasoline legacy continues to plague us.

Chained to Oil

America and all industrial nations of the world no longer "drink" from an unpolluted spring; we haven't for more than a half century. We are addicted to using fuel that robs each one of us of our hard-earned money. What's more, the over-hyped, worrisome economic or political news about oil, and the greedy cycle of oil commodity speculation, keeps all segments of the financial markets teetering on the brink of disaster. The richest and most wonderful country in the world has become just another debtor-nation. Fifty percent or more of our national debt is owed to other countries to pay for the oil that we helped discover, subsidize, and then protect through two world wars and several regional conflicts. And we are forced to fund terrorist regimes who would like to see us dead.

We willingly support an industry that has given us mass-murdering billionaires. Every gallon of gasoline or diesel fuel that we buy props up bad actors, such as the likes of Hugo Chavez, Mahmoud Ahmadinejad, Saddam Hussein, Muammar Gaddafi, Bashar Assad, Osama bin Laden, the so-called Saudi royal family, and even Vladimir Putin.

Oil leaders rogues gallery.

It doesn't matter that some of these men are now dead; the system will continue to churn out new monsters like a well-oiled machine. Their wealth and power are directly derived from our use of gasoline, whether we buy their respective oil or not. The oil and gasoline market is not an open, supply-and-demand marketplace. It is a tightly closed, tightly controlled, fixed enterprise that doesn't permit rogue competitors or independent action. Boycotts or restrictions on the purchase of oil from especially insidious dictatorships are irrelevant weapons; threatening such action is perceived as a joke. If we refuse to buy oil from Iran, for instance, we still must buy it from someplace. As we pull oil from the new source, a void opens that must be filled. Iranian oil fills that void, and another country (customer) that doesn't share our sense of moral outrage rushes in to buy the Iranian oil—at the same global market price. The other trick oil producers use is to sell and/or ship their oil to a middleman country, such as Canada or India, where it is re-labeled, invoiced and sent to us. So, we, the public, are the only ones negatively affected.

When General Motors lost their way ninety years ago, the United States lost its economic and energy independence.

Notes

[1] My reason for calling Steven Rattner a "know-nothing" is best exemplified by the criticism he leveled at Mitt Romney, after Romney criticized the GM bailout. Rattner said, "If Mr. Romney disagrees, he should come forward with specific names of willing investors in place of empty rhetoric," Rattner wrote. "I predict that he won't be able to, because there aren't any."

While I agree that there were probably no willing investors to bailout GM (they would have had to be insane to do so), Rattner, via Obama, should have forced the oil industry to come to GM and Chrysler's rescue. How would he have forced them to do it? Simple, he would have said the following: "Either you provide the funds, or we withdraw our military forces from around your foreign oil installations, and you have to provide your own protection for your oil fields and tanker shipments around the world. In addition, I will give immediate emergency orders to jumpstart a serious alternative fuel program in the U.S. and we will phase gasoline out forever."

If this option was ever explored, we should have been told about it. If the option was explored and the oil industry refused, we should have been told that, too. If Obama appointed someone as "Car Czar" who didn't have the imagination to make a suggestion like this, he never should have been appointed in the first place. But what should we have expected from a president who was without any experience or imagination of his own?

Our 'Opium War' with Gasoline

Originally published October 23, 2010

The United States and most nations we might include as part of democratic industrialized society are at war. In effect, it is the Grand World War. The "hot zone" battlefields are currently in Iraq and Afghanistan. It is brought to life every day on our TV and computer screens, in our newspapers and magazines, and on the radio. This war has been going on since 1914 at least. Over the years, the battlefields have moved, covering much of Europe, Asia, the Middle East and America's Pacific Ocean possessions.

However, the main battleground in this war is not in the hot zones where bullets are fired; the main battleground is anywhere where there is a gasoline filling station: along our streets and boulevards, our highways and retail shopping strips.

We are fighting for our economic lives. Our physical health is also under constant attack, even during this time of environmental enlightenment.

The enemy is already here; indeed, the enemy was pretty well born here, in Pennsylvania. Today the enemy makes its headquarters in cities like Riyadh, Tehran, Baghdad, Kuwait City, Abu Dhabi, Caracas, Moscow, Tripoli, Geneva, Houston and Ottawa.

The Enemy's Deadly Weapon is Gasoline

That sounds quite melodramatic; I intended it to be. I wrote this to be inflammatory. However, while this may be overly dramatic, that doesn't mean it's not true. If it didn't incite intense emotion, it wouldn't have captured my, and my business partner's attention.

We are in a war where nothing is sacred—not family, not religion, not concern for the health of our families or fellow men—not even national pride or patriotic spirit. No humane sentiment is of concern to this enemy; all kindly matters are ignored.

On several occasions, the world's super-powers (whoever they were at the time) have gone to war to force or defend trading rights. As Anne Korin and Gal Luft point out in their books* and public speeches, for centuries common table salt was the reason for war.

In the 1800s, Britain and France went to war in Asia to force entire populations to use and become addicted to opium, produced and sold by their respective commercial entities under "Royal Charter." When the indigenous governments tried to interfere and protect their citizens, Britain and France used their national armies to force the issue at gunpoint. The opium trade was one of Britain's most profitable. To be fair, if there is a fairness element in this history, Britain and France weren't the only countries involved in the opium trade; and they weren't the only ones to use military force—but they were the two most egregious offenders.

Britain defends right to force people to use opium.

The important point here is that Britain is typically portrayed as being a "compassionate nation." So, if a compassionate nation is willing to kill people to force them to become drug addicts (to drugs that they wouldn't let their own citizens freely use), imagine what uncompassionate regimes, cartels and business groups would do to ensure the uninterrupted addiction and use of their "drugs." Think of the Mafia, on a huge scale.

I will not mince words or hide behind metaphors. We are being manipulated by external and internal entities to use gasoline. Our politicians are bought off to ensure that the addiction never ends. Media personalities we trust on all sides of the political spectrum are likewise bought off (in one manner or another) to spread lies and confuse their audience as to the benefits and necessity of our gasoline addiction—not just the addiction to foreign-oil gasoline, but to all gasoline. Regardless of where the petroleum oil originates or where the gasoline is refined—because the industry is unified in global pricing and supply, every gallon of gasoline we buy props up the wealth and power of our enemies. We must figuratively disembowel and behead them by eliminating their drug. As Korin and Luft advise, we must turn "oil into salt" and strip away oil's strategic importance.

We do not need gasoline to power our vehicles. We do not need petroleum oil diesel to power our trucks and heavy machinery. There are alternatives that are available right now that are, and can be, domestically produced. What's more, you can make your own engine fuel right in your own backyard.

When we at The Auto Channel began our investigation into alternative fuels several years ago, among the issues we had to consider was why and how we got ourselves into this state of gasoline addiction. Why would auto manufacturers not want diverse sources of fuel to power their vehicles? Didn't they want to encourage personal mobility? What do they stand to gain from poor fuel economy? How did America and American companies lose their position as the dominant force in the oil industry and become the puppets of a small group of nomadic highwaymen who claim to rule under the preposterous notion of Divine Right?

There is not now, nor was there ever a good technical reason for gasoline to be the world's primary engine-fuel of choice. It was not because of greater availability—in fact, the fear that the world's supply of petroleum oil would end has been a serious concern since the late 1800s, when early oil fields in Pennsylvania, Ohio and Oklahoma began petering out. (Only in recent years have we come to learn that the world is not in danger of running out of petroleum-based sources to make gasoline.)

The reason gasoline became the dominant engine-fuel is an economic one. Unfortunately, it wasn't because it was economically beneficial to consumers—which would have been justifiable.

A hundred years ago, alcohol (ethanol) was the engine fuel of choice for unbiased persons-in-the-know. This was true in America and anywhere in the world that was impacted by that new-fangled invention, the automobile. For alcohol to be preferred over gasoline was rather remarkable, considering that it had to overcome the oil industry's 40-year head start in developing sophisticated production facilities and distribution outlets.

Moreover, alcohol had to overcome an onerous $2.08 tax on every gallon, not to mention the significant marketing war chest of John D. Rockefeller and Standard Oil.

Even Prohibition, with its severe criminal penalties for producing alcohol, which was funded by Rockefeller's multi-million-dollar donations to politicians, failed to quench the thirst for ethanol in business and technology leaders who knew its advantages.

The economic reason—the game-changer—was General Motor's discovery of lead as the gasoline additive that silenced annoying engine knock, while allowing GM to build better performing engines (engines had to be dumbed-down to accommodate gasoline). By comparison, alcohol did not cause knock and could be used to fuel higher compression engines. GM's patented process (and their subsequent teaming with Standard Oil) meant that GM would earn about three cents per gallon for almost every gallon of leaded "ethyl" gasoline sold. They estimated that profits would be in the billions by the time their patents expired; profits they would receive regardless of a vehicle's manufacturer, as long as the consumer used leaded gasoline. Alcohol distillation didn't offer comparable unique processes that could be patented.

With their attention riveted on this pot of gold, there was nothing to stop General Motors—not the health risks to the general population and not even the deaths of dozens of employees who worked with the lead formulas. It was full speed ahead for Standard Oil, too. It now had a gasoline-committed partner/customer that just happened to be the world's number one automaker. (Although GM and Standard Oil would never be charged for collusion in their lead-gasoline venture, they were found guilty of conspiracy in Federal Court in what became known as the "Great American Streetcar Scandal.")

However, one thing stood in the way of gasoline's complete dominance as an engine fuel: Alcohol was better in every way (power, environment, cost, safety, renewable availability, and no reliance ever on foreign sources). Considering that America was the only industrialized nation with Prohibition, there was still plenty of potential competition from alcohol. And, as it was possible that Prohibition could eventually be overturned (which it was), alcohol and all its advantages would be back to challenge the monopoly of gasoline and its billions of dollars in profits. There was only one thing car companies and oil producers could do—fabricate lies about alcohol.

Nearly all the lies and misconceptions dreamed up by the oil/gasoline lobby to fight ethanol in the early days are still with us today. The lies are the direct opposite of the actual truths.

Betraying the American People

The motive was profit. That's not so bad considering that GM and Standard were public companies, with many American employees, so at least a small portion of the profits did go to shareholders and workers. That's what's supposed to happen, isn't it?

Yes, sort of—as long as in doing so our birthright is not given away or sold out from under us to people who don't care about the United States or their own country and people. This issue is not just a concern for America, it affects all the aforementioned democratic industrialized nations. At the beginning of the last century, as it became clear that the domestic oil supply wouldn't last forever, a worldwide search was on for new deposits. Oil in the Middle East was not discovered by Iranians or Iraqis or Kuwaitis or Saudi Arabians; such people didn't even exist. There was no Iran or Iraq or Kuwait or Saudi Arabia or United Arab Emirates. Those names and geographical designations didn't come into use until well after a British businessman backed by his government made the initial discoveries. America's entry (defined as the huge expenditure of taxpayer dollars and American lives) into World War I secured Britain's oil interests and led the way to taking away adjacent oil-rich territory from the Turks (part of which is now known as Iraq).

Later, just prior to World War II, America's Standard Oil company discovered oil in the Persian Gulf and peninsula. This discovery rescued the so-called Saudi royal family from complete financial ruin. The development of all oil production was financed by Standard Oil, along with other commercial and government assistance. These oil discoveries and holdings were then secured through the United States' entry into WWII (defined as the enormous expenditure of financial and human resources).

So, we, the American people, paid for the development and protection of the oil fields in the Middle East; then, we gave it all away.

There are reasons we gave it away—excuses for why it could be said that there is a benefit to giving some of the indigenous people part of the pie. The problem is that we, the American people, not only gave up the entire pie, we now pay premium rates for every additional piece of the pie, in addition to continuing to pay for the protection of the pie-making process.

Look at it this way: you start an ice cream business in another town. You need help running the store because you have other interests and don't live in the area. Your son-in-law, who is currently out of work needs something to do, and you want to help your only daughter and her kids, so you bring your son-in-law into the business. You pay for his family's relocation, you give him a salary and a title, set up distribution for the ice cream in other cities, and you continue to finance the business. Sometime later you find out that your son-in-law is skimming from the receipts. Alright, you let it pass, figuring that it's okay since it's going to support your daughter and grandchildren.

But then, on a visit to the ice cream store, when you want a double-scoop of chocolate fudge swirl, your son-in-law asks you to pay full retail price for the ice cream cone. Hey, it makes you giggle a little—he's just trying to protect the business, your business. The only thing is that when you do the year-end accounting you find that there's no money in the bank and that your son-in-law has a girlfriend on the side (and that the trip he took to the ice cream industry convention in Las Vegas was really a sex-junket for him and the girlfriend). To add insult to injury, your son-in-law then changes the name of the business, changes the locks on the door, and files for divorce from your daughter.

If this happened to you, you would take some kind of action. Neither the American government, the British government, the French government, nor any of the other leaders of the countries that paid for the discovery and protection of Middle East oil did anything. Instead we pay more and more and more for the "ice cream" that we should at least be getting at a wholesale price.

And our political leaders, who should be helping to get us off our reliance to gasoline, do nothing to end the addiction.

Why? How? The answer is "follow the money." The oil companies and politicians gave away the oil deposits, but kept the shipping and distribution, the refineries and the retail outlets. And perhaps most significantly, by installing their brothers-in-law (the nomadic highwaymen) as the groundskeepers, they created a ready-made excuse to drive prices up and keep them up to ensure their intake of trillions of dollars.

Political corruption did not end with the demise of Tammany Hall; it did not end with the deaths of Boss Tweed or Jimmy Walker or Richard Daley or Spiro Agnew or Huey Long. It is alive and well in every significant state and federal elected official's office. It is rampant throughout the world. The oil and gasoline industry have bought our politicians and pay with our money and lives.

Instead of the world's carmakers going to the gasoline industry to demand their financial bailout (in order to continue producing inefficient gasoline-powered vehicles), they went to our corrupt politicians to get us to pay again for vehicles we already paid for.

The production of gasoline-powered vehicles must end. This gasoline-opium war must be terminated in the peoples' favor. As we have stated in separate editorials on TheAutoChannel.

com, we believe electric vehicles are just a diversion to keep us globally addicted to gasoline. Electric vehicles might be the answer someday; but according to all major carmakers, that 'someday' is decades away. The immediate answer, the near-time solution, is domestic ethanol. The cure for the current financial depression is the swift and vigorous development of domestically produced alternative fuels and alternative energy technologies. This is true for the United States and for all other democratic industrialized nations.

Notes

See also: Bob Gordon's "Alcohol and Driving Do Mix"
https://www.theautochannel.com/news/2009/10/04/480130.html.

To watch Ann Korin's complete presentation at 2010 American Coalition of Ethanol conference at:
https://www.youtube.com/watch?v=06LdZ3EPvTM
https://www.youtube.com/watch?v=mC0rJ_-EADY

Anne Korin and Gal Luft co-authored "Turning Oil into Salt: Energy Independence Through Fuel Choice" in 2009

The Gasoline Companies Should Fund the Big Three Bailout

Making the Case for Not Bailing Out the U. S. Car Companies

Is America Allowing the Big Three to Fail or Have the Big Three Failed America?

Originally published November 17, 2008

In November 2008, Congress approved a $25 billion aid package to the American car companies to help them design and build the vehicles of the future; cars and trucks that would use alternative fuel propulsion systems.

Around the same time, they agreed to provide this aid package to the carmakers, Congress debated and approved two other aid packages (one to the insurance company AIG, the other to the banking and Wall Street firms that made such a mess of the financial and housing markets). These two bailout packages amounted to about $800 billion.

Then there was a movement afoot to allow the carmakers to shift the focus of the $25 billion alt-fuel vehicle aid package from helping them create the next generation of alt-fuel vehicles to using the funds to shore up their poor general financial condition, and to ask the Federal government to provide at least an additional $25 billion in aid for the same cause.

As I was leaving Detroit on a Friday evening, after spending two days at the Propane Engine Fuel Summit, I listened to a local radio talk show host harangue his audience about those (in and out of Congress) who are against allowing the car companies to get off the hook (again) to develop cars and trucks that don't rely on gasoline, and to award them the additional $25 billion for their financial debacle. "Is America going to just let the Big Three fail?" he screamed. "Don't you stupid people realize what will happen to this country?"

When I arrived at the airport, I went online and read an urgent email message sent to automotive journalists by Chrysler entitled, "Making the Case for Assistance." The email also directed readers to a video message. *(https://www.youtube.com/watch?v=bjctwIsTzgg)*

The video presents the importance of the auto industry in America and cites numerous significant statistics that add up to this: if the auto industry fails, America will fail; basically, the same point that the radio host was trying to make, but without the yelling.

The impressive stats presented by Chrysler are correct. The automotive industry is America's largest and most important industry. (In fact, the auto industry is not just the most important business sector in the United States; it's most important in every industrialized country in the world.)

The greater auto industry includes the three U.S. carmakers that are collectively still known as the "Big Three," as well as tens of thousands of supportive dealerships, fabricators, service firms and financial institutions. In all, as the video points out, millions of Americans earn their living from one part or another of the auto industry.

However, there is a giant disconnect in the argument presented by the talk show host, Chrysler, General Motors or Ford that if we (the American people) don't ante up $50 billion to save the U.S. car companies from bankruptcy, we (the American people) are allowing them (and the country) to fail.

To begin with, even if the Big Three went bankrupt and eventually ceased to exist, the automotive industry in America would continue. It's not like we would suddenly be stranded on empty highways without any 'wheels,' or left with broken down cars and no place to go to get them repaired. There are, after all, other automakers manufacturing cars and trucks in the U.S. and others that sell their foreign-built vehicles in this country. These plants and dealerships employ American workers, who pay taxes to our local and federal government. There would still be ample opportunity for service facilities, fabricators, aftermarket products, etc., to prosper. Any vacuum left by the disappearance of Big Three products would be filled by other manufacturers. If we want to ensure the continuance of a strong domestic assembly-worker base, Congress could simply mandate that all or most vehicles sold in America be manufactured in America. The industry would be changed, but far from shuttered.

Secondly, approval of the $50 billion aid package will do nothing to solve the immediate financial crisis. It will not re-open any closed plants, it will not stimulate vehicle sales, and by shifting the focus of the first $25 billion away from helping to build alt-fuel vehicles, it will certainly not create a new fleet of desirable new technology cars and trucks. In short, giving the car companies a $50 billion gift is throwing money down the toilet. There would be no reason for the American public to buy new American vehicles and get themselves further in debt.

Rejection of a bailout plan that allows GM, Ford and Chrysler to return to business-as-usual is not a matter of the public allowing them to fail. It is the American car companies that have failed the American people. If a sacred debt ever existed between the U.S. carmakers and the American public, it's not that the public owes them; consumers have already paid for the vehicles produced and sold by Detroit. The cars and trucks weren't given away free. And if you factor in the finance charges that most of the purchases accrued, then the public not only paid for the vehicles once, it may be said that they paid for the vehicles two or three times over. Using the fear of the demise of America as the reason for granting $50 billion in aid is essentially asking consumers to pay for the cars and trucks all over again, while the carmakers return to manufacturing relatively uninteresting gasoline-powered vehicles.

The American people had already bailed out the Big Three in 1993 during the Clinton Administration by providing government loans, tax incentives/credits and favorable environmental regulations. Instead of working in earnest to fulfill their part of the deal, they chose to make and sell gasoline-guzzling vehicles that took advantage of semantic loopholes to shine us on. While these actions resulted in Detroit experiencing several of the best financial years they ever had, today we're left waiting for a token alt-fuel vehicle effort from the Big Three that isn't really even worth considering, vis-à-vis the request for a taxpayer-provided $50 billion aid package: GM will have one electric vehicle ready for sale by late 2010 (the Volt), one electric Chrysler model ready for sale by mid-2010, and no real commitment by Ford, except that they think they'll be ready to mass market electric vehicles by 2020 to 2025 (twelve to seventeen years from now).

These efforts aren't enough. If GM has a good electric propulsion system, that same propulsion system should be slated for use in every GM model, whether it's a Chevrolet, or a Buick, or a Saturn, or a Saab, or a Pontiac, or a Cadillac. For example, if the new gasoline-powered Camaro is slated for 2010 delivery, why isn't GM planning for a Camaro Volt? Chrysler is looking at three potential electric models but will only select one to manufacture.

Sorry, that is simply not good enough. Every Chrysler model on sale should offer an electric variant. If the best that Ford can promise is more than a decade away, then they should do a 'Willie Wonka' and close up shop until they're ready to re-enter the auto business with relevant products.

And with all the misery, there's still no discussion about the Big Three selling energy efficient, low polluting CNG and propane vehicles in the U.S., even though they make and sell them in other parts of the world. The Auto Channel has reported extensively on this issue. There is a great opportunity for the Big Three to salvage what they can of existing models by converting them to use these fuels. This might actually add some sizzle to their lackluster product line-ups.

If all that GM, Ford and Chrysler offer in exchange for a $50 billion aid package is business-as-usual, then they should get the aid from the people that benefit the most from their gasoline-powered products: the gasoline companies. If I recall correctly,

Exxon/Mobil earned a $40 billion-plus profit this past year, and that's just one gasoline company. The combined oil and gasoline companies could easily provide the aid. Considering that the American people have paid hundreds of billions of dollars for two wars that have done nothing more than protect the earnings capabilities of the OPEC members and the gasoline companies, we've already done more than our share.

Notes

Bailout Details

Company	Invested	Sold For	Profit/Loss	Bailout Ended
GM	$51.0B	$39.7B	-$11.3B	Dec. 9, 2013
GMAC (Ally)	$17.2B	$19.6B	+$ 2.4B	Dec. 18, 2014
Chrysler	$12.5B	$11.2B	-$ 1.3B	May 2011
TOTAL	$80.7B	$70.5B	- $10.2B	

(Source: "Key Facts," U.S. Department of the Treasury.) *https://www.thebalance.com/auto-industry-bailout-gm-ford-chrysler-3305670)*

2008 LA Auto Show:
Is this the start of the big three farewell tour?

Originally published november 19, 2008

Los Angeles, San Francisco, Detroit, Chicago, Philadelphia, New York... sounds like the city stops along a rock band's national tour; and it could be. But in this case, these are major cities that are part of the 2009 model-year auto show exposition tour that car companies will follow to display their products to the american public. For the big three—General Motors, Ford, and Chrysler, it could be their farewell tour.

While nearly everyone hopes this won't be the situation, it doesn't look particularly promising judging from what the U.S. carmakers unveiled (or more pointedly, did not unveil) this week at the Los Angeles auto show.

Last Tuesday night, Ford debuted its latest muscle car, the 2010 Mustang. Yes, the Mustang has always been an American favorite, and the unveiling is obviously intended to blunt Chevrolet's intended 2010 relaunch of the Mustang's longtime competitor, the Camaro, but is this the right time to be talking about gasoline-guzzling muscle cars—regardless of how relatively fuel-efficient the new Mustang may be?

By comparison, that same evening, BMW unveiled its electric Mini E concept, a seemingly smarter move for these times. Even in its gasoline-engine version, the popular Mini could be said to be an energy efficient homerun. With an electric propulsion system, it would be a grand slam: a desirable car with near-zero harmful emissions.

On Wednesday, the first day of the LA auto show's two press-introduction days, when journalists would normally expect to see major presentations by General Motors, Ford and Chrysler, GM and Chrysler were truant. Ford staged a modest duo event in which they presented a 2010 Lincoln MKZ and

2010 Ford Fusion—two vehicles that will probably have, at best, lukewarm acceptance by American consumers. Keep in mind that in past years, the big three would not have just one overall brand presentation each; they would usually have individual presentations for their respective specific makes, i.e., Chevrolet, Cadillac, Buick, Saturn, Saab, Ford, Mercury, Volvo, Chrysler, Jeep, and so on. Not having any unveilings by two of the big three at North America's second most important car show (in the number one regional car market) spoke volumes.

Yes, it is true that the heads of the big three were quite busy on Capitol Hill at about the same time as the press-intro days. And certainly, any extravagant presentation in Los Angeles would have been criticized as wildly inappropriate and too AIG-like, given the reason why Wagoner, Mulally and Nardelli were sitting before a congressional sub-committee. But instead of playing 'poor-little-us,' it might have been a grand opportunity to display the responsible action that America (and some members of Congress) had been anxiously waiting for: re-seizing the initiative from foreign automakers to deliver products that we need and want, in a timely and meaningful way (production numbers that are impactful), and work toward the elimination of the need for gasoline. Just one month ago, General Motors did stage multiple extravagant product unveilings at the Paris Motor Show. Surely, they knew of their dire financial situation at that time. One item that was presented by GM in Paris that could have been talked about in Los Angeles was natural gas-powered vehicles: an environmentally friendly, less expensive solution that is being offered in other parts of the world. Unfortunately, GM, along with Ford and Chrysler, continue to deny this proven alternative to America.

Meanwhile, although non-U.S. Car companies all spoke of the past year's poor financial results and the bleak outlook for the near-term future, they carried on with business-as-usual product premieres and high-tech alternative-fuel innovations. And, as troubling as it is to say, at the end of the two days, other than a general nostalgic sadness or valid concerns for the domino effect that could devastate the American work force and its army of mom-and-pop suppliers, I didn't talk to one journalist who didn't express the opinion that the auto industry and the consumer world could not thrive without the big three.

Notes

The Auto Channel has presented extensive coverage (including tons of video) of several major alternative fuel and vehicle events, which can help you understand all the issues and all the options that are available. We suggest you start with the following three links and then simply enter "alt fuel" or "alt vehicles" in our search box at the top of any page on TheAutoChannel.com for hundreds of additional individual stories and videos.

2008 Alternative Fuels and Vehicles Conference in Las Vegas:

https://www.theautochannel.com/news/2008/05/23/087900.html
ALT CAR EXPO in Santa Monica:
https://www.theautochannel.com/news/2006/12/14/031415.html
Plug-In 2008 Conference and Exposition in San Jose:
https://www.theautochannel.com/news/2008/08/03/095252.html

Why It's Essential for Congress to Provide Financial Stimulus to the U.S. Auto Industry

The Fate of the Industrialized World Hangs in the Balance

This Is Not Merely an Issue of Protecting a Segment of the Work Force; It's an Issue of Societal Progress

Originally published November 26, 2008

Innovation has almost always been the stimulus to advance human civilization. Sometimes the innovations were originally intended for military use and then adapted for everyday life; sometimes they were specifically meant to improve the human condition. But as far back as the controlled use of fire and the first application of the wheel, to gunpowder, sails, moveable type, steam engines, telegraph, radio, and electric lights, innovations took our primitive species and turned us into relatively sophisticated creatures on the verge of serious interplanetary exploration.

Along the way, the innovations resulted in sky-rocketing economic bubbles and the inevitable financial busts when the booms petered out or failed to meet the over-hyped goals. The last worldwide bubble, the Internet bubble, sent all economies to new heights, and the innovation will undoubtedly change mankind forever. Unfortunately, the excesses of this unprecedented boom and its subsequent effects left us with the worst global economy ever experienced. In time, as politicians and economists come out from under the covers of denial, they will admit that this bust period was more devastating than even the Great Depression in 1929.

That the world will eventually recover is indisputable. But the questions that must be answered are what will be the stimulus to recovery and how long it will take?

As we have long stated on TheAutoChannel.com, it's our belief that the next stimulus to economic well-being is what we all refer to today as "alternative-energy" and "alternative-fuel vehicles." When we are finally able to take the giant step for mankind that allows us to exchange the word "alternative" for "modern-day" or "present" when describing energy and fuel and transportation propulsion systems, we will advance to better financial health and wealth. And we believe that this new boom period won't last for just three or four years, but for decades.

Furthermore, we believe that American economic success from the adoption of new fuels and energy technology will spill over to the rest of the world and alleviate their economic malaise, as well.

To go down the path that leads to the new boom, America must shake off the gasoline yoke now, not in ten or fifteen or twenty-five years. Not only does this mean that American carmakers must get out of the gasoline-engine making business, they have to reign in their attack-dog lobbyists and political henchmen from hindering change on both national and local levels.

By themselves, the three U.S. car companies are incapable of making the necessary transformation. Bogged down by constipated corporate philosophies and worker encumbrances, they lack the resolve and clearly lack the financial resources.

(It's interesting to note that the heads of the Big Three car companies didn't go before the British Parliament to beg for billions of dollars to save the American auto industry, nor to the European Union's Council, nor the Russian Federation Assembly, nor China's National People's Congress. They didn't even appear before an OPEC meeting to ask for the $50 billion loan (although this group is the most likely to benefit from the Big Three's continuance of gasoline-powered vehicles). They went to the United States Congress in Washington, DC. So regardless of how international GM and Ford and Chrysler pretend to be; they are still just American companies, not global companies. And because they are seeking a bailout from all the American people, not from American union pension funds, it's the American people—all the people, not just auto workers—that they must serve. They have a responsibility to do what is right for America, not what is expedient for the oil producers or unions.

This is where Congressional intervention comes in: The United States Congress can provide the financial stimulus and enact laws that mandate the necessary changes, and in the process possibly redefine what "American auto industry" means in the 21st Century (a financially viable American auto industry doesn't necessarily have to include American brands, or American brands as we know them today.)

If Congress is willing to bailout a mismanaged insurance company and banking group, whose significance to our national well-being pales by comparison to the energy-fuel-transportation issue, then they can and should put the American auto industry back on track.

But Congress' own resolve must match the level of the financial commitment that Detroit requires and the importance of the intended result. For Congressmen to truly be helpful they must overcome their own personal real-world incompetence, selfish pork-barrel entanglements, and addiction to oil company inducements.

The specific dollar amount provided to the auto industry is less important than the stipulations that must be put in place to ensure that the funds are used appropriately; and the dollar amount and the stipulations are less important than the clear laws that must be swiftly enacted to ensure that there is no misinterpretation of what the goal is.

For example:

No loophole laws to permit the continued production of grossly inefficient gasoline-guzzling vehicles under a deceptive fuel-efficiency averaging formula. Each vehicle must stand on its own as being fuel-efficient and operate without gasoline.

If any vehicle manufacturer wants to sell their products in the U.S., then the vehicles must be manufactured or assembled in the U.S., with the requirement that a "reasonable" portion of the parts required for domestic assembly be manufactured in the U.S.

If an automaker accepts financial assistance from public funds, executive compensation (including benefit package) should be severely restricted, and employee compensation (including benefit package) should not exceed their local community private-sector standards.

If Congress can take the bold steps necessary to protect whatever bailout is ultimately decided upon, then we have the chance to pull out of this depression in the next two to four years.

The U.S.

If an automaker accepts financial assistance from public funds, executive compensation (including benefit package) should be severely restricted, and employee compensation (including benefit package) should not exceed their local community private-sector standards.

If Congress can take the bold steps necessary to protect whatever bailout is ultimately decided upon, then we have the chance to pull out of this depression in the next two to four years.

The Ethanol Dividend

"Unquestionably alcohol is the fuel of the future..." Thomas Midgley, Jr., General Motors scientist, co-inventor of 'leaded' gasoline

Originally published August 6, 2011

A hundred years ago, ethanol (alcohol fuel) was the engine fuel of choice for objective persons-in-the-know. This was true in America and anywhere in the world that was starting to be impacted by that new-fangled invention, the automobile. For alcohol to be preferred over gasoline was quite amazing considering that it had to overcome the oil industry's 40-year head start in developing sophisticated production facilities and distribution outlets. Moreover, alcohol had to overcome an onerous $2.08 tax on every gallon, as well as the significant marketing war chest of John D. Rockefeller and Standard Oil.

Even the 18th Amendment to the U.S. Constitution (Prohibition), which was funded by Rockefeller's multi-million dollar "donations" to politicians, failed to quench the thirst for ethanol in objective business and technology leaders who knew its advantages.

The event that changed everything was General Motor's discovery of tetraethyl lead as the gasoline additive to silence annoying engine knock, while permitting GM to build better performing high compression engines. Conversely, alcohol fuel did not cause knock and could be used in higher compression engines. Engines had to be dumbed down to accommodate early gasoline, which was more readily available than alcohol. GM's patented process was the game changer, because with the patent and their subsequent teaming with Standard Oil, it meant that GM could earn about three cents per gallon for almost every gallon of leaded "ethyl" gasoline sold. They estimated that profits would be in the billions by the time their patents expired. Keep in mind, these were profits that required almost no GM personnel, factory allocations or product design and development funding (past the initial exploratory phase). It was pure profit, profit they would receive regardless of a vehicle's manufacturer, as long as the consumer used leaded gasoline. Alcohol distillation didn't offer comparable unique processes that could be patented.

Armed with the newfound (potential) wealth, there was nothing that was going to stop General Motors, not even the deaths of dozens of employees as a result of working with the lead formulas. Likewise, it was full speed ahead for Standard Oil, who now had a gasoline-committed partner/customer that just happened to be the world's number one automaker. (Although GM and Standard Oil would never be charged for collusion or conspiracy regarding their lead gasoline venture, they were found guilty of conspiracy in Federal Court in what became known as the "Great American Streetcar Scandal.")

However, there was one thing that stood in the way of gasoline's complete dominance as an engine fuel: ethanol. Ethanol was better than gasoline in every way (power, environmental issues, cost, safety, renewable availability, and no reliance on foreign sources). Considering that America was the only industrialized country in the western hemisphere or Europe with alcohol prohibition, there was still plenty of potential competition from alcohol. Additionally, as it was possible that Prohibition could eventually be overturned (which it was), alcohol and all its advantages would be back to challenge the monopoly of gasoline and billions of dollars in extra profits. There was only one thing to do—lie about alcohol. Yes, good old propaganda.

Amazingly, nearly all the lies and misconceptions dreamed up by the oil/gasoline lobby in the teens and twenties to fight ethanol are still with us today. Ironically, most of the lies are the direct opposite of the actual truths. For example, the most common lies told about ethanol are:

- Ethanol damages engines
- Ethanol produces less power than gasoline
- America can't produce enough ethanol to replace gasoline

The most pernicious misconceptions ascribed to ethanol are:

- Using corn for ethanol causes food prices to rise and takes food away from the world's starving peoples
- Ethanol is environmentally harmful
- It takes more energy to produce a gallon of ethanol than the gallon of ethanol puts out
- Ethanol gets unfair, wasteful government subsidies

But the truth is that:

- Alcohol cleans engines, and they burn cooler. An engine that uses ethanol can last two or more times longer an engine that uses gasoline.
- Alcohol fuel is used in race cars that want to get greater power and performance. In an engine that is optimized to run on gasoline, it is natural to expect that any other fuel might not perform as well. However, if the same engine is then optimized to use alcohol, it will perform as well or better.
- If 100% of all corn grown in America were used to make ethanol, then we could produce enough ethanol to replace all gasoline...and we would still be able to get the dried distillers' grains [DDG]for needed feedstock. In reality, other crops would feature in the mix, some that don't require existing crop or farmland and very little water. The alcohol yield from these plants is significantly higher than corn. Seaweed harvested from the Gulf of Mexico alone could provide all the ethanol needed to replace gasoline in America.
- The food-for-fuel argument is completely false. It has been debunked from every angle and the World Bank has retracted its original accusation that corn used for ethanol caused food price increases and food shortage.
- Ethanol critics deride ethanol production if fossil fuels are used anywhere in the process of growing, harvesting, distilling and shipping alcohol. These charges are presented as if oil and gasoline weren't guilty of the same criticisms. In the worst-case situation, at least it could be said that the final ethanol

product vs. gasoline is far more environmentally friendly. Gasoline has no advantages whatsoever. In truth, however, fossil fuels do not have to be used to produce ethanol. In some instances, natural, on-site nutrients can be used as fertilizer. Ethanol or other biofuels can be used to operate harvesting equipment, and non-fossil fuel energy sources can be used to power the distillation process.

- Critics make it sound as though subsidies are exclusively given for ethanol production. However, the oil industry has received more government subsidies, allocations, inducements, and tax breaks than any other energy technology, energy source or fuel. Every subsidy ever granted to oil/gasoline in the last hundred years is still in effect. In addition, untold billions of dollars, and tens of thousands of American lives have been spent on wars related to protecting foreign oil. No dollar amount of subsidies to produce domestic ethanol will ever be too expensive compared to our servicemen's lives.
- Using arcane and irrelevant methods of measuring how much energy is expended to produce ethanol (such as the energy required to build brand new tractors), critics like David Pimentel don't take into account comparable measurements in evaluating EROEI (energy returned on energy invested) for gasoline.

But what they forget or didn't understand in the first place is the "ethanol dividend."

The Obama Administration, or any federal administration, pumps in billions and trillions of dollars to stimulate the economy. If those stimulus dollars are not surplus, the government has to print more dollars, thereby devaluing all dollars. It's said that one billion dollars leaves America every day to pay for foreign oil. These dollars may never make their way back to America. If we can avoid sending the $1 billion a day ($360 billion per year) out of the country, and instead use those dollars here to pay salaries and buy domestically produced goods, those dollars become stimulus for the economy without having to print additional dollars and devalue those already in circulation. Then, if those dollars are spent on American workers and American products, we actually wind up recycling the dollars and multiplying its stimulus effect severalfold. It could be possible to recycle the $360 billion 3, 4, 5, 6 times, giving us a trillion dollar-plus stimulus every year for every year that we do without gasoline.

Consequently, even if the cost or EROEI for ethanol is initially upside-down, the "ethanol dividend" would make using ethanol wildly advantageous and far safer for our military personnel.

Notes

Abbreviated list of links related to oil/gasoline subsidies

http://www.nytimes.com/2010/07/04/business/04bptax.html
http://www.cato.org/pubs/pas/pa390.pdf
http://www.elistore.org/reports_detail.asp?ID=11358
http://www.imf.org/external/pubs/ft/spn/2010/spn1005.pdf
http://en.wikipedia.org/wiki/Energy_subsidies
http://blumenauer.house.gov/index.php?option=com_content&view=article&id=1672:blumenauer-introduces-bill-to-close-billions-of-dollars-in-oil-industry-tax-loopholes&catid=48:2010-press-releases-new
http://merkley.senate.gov/newsroom/press/release/?id=F68FED0D-08C0-4DC4-9035-211E28DD9B7E
http://go.ucsusa.org/publications/report.cfm?publicationid=149
http://en.wikipedia.org/wiki/Petroleum_industry
http://investmentwatchblog.com/oil-subsidies-for-the-five-largest-oil-companies-jesus/
http://www.energetixllc.com/files_uploaded/e7e1c329322b3712b2c8e3aa0b7e78be.pdf
http://www.progress.org/2003/energy22.htm
http://www.huffingtonpost.com/2011/02/01/obama-moves-to-cut-oil-subsidies_n_816787.html

http://themoderatevoice.com/78678/unemployment-benefits-and-oil-subsidies/
http://www.its.ucdavis.edu/publications/2004/UCD-ITS-RR-96-03(15)_rev1.pdf
http://agfax.com/Content/dtn-biofuel-subsidies-10132010.aspx
http://www.americanprogress.org/issues/2010/05/oil_company_subsidies.html
http://energydeals.wordpress.com/2011/02/02/obama%E2%80%99s-bid-to-end-oil-subsidies-revives-debate/
http://iipdigital.usembassy.gov/st/english/article/2008/08/20080814172401saikceinawz0.4823267.html#axzz4GpxqWBh7
http://ilsr.org/wp-content/uploads/2012/05/Oil-Slickers.pdf

Why Is Getting US Off Gasoline So Important to The Auto Channel?

Originally published November 10, 2010

Engine fuel and energy technology doesn't just power automobiles, it powers our entire economy. I take that back—it powers the world's economy.

The automobile industry is the world's largest and most important industry. If we, the people in all democratic countries, can find a new and better way to fuel vehicles, something that saves money and keeps the money in the domestic "purses" of each country, then it gives everyone in that country a boost. If we can find alternatives that don't damage the environment, then we're all healthier and happier. And if we can find alternatives that cannot be used as strategic weapons by one country against another, then we have a freer global society. And if we can do all of this, then people buy more cars and trucks, which increases job opportunities, corporate revenues, taxes, and enjoyment of life.

The bottom line is that we support all alternative fuels; particularly those that we can produce here in America from American raw materials. We love CNG, we love propane, and we love algae-based biofuels. But we especially love ethanol because it can be produced from so many different types of abundant raw materials right now, today, and the use of ethanol doesn't necessarily require any engine conversion. Depending upon your vehicle, you can put a rather high percentage of ethanol in your fuel tank right now. And the only effect you will experience is that you will save money, up to $1.00 or more on every gallon of ethanol you use. You don't require government approval or EPA sanctioning to do it. You just drive up to an ethanol pump and fill 'er up. You can even make your own ethanol in your backyard. You can't do that with CNG, propane or algae biofuels.

(Please see: "Save 20%* on Gasoline Immediately...Guaranteed!" for more information at *https://www.theauto channel.com/news/2010/09/27/498407.html*)

We also love the idea of electric vehicles, but we're very skeptical of electric's potential to help us out of our present economic depression. Electric cars may just not be a significant factor until the end of this century, the 21st century. That means that not only will you be dead and buried by the time of an electric vehicle revolution; but your children, if they are above the age of puberty, may not see it either. We feel that electric gets the support they do from the gasoline companies because the gasoline companies know that electric isn't a real threat. They use electric as a carrot before the mule: "Just keep on using gasoline; we'll get to that there electric soon enough." But soon never comes.

As a matter of interest, we've been approached by Brazilian ethanol entities to help promote the sale of Brazilian ethanol in America, but we don't do it. We're happy that the Brazilians have developed such a strong ethanol industry; and we're certainly thankful for what the Brazilians have done to show the contemporary world that ethanol is still as great an engine fuel as it was considered nearly 100 years ago, when Henry Ford and the top engineers at General Motors thought ethanol was the best engine fuel. But we don't want to replace our dependence on Middle East or Venezuela or Russian petroleum oil with Brazilian ethanol. We must stand on our own and every other country should do the same.

In any event, what this all gets down to is building new factories, creating new supportive businesses, putting people to work, and keeping the money here to be used to buy more domestic products that are built by our own citizens. The petroleum oil and gasoline industry just suck us dry. It's great for the Arab sheiks and dictators so that they can live fantasy lifestyles and pay terrorists to wreak havoc around the world, but it does nothing for us. This is why the issue of alternative fuel is so important to The Auto Channel.

No New Gasoline-Powered Vehicles In The U.S. By 2014...Can It Be Done?

Originally Published on June 23, 2008...
If only someone would have paid attention

The world seems to be in the midst of an energy crisis, fueled (if you'll pardon the pun) by issues related to petroleum oil and gasoline. Among the consequences of the way these issues play out are soaring prices of other consumer products and services, limited freedom of personal movement, erosion of individual wealth, funding of global terrorism, and the granddaddy of all oil/gasoline consequences, pollution.

Assuming that there are no great universal conspiracies in play designed to keep individuals from acquiring personal wealth or to curtail personal freedom of movement or to keep terrorism alive and well for the benefit of the 'military-industrial complex,' the remedy may be as simple as the elimination of gasoline-powered vehicles.

Can it be done? Can it be done quickly and without the creation of even more drastic consequences to consumer product pricing, personal freedoms, individual wealth, societal safety and the environment?

I've initiated a multi-part investigation to determine if and how we could accomplish this herculean feat. In a concurrent article published on The Auto Channel, I presented that information. You can find that piece in the NOTES section

My partner, Bob Gordon, and I have often mused that to solve the seeming energy crisis we need a government mandate equivalent to the Manhattan Project or Kennedy's effort to land a man on the moon. Although a unanimous government mandate may not be necessary to accomplish the task (maybe a spontaneous unanimous popular movement could erupt), the assistance would be most helpful in order to overcome the myriad legal and licensing requirements that are sure to arise.

So as not to appear too Polyanna-ish, I do want you to know that I have taken into consideration the financial devastation of the existing oil/gasoline industry (and its collateral distribution and retail components) that would result from the elimination of gasoline. That is why my premise proposes the idea of the elimination of NEW gasoline-powered vehicles by 2014, and not the total elimination of all gasoline-powered vehicles and machinery by 2014. Simply put, with the base of existing gasoline-powered vehicles and stationary machinery that now exists, the oil/gasoline industry will continue to thrive in the United States for at least a couple of decades. In the meantime, those oil/gasoline producing companies could turn their efforts to seriously producing alternative fuels, and the associated industry components would adapt to distributing and selling the alternative fuels. As for those companies and individuals who do not want to adapt to the new alternative fuel, they would go the way of the buggy whip industry or do business with other nations that continue to rely on gasoline.

The Underlying Issue

In referring to the gasoline crisis, I say "seem" because there is evidence to suggest that the high cost of gasoline at the pump is nothing more than the contrived gasoline-shortage-scare of the 1970s. Case in point: last year, when the barrel price of oil was in the $60 range, gasoline at the pump in the U.S. was in the low to mid $3 per gallon. Today, oil is running around $130 per barrel, yet gasoline is "only" low to mid $4 per gallon. Why isn't the price $6 or $7 per gallon, or better yet why wasn't the price last year in the $2 range?

While I understand the arithmetic tied to the issues of "oil lifting," production and refinement facilities, and increased world demand for petroleum and gasoline, I believe the real culprits behind today's pump price are commodity speculators, OPEC, and the oil companies.

It's said that it costs the Saudis and others about $2 to "lift" a barrel of oil. There's a wide gap between $2 and $130. It would be one thing if the $130 per barrel price included all the eventual shipping, refining, and marketing costs related to taking the oil from underground and then transforming it from its raw state to the product you pump into your vehicle, but it doesn't. Those costs are additional. Therefore, the $130 price is not predicated on arcane consumer supply & demand economic theories. It's because of greed.

I pretty much discount the notion that the world is running out of oil; that ain't happening in my (or Al Gore's) lifetime, and it probably won't happen in my grandchildren's lifetime, if I had grandchildren.

Moreover, the price of gasoline has only risen in concert with prices of most other items such as milk and bread and a slice of pizza. A great case could be made for saying that all the news about oil production and gasoline prices is nothing more than overly hyped excuses for raising gasoline prices to match the retail prices of everything else. (In the mid-1960s, when gasoline in New York was around 30 cents per gallon, I remember paying about 25 cents for a loaf of bread or quart of milk and 10 or 15 cents for a slice of pizza.)

Ironically, if you consider all the greed exhibited to date has only pushed gasoline prices in the U.S. to the point where it is level with other consumer items, then maybe it could be said that $4 a gallon gasoline is a great deal.

The reason why there's such a great uproar over the price of gasoline, versus the price of milk or bread, is that no individual consumer spends $50+ per week to fill up on milk or bread or pizza (with some multi-vehicle households spending two or three times the $50+ per week). So, perhaps the gasoline crisis is simply a crisis of perception, not reality.

Either way, knowing that the price of gasoline is where it should be (by standards of inflation), doesn't make any of us feel better, unless you're an oil company executive. And knowing that there's no real end to the world's supply of oil doesn't mean that temporary shortages cannot occur if oil production levels don't meet real demand or if there are failures in the pumping, refining and distribution system. Understanding that $130 a barrel oil price is determined by greedy commodity speculators doesn't mean they can't push the barrel price up to an amount that makes the per gallon price double or triple what it is now.

The Real Reasons We Need Alternative Fuel Sources

To me, the reasons to find alternative fuel sources are:

1. To take the weapon of oil production and supply out of the hands of fear mongers or greedy speculators and those peoples who would like to use it against us democracy-loving consumers.
2. To improve the environment (not because of a man-made global warming fantasy, but because clean air is just a whole lot nicer than filthy air). And, of course, if we can do all of this at a lower price than gasoline costs, that's even better.

Requisites for a Gasoline-Alternative Future

1. We need an ample supply of vehicles that can reliably use an alternative fuel.
2. We need ample distribution and supply outlets for the alt-fuel(s).
3. We need abundant resources (preferably located on the North American continent) to ensure a continuous supply of the raw material to produce the alt-fuel.
4. We need available, plentiful, and affordable environment-friendly fuel solutions to the gasoline predicament (regardless of how or why the crisis exists and which conspiracy theory you subscribe to).

The Contenders

The potential alternate solutions are:

- Ethanol
- Electric (straight plug-in battery or fuel-cell related)

- Hydrogen - internal combustion engine
- Solar
- Biodiesel
- Compressed Air
- Compressed Natural Gas (CNG)
- Propane

No new gasoline-powered vehicles in the U.S. by 2014...How it can be done

Originally published June 23, 2008 and updated January 15, 2011

A couple of weeks ago I had the opportunity to sit down with Roy W. Willis, President and CEO of the Propane Education & Research Council, to talk about propane as a vehicle fuel. Roy and I were at the Hyatt Regency in Washington, DC; the site of the 2008 Propane Days National Conference. This article features the video recording that I made of that conversation. Although I originally planned to only use the video as a reference source, I found that it contained too much great information to be transcribed. Fortunately, Roy Willis generously consented to allowing The Auto Channel to use the video in its entirety. The video is presented in two parts.

There are, right now in America, two very workable fuel solutions to the on-going energy crisis that we seem to be facing that almost every homeowner and apartment-dweller uses on at least a semi-regular basis: CNG (compressed natural gas) and propane.

In the overview of this "No New Gasoline-Powered Vehicles" challenge, I set forth the requisites for a gasoline-alternative future. To recap, we need an ample supply of vehicles that can reliably utilize an alternative fuel; we need ample distribution and supply outlets for the alt fuel(s); and we need abundant resources (preferably located on the North American continent) to ensure a continuous supply of the raw material to produce the alt fuel.

Next, we need available, plentiful, and affordable environment-friendly fuel solutions to the gasoline predicament (regardless of how or why the crisis exists and which conspiracy theory you subscribe to.

THE CONTENDERS for the fuel solutions are:

- Electric (straight plug-in battery or fuel-cell related)
- Hydrogen - internal combustion engine
- Solar
- Ethanol
- Biodiesel
- Compressed air
- Compressed Natural Gas (CNG)
- Propane

A LOOK AT THE RUNNER-UPS

• Batteries are not yet a reality for highway-speed driving and distant trips, although they may be by the end of this year if Phoenix Motor Cars, Miles Electric, Tesla Motors and Think Electric Cars actually deliver on their promises of delivering their respective cars and trucks. However, none of these auto manufacturers are major auto producers and their ability to consistently deliver reliable vehicles in volume are as uncertain as the battery technology itself. Because hydrogen fuel-cell vehicles also rely on batteries, this is also only a projection for the future. At some point batteries will undoubtedly figure in as one of, if not the main method to power vehicles, assuming that an electric generation "crisis" doesn't replace the perceived gasoline crisis.

• Internal combustion hydrogen engines, my personal long-range favorite (and Jay Leno's too), appears to be a dead issue since none of the other major manufacturers want to support this BMW led initiative. Granted, there are also significant issues related to the current hydrogen distribution infrastructure. So, this may never be a workable solution or part of the mixed bag of solutions.

• Solar power for vehicles, while the sexiest of all solutions, is just a novelty and will remain so until solar cell technology and battery technology substantially improve.

• Biodiesel? Imagine creating diesel fuel from algae. That's just what some people, like Valcent Products (VCTPF.OB) and Solazyme are doing, and it will be fantastic if it's proven and can be delivered. Biodiesel from plants we don't eat, like ethanol, can be great once production gets ramped up, but it's not there yet.

• Compressed air engine technology is sort of the latecomer to the party. It's as sexy a solution as solar, but not as well known. India's Tata Motors and Spain's Air Car Factories are currently working on air-powered vehicles for public release, so they say, by 4th quarter 2008. We hope to visit the Spanish group this Fall for a look-see, so stay tuned. For now, who knows? But it is a sexy idea.

The Finalists

Only three of the contending solutions mentioned above can provide the answer we need now to start the revolution: ethanol, CNG and propane.

Of the three, ethanol is my hands-down personal favorite for a solution that is the closest we will ever get to a single-bullet alternative to gasoline and diesel. However, because of the enormous anti-ethanol efforts paid for by the petroleum oil industry, ethanol may never be as widely accepted as it should be. The oil industry is particularly frightened by ethanol because they can't control it. Ethanol can be produced very simply and easily from a wide variety of commonly grown items such as corn, sugar, cattails, switchgrass, buffalo gourds and seaweed.

On the other hand, CNG and propane, which can also be produced exclusively from domestic sources, does require oil industry production facilities and distribution, and therefore can (or might not) be oil-industry controlled. In fact, the primary negative aspect of CNG and propane is that their success as an engine fuel substitute will ensure the continued success of the same domestic oil/gasoline entities that are responsible for creating the dire energy situation we are now in.

This specific report will cover CNG and propane. Ethanol is discussed in great detail throughout this book and on TheAutoChannel.com.

Although very similar in nature and in how they are found, CNG and propane have distinct advantages that separate them.

CNG has a lower price per gallon (from about $1.25 to $2.75) and greater domestic reserves. Propane provides greater energy output per unit of measurement and offers more existing consumer outlets. (There are literally thousands of propane dealers located throughout the country, in virtually every city and town.)

CNG and Propane Vehicles

The prime requisite to make No New Gasoline-Powered Vehicles By 2014 a reality is to manufacture vehicles that can reliably use an alternative fuel source. Unbeknownst to a great many Americans, most—if not all major—automakers already produce vehicles that use CNG and/or propane, and they've been doing so for years. The list includes GM, Ford and Chrysler. Perhaps even more surprising is that many of the models, including Honda's Civic GX are manufactured in North America. So obviously the technology is proven, and the car companies know how to make them.

Unfortunately, nearly all CNG and propane cars are shipped overseas, and the domestic producers of these vehicles are extremely reluctant to increase or shift allocation to their fellow Americans. While no one from the car companies will give you a straight answer as to why this is so considering the burgeoning U.S. demand for these vehicles, it seems they don't want to disrupt the efforts they've been expending to introduce forthcoming gasoline/electric hybrid or electric fuel-cell models. Consequently, we're left in the lurch while underused production facilities, that could be building more CNG and propane vehicles, are laying off American workers. A flood of phone calls, emails, faxes and letters to Congress may help.

A Closer Detailed Look at Propane as a Vehicle Fuel Source

As mentioned, I had the opportunity to talk with one of the propane industry's leaders, Roy W. Willis of the Propane Education & Research Council. The complete conversation can be watched by clicking on the two images below.

https://www.youtube.com/watch?v=54ctiDfPpTw
https://www.youtube.com/watch?v=rGL2DMzrfM0

The following is exclusive video coverage of the propane presentation given at the 2008 Alternative Car and Transportation Expo in Santa Monica, California. The presentation was made by Leslie Brown Garland from Western Propane Gas Association. Leslie provides additional in-depth information and statistics about the use of propane as a vehicle fuel.

https://www.youtube.com/watch?v=SDmINMWL_PA

Propane Facts

- Outside of North America, liquefied propane vehicle fuel is known as autogas, and it enjoys great popularity in Australia, throughout Europe and the United Kingdom, Hong Kong, India, South Korea, Serbia, the Philippines and Turkey. In Australia, autogas is about half the cost of gasoline in urban areas.
- Propane-powered vehicles are currently used in full-size vehicle fleets (buses and passenger cars) throughout the United States, including significant usage numbers in California, Texas, Florida, New York, Arizona, and Pennsylvania.
- Existing conventional gasoline-powered cars and trucks can be converted to use propane. Federal and State tax incentives can offset the cost of conversion, depending upon your state.
- Propane fueled vehicles may also qualify for use in your state's carpool (HOV) lanes and may be exempted from highway or bridge tolls.
- Propane is on the Environmental Protection Agency's

(EPA) list of federally approved clean-burning fuels. Furthermore, propane is always contained in a sealed, pressure-tight system and there are no evaporative emissions. This eliminates a significant source of secondary pollution.

• Propane is clean burning. Tests conducted by the EPA show that propane engines produce 60 percent less ozone-forming emissions than reformulated gasoline. Recent tailpipe emissions tests performed on Orange County (California) Transit Authority's propane buses showed they emitted 87 percent less total hydrocarbons, 50 percent less nitrogen oxides, and 40 percent less particulate matter than gasoline-fueled buses.

• More than 90% of all propane used in the US is produced domestically.

For more information about propane and the Propane Education & Research Council visit: *http://www.propane council.org*

The Hypocrisy of Big Oil and API

hy•poc•ri•sy
/he´ päkresē/
(alternate pronunciation /big oil/)
noun 1.
the practice of claiming standards or beliefs to which one's own behavior does not conform.
synonyms: deceit, dishonesty, mendacity

ALERT... Prepare for a Shocking Information Discovery

Originally published December 20, 2017

I came across a website titled www.*FillUpOnFacts.com*, sponsored by the American Petroleum Institute (API) to bash ethanol. The hypocrisy begins with the website's name; the website is filled with falsehoods.

It lists 13 categories, and then links to details API pretends are facts to support their bashing of ethanol. I provide a list of these categories below, along with my own previously published responses to explain why API is wrong on these issues.

Without having to go through the 13 categories and my responses, there's a far easier way to prove the duplicity of Big Oil's attacks on ethanol.

For decades, while Standard Oil and other gasoline companies vehemently protested against ethanol fuels in the United States, Standard Oil, Cities Service (CITGO), and other American-based oil companies, sold ethanol-gasoline blends in Great Britain, where they marketed these fuels as being more powerful, cleaner, and cooler running.

In the 1920s, a consortium of scotch whisky distillers began producing and marketing an ethanol-gasoline blend for automobiles called "Discol." A few years later, the Discol name was licensed by another British company (Cleveland Petroleum Products) to sell an ethanol-gasoline blend only if Cleveland

CLEVELAND
DISCOL
CLEVELAND DISCOL HAS THE HIGHEST REPUTATION FOR SMOOTH POWER PRODUCTION, ECONOMY, AND CARBON-FREE ENGINES ON THE ROAD: THE SAME ADVANTAGES ARE YOURS IN THE AIR
ALCOHOL FOR FLYING HORSE POWER
"Cleveland" is the trade mark name of Petroleum Storage and Finance Corporation, Ltd., and "Discol" the trade mark name of The Distillers Company Ltd.

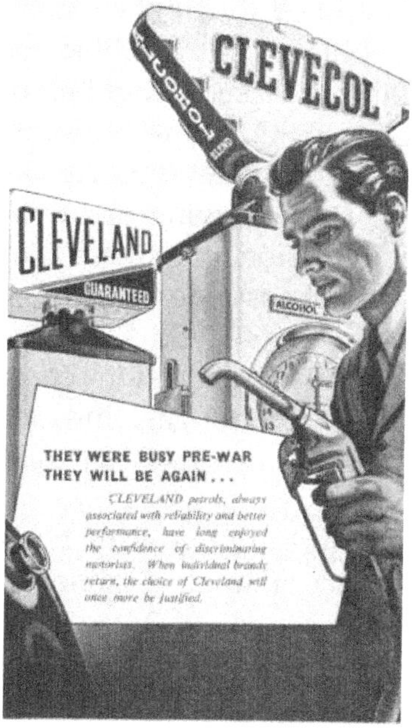
CLEVECOL
CLEVELAND
GUARANTEED
ALCOHOL
THEY WERE BUSY PRE-WAR
THEY WILL BE AGAIN . . .
CLEVELAND petrols, always associated with reliability and better performance, have long enjoyed the confidence of discriminating motorists. When individual brands return, the choice of Cleveland will once more be justified.

STILL BETTER IN WINTER
CLEVECOL
SPECIAL
WITH
ALCOHOL
WINTER GRADE
CLEVECOL
SPECIAL
WITH ALCOHOL
ALCOHOL has long been used to PREVENT CARBURETTOR ICING in aircraft and motor engines
ALCOHOL FOR
1. Immediate Winter starting
2. Amazing acceleration and smoother running
3. Clean engines and fuel lines
4. More top gear miles, and therefore more miles per gallon
5. Highest anti-knock value available, and therefore no pinking
RESULTS ARE IMMEDIATE
The Best Petrol that money can buy
CLEVELAND
The Specialists in Motor Fuels

A remarkable petrol
combining
OUTSTANDING PERFORMANCE
and
EXTREMELY GOOD MILEAGE
CLEVELAND
DISCOL
THE ALCOHOL FUEL
Why remarkable? Because only Cleveland Discol, of all today's petrols, has added alcohol. Alcohol makes engines run smoother, cooler and cleaner, improving engine performance and miles per gallon.
A trial tankful should certainly demonstrate these facts. Fill up with Cleveland Discol next time.

continued to use the alcohol produced by its distillers for the fuel blend. In the 1930s, Standard Oil (ESSO/Anglo-American Oil) purchased about half of Cleveland Petroleum Products. In 1958, ESSO (formerly Standard Oil and known today as Exxon/Mobil) purchased the remaining shares of Cleveland Petroleum.

The following are some of the print advertisements and a television commercial used to promote Cleveland Discol ethanol-gasoline fuel. The ads span the 1930s through the 1960s.

British television commercial from 1966 for Cleveland Discol ethanol-gasoline fuel:

https://www.youtube.com/watch?v=4uEqqlqqf4Q

Here are two British newspaper ads in the 1930s for Cities Service ethanol-gasoline blend called "Koolmotor."

THE YORKSHIRE EVENING POST, WEDNESDAY, JANUARY 31, 1934.

WHY YOU SHOULD USE
KOOLMOTOR ALCOHOL
THE ORIGINAL BLEND OF ALCOHOL AND MOTOR SPIRIT

KOOLMOTOR ALCOHOL

CITIES SERVICE OIL CO., LTD.

LOOK FOR THE GLOBE

APPLEYARD OFFERS PROOF OF SATISFACTION.

5 REASONS FOR BUYING

APPLEYARD O.K. CARS.

SPECIALLY SELECTED.

THE OPEN ROAD

The Campaign Opens.

Glaring Headlights.

DEATH OF TALLEST BISHOP.

DR. TALBOT'S LEEDS ASSOCIATIONS.

How He Ended a Gas Strike.

THE CARS WITH 'the easiest gear control of all' ARE HERE

BARGAIN WEEK
USED CARS
AT
Rossleigh

THE YORKSHIRE EVENING POST, FRIDAY, JULY 8, 1938

FOR TWO WEEKS ONLY!

Tate's GREAT CAR Sale

TAX & YEAR END

BUY BRITISH
Rossleigh
STANDARD
FLYING STANDARD
STANDARD FLYING
WE SELL THEM

ROSSLEIGH LIMITED

SUMMER SALE
Slashing Reductions

CENTRAL GARAGE
BRADFORD

Pointing
5, NEW YORK ROAD and 7D, ALBION STREET
LEEDS

Call and see our good selection of
USED CARS
From £30
THE ASHTON GARAGE CO.
TRIAL RUNS WITHOUT OBLIGATION

KOOLMOTOR
ALCOHOL PETROL
CITIES SERVICE OIL CO. LIMITED
STILL 1/7 PER GALL.

MORRIS

As you see, the message told to the consumers in Great Britain about ethanol was far different than the story told to the American people, then and now. The benefits mentioned in the Discol ads include:

- Ethanol (alcohol) contributes to a brilliant performance and better mileage
- Engines run cooler and cleaner
- Ethanol has long been used to prevent icing in aircraft and car engines
- Amazing acceleration and smoother running
- More miles per gallon
- Highest anti-knock value
- Its thermal efficiency is higher than non-ethanol gasoline (BTU irrelevant)
- Less carbon is produced, ethanol removes carbon deposits

Great Britain wasn't the only place in Europe that sold ethanol or ethanol-gasoline blended fuels; alcohol was used extensively throughout Europe, South America, and Asia. Benzacool in Italy, Moltaco in Hungary, Lattybentyl in Sweden, South Africa had Natelite, Brazil's big brand was Alcool, and in the late 1930s America had an ethanol-gasoline blend called Agrol.

The Forbidden Fuel, a book co-authored by William Kovarik, Hal Bernton, and Scott Sklar, details the use of ethanol throughout the world. Bill Kovarik expands on all this through his Internet presentations, such as his "History of Biofuels" at: *http://www.environmentalhistory.org/billkovarik/about-bk/research/cabi/*

In post-Prohibition America, ethanol-gasoline fuels were often sold without all the boogeyman histrionics used by API members in recent years. For example, in the 1970s, Texaco produced an ethanol-gasoline blend and sold it in the United States. Here's a TV commercial from 1979 featuring Bob Hope promoting corn alcohol blended with gasoline.

https://www.youtube.com/watch?v=IWr0avK1Z7I

Canadians don't seem to know much about ethanol-gasoline blends. But here are two 1980's TV commercials from Mohawk gasoline stations in Canada. They sold an E10 ethanol-gasoline blend. It gave consumers "faster starts, more power, and cleaner emissions."

https://www.youtube.com/watch?v=RNo1HrV7U-c
https://www.youtube.com/watch?v=I_ivQ_gyyaQ

The great significance in all this is that prior to our modern Ethanol Renaissance (and production of vehicles with "ethanol-compatible components" starting in the 1990s) tens of thousands of vehicles—sorry, make that hundreds of thousands of vehicles—no, I mean millions of vehicles—In Europe, the Americas, and Asia were powered by fuels that contained ethanol. On top of that, we should assume other internal combustion engine devices (lawn mowers, chain saws, stationary machines) throughout the world used ethanol-gasoline blends at one time or another.

With all these millions of engines using ethanol-gasoline blends, where are all the newspaper/magazine reports of ethanol causing engine damage? Where are all the stories from the 1920s, '30s, '40s, '50s, and '60s screaming about ethanol corroding engine parts? Where are all the stories about ethanol causing less power and fewer miles per gallon? How could there not be hundreds, thousands, tens of thousands of published complaints that singled out ethanol as the evildoer?

Of course, there wouldn't have been such stories sponsored by the oil companies while they were promoting the ethanol-gasoline blends as a superior fuel. But what about all the know-it-all ethanol-haters who nowadays haunt and troll every publication and media outlet they can find with ethanol-horror stories?

Surely there must have been empty-headed know-it-alls in those days insisting that because of ethanol's lower BTU values that it must deliver less power. Where are all the published stories about how ethanol corroded the rubber, plastic, and metals used in the engines and fuel systems?[1] (Incidentally, in all the years up to the 1980s when petroleum oil companies sold ethanol-gasoline, they weren't "forced" to; there were no "ethanol mandates." They did it because it was the right thing to do. They did it because they knew that ethanol added power, cleaned the engines, emitted less pollution, and was the least expensive way to mitigate engine knock.)

A Remarkable Revelation

A newly published report from the National Renewable Energy Laboratory (NREL) sheds some light on this subject, and at the same time throws some shade (as the young people say) on a possible suspect for the modern complaints against ethanol. Can you guess who that suspect is?

If you said "petroleum oil producers and blenders" you win a kewpie doll. Imagine, the oil industry causing the very same problems that they blame on ethanol.

The report is titled, "Sulfate Salts in Gasoline and Ethanol Fuels—Historical Perspective and Analysis of Available Data." It was published in September 2017 and can be found at *https://www.nrel.gov/docs/fy17osti/69001.pdf*

As I often say during my public speaking appearances, all liquids are corrosive, and one of the most corrosive liquids in the world is water. Water is so corrosive that if metals aren't treated or cared for, it could cause our largest aircraft carriers and biggest passenger jets to fall apart. Yet, we drink water, we wash in water, we even bathe newborn babies in water. But because the corrosive characteristics of water can be overcome with appropriate materials, babies can go about 60 or 70 years before they start to corrode (yes, you're supposed to laugh at this point).

Gasoline is corrosive, too. Inhibitors are added to gasoline to reduce corrosion. Many, people think that non-ethanol gasoline is a benign fluid gently caressing engine components and protecting the metal and rubber parts in an on-going mechanical love ballet. Not hardly. Gasoline is a toxic, caustic, corrosive liquid.

Unbeknownst to most people, especially those ethanol-bashers who don't want to learn facts, prior to the invention of leaded gasoline, ethanol was considered the best substance to put in your vehicle's engine during long periods of inactivity. That's right, it was recommended to fill your vehicle's fuel tank and fuel system with ethanol fuel. In his 1919 maintenance manual for the Ford Model T (the world's number one selling automobile for almost two decades), Victor Page wrote, "Denatured alcohol (ethanol fuel) is without doubt the best substance to use as it does not have any destructive action on the metals and rubber hose, will not form deposits of foreign matter, and has no electrolytic effect." No literature of the period was ever published to refute that recommendation.

The above quote is located on the top of page 156 in Victor Page's manual at *https://books.google.com/books?id=dqEgAAAAMAAJ*

When tetra-ethyl lead was added to gasoline by the General Motors scientists in the early 1920s, it worked like ethanol in quieting engine knock. However, it was highly corrosive, so they had to add ethylene dibromide to the gasoline-tetraethyl lead fuel mixture. Now they had a poison (gasoline), mixed with another poison (tetraethyl lead) in order to stop the engine knock caused by gasoline, with yet another poison (ethylene dibromide) to stop corrosion caused by tetraethyl lead. It's like a triple-threat poisonous concoction unleashed on the public by a science-fiction madman—except it really happened.

The advent of leaded gasoline also led to changes in the design and manufacture of engine components to help handle this fuel. These changes probably were at least part of the reason why ethanol may have gone from being the harmless substance described by Victor Page to a potential corrosive fuel.

However, the new report released by National Renewal Energy Laboratory [NREL] introduces the distinct possibility that the modern complaints of ethanol corrosion are the result of intentional or accidental monkeying with the ethanol-gasoline blends by the oil industry (or cronies of the oil industry) to denigrate ethanol.

Could the oil industry or its minions be responsible for this? Would they stoop to doing this?

Of course, they would, trillions of dollars are at stake. The petroleum oil industry colluded with the Nazis before and during World War II in order to ensure their position of dominance after the war in case the Allies lost. They're responsible for unleashing terrible poisons into our air and water. Oil wars have killed millions of Americans, and untold tens of millions of people around the world. Oil industry carelessness has killed millions of non-human animals of all types.

On the other hand, perhaps it's all just a coincidence? Perhaps it comes from the sheer competitive desire by the oil industry to keep their engine fuels dominant, even though there is a superior fuel (ethanol) at hand. This attitude lends itself to circumstantial conjecture.

I can go with circumstantial conjecture. I don't think I need to paint a picture of premeditated death and destruction, as long as everyone understands that everything else I've written isn't coincidence or circumstantial conjecture. Ethanol was the best fuel to use in internal combustion engines; it was the safest fuel to use in internal combustion engines; it provided the most power for internal combustion engines; and ethanol is still the best, safest, most powerful fuel to use in internal combustion engines (spark-ignited and compression-ignited engines).

And now I'll return to the lies posed by the API website

FillUpOnFacts.com indexes the following 13 issues. My replies can be found under the headings

The Renewal Fuel Standard Needs Reform

"Reform" suggests an evil act has been perpetrated. No one has committed more evil acts than the oil industry. Read "Why Do the AAA and Big Oil Feel They Must Lie to America?"

https://www.theautochannel.com/news/2013/05/23/077318-why-does-aaa-feel-it-must-lie-to-america.html

The Cost of Mandates

The U.S. and our fellow citizens have been kept chained to oil for a century and a half via onerous taxation, an unpopular law of prohibition, military conscription, and never-ending taxpayer subsidies. Read all about oil subsidies in Part 4 of my full-length report "Truth About Ethanol."

https://www.theautochannel.com/news/2013/06/11/082056-gusher-lies-book-review-and-reply-to-robert-bryce-pt.html

The Reality of America's Energy Revolution

API is complaining that cellulosic ethanol hasn't yet come to fruition. Their argument is a ruse. They're looking for something to complain about, like Chico Marx's nonsensical add-on complaint in "Night at The Opera." He said, "Yeah, and your ice isn't cold enough."

It's not important that cellulosic ethanol isn't being used more; we have no shortage of ethanol or ethanol-crops. Here's my response to a Big Oil shill who complained that things are moving too slowly:

https://www.theautochannel.com/news/2016/07/30/266179-biofuel-blunder-renewable-fuel-mandate-having-negative-effect-home-run.html

Outdated Supply and Demand Projections

API has built many of its anti-ethanol arguments on outdated information and production techniques. The oil industry funded studies conducted by David Pimentel and Tad Patzek are rife with rotten data. Read all about the Pimentel-Patzek sham studies in Part 5 of my full-length report "Truth About Ethanol."

https://www.theautochannel.com/news/2013/06/11/082057-gusher-lies-book-review-and-reply-to-robert-bryce-pt.html.

Disconnect Between EPA Mandate and Technology

Read about the real disconnect. Taking On EPA, CAL Air Resources Board, Obama, and Big Oil.

https://www.theautochannel.com/news/2017/03/16/357694-taking-epa-cal-air-resources-board-and-obama.html

What is the Blend Wall?

The Figurative Ethanol Blend Wall is a Fictional Ethanol Blend Wall.

https://www.theautochannel.com/news/2014/06/13/111547-figurative-blend-wall-is-fictional-blend-wall.html

E15 and Your Car

Big Oil E15 horror stories show up in the most remote places. To read about the debunking I did of one story:

https://www.theautochannel.com/news/2015/08/28/140112-lauren-fix-takes-ethanol-opposition-to-new-level-stupidity.html

Fairytale Thinking on E85 and Ethanol Mandates

The fairytale thinking is in believing that E85 is limited to certain vehicles. All gasoline-powered automobiles can use E85, not just flex-fuel vehicles. Consumers can 'splash-blend' their own blend levels to save money and get better MPG results.

Read, "Using E85 in Non-Flex Fuel Vehicles Can Save Almost $1 per Gallon and Fight Terrorism," at:

https://www.theautochannel.com/news/2011/05/09/531654-using-e85-non-flex-fuel-vehicle-can-save-almost-1.html

The RFS is Bad Policy

We use as much corn for ethanol as we do because we grow it for ethanol production. If the ethanol market didn't exist, farmers wouldn't grow as much corn as they do. Even with the amount of corn that used for ethanol, we have virtually as much corn for human consumption as we did in the past. There is no shortage of popcorn, corn-on-the-cob for picnics, tortilla chips, or canned corn. Read "Biofuels Are Worth Every Effort and Cost" for more details.

https://www.theautochannel.com/news/2012/07/24/043980-biofuels-are-worth-any-effort-or-cost.html

A Bumpy Ride for Motorcyclists

Motorcycles and Ethanol (E10) – The Shocking Truth.

https://www.theautochannel.com/news/2017/04/24/372824-motorcycles-and-ethanol-e10-shocking-truth.html

Rising Ethanol Blends Don't Float All Boats

Mercury Marine, the world's largest manufacturer of marine

engines likes E10 and thinks it's a better fuel than ethanol-free gasoline. Read/watch Mercury's ethanol webinar at *https://www.youtube.com/watch?v=GbkmY7TMKMg.*

Outdoor Equipment and the Renewable Fuel Standard (RFS)

One of the oil industry's most pernicious myths about ethanol concerns small engines. Read my take on ethanol and small engines: *https://www.theautochannel.com/news/2015/09/01/140446-why-do-small-engines-suffer-from-ethanol-problems-video.html*

How Has the RFS Evolved Over Time?

You want to know how the RFS has evolved over time? Go to Los Angeles Airport and look north. On many, many days you can see the Hollywood Hills. In Pasadena you can look to the west and south west and see the ocean. This wasn't true in the late 1980s and '90s.

On February 14, 2018 I had the opportunity to present this paper in speech form at the Renewable Fuels Association National Ethanol Conference in San Antonio, Texas.

See a complete video of that presentation at: *https://www.youtube.com/watch?v=sdP4w07ttYc*

Notes

[1] The rubber, plastic, and metal used in those vehicles wasn't susceptible to ethanol; they were however susceptible to gasoline and aromatics. See "Why Do Small Engines Suffer from Ethanol Problems?" included in this book.

BREAKING NEWS:

No American Servicemen Die Defending Ethanol — Again

Updated August 1, 2018

AUTO CENTRAL - The Auto Channel was able to confirm this morning that no American military personnel have died or been wounded defending the production, distribution or sale of ethanol in America. And while we're still awaiting confirmation that no hostages have been taken at any ethanol plants throughout the world, TACH's (The Auto Channel) management is confident that once reports are finalized, that will be confirmed.

Counting today, it appears, unofficially at least, that this is the 16,305th consecutive day since the October 1973 oil crisis that no hostages have been abducted by any militant group from any ethanol plant, no journalists have been beheaded, and no one has been injured as a result of a car bombing or sniper attack. In May 2003, there had been a report that Mary Bruener, a salesclerk at an Iowa-based ethanol plant was taken hostage, but it turned out it was just her friends who were taking her to the local Applebee's for a birthday celebration.

Today's report means that, once again, no lives were lost, no U.S. combat troops were dispatched, no naval assets were deployed, no Tomahawk missiles were fired, no helicopters were shot down, and no drones were needed to go behind enemy lines in order to protect any ethanol production facilities or the flow of ethanol to American service stations.

In a related story, there are no televised reports showing that service stations around America are experiencing long waits for motorists to fill up on any fuel.

© Iculig

MYTHS & LEGENDS

Examines and explains the un-science that has been used to discredit ethanol and other alternative energy solutions for a century and more.

Ethanol Does NOT Suck Water Out of the Air

Originally published March 18, 2016

Among the many ignorant and insidious claims spread by the oil industry to denigrate ethanol is that ethanol sucks water right out of the air. Sometimes it's phrased as "it absorbs water right out of the atmosphere." You see this claim on lots of websites (often on the websites that are trying to sell some kind of whizzbang engine treatment product); you hear it on radio and TV, and you see it all over the place in the comment sections of anti-ethanol news stories and blogs. Even comedian-auto enthusiast Jay Leno made this claim last year in a story he wrote for *AutoWeek* magazine.

Well, it's a lie: Ethanol does not suck or absorb water right out of the air. The only thing that sucks is that the oil industry has been allowed to propagate lies like this for so long.

The origin of the claim is that ethanol is classified as a "hygroscopic" substance. Please note the spelling: h-y-g-r-o-s-c-o-p-i-c (we've had people send us emails that say "ethanol is hydroscopic…"). It is HYGRO, with the letter "G."

It seems that many years ago some clever oil industry person must have learned that ethanol (alcohol) is a hygroscopic substance, and that the general dictionary definition for a hygroscopic substance is that it can attract moisture from its environment.

What the oil industry wag then did was to play a semantic word game by substituting the word "attract" with "absorb," and "air" for "environment." The words attract and absorb are two different things, with two different meanings:

Attract: to pull to or draw toward oneself or itself (as a magnet attracts iron).

Absorb: to take in and make part of an existent whole (as a sponge absorbs water).

And the words environment and air have different contexts:

Environment: the immediate adjacent surroundings (as in the type of environment in which you live).

Air: the mixture of invisible odorless tasteless gases that surrounds the Earth.

How do we know that a semantic word game was played, and that alcohol will not absorb water right out of the thin or ambient air? It's simple: fill any open container halfway with alcohol and place it on your kitchen countertop. Allow it to sit for one or more days. If alcohol absorbs water right out of the air, then when you check the level of liquid in the ensuing days you should find that the volume of liquid has increased. Try it and see what happens.

Incidentally, cotton is also a hygroscopic substance. To additionally prove that a hygroscopic substance doesn't absorb water right out of the air, place a cotton ball on the other side of your kitchen counter and see if it gets saturated with water from just sitting out in the open. Spoiler alert! The cotton ball will not get saturated with moisture. However, if you want to see a hygroscopic substance in action, place the cotton ball immediately adjacent to some water, so that the cotton ball is in an environment that includes liquid water. The cotton ball "wicks" up the water. If you place a piece of glass next to the water, instead of the cotton ball, the glass does not wick up the water because glass is not a hygroscopic substance.

Other than by a dumb accident, water will form in your fuel system because of condensation, but condensation has nothing to do with hygroscopic action. For example, if you have a glass-top outdoor table you have probably noticed that moisture formed on the glass after a cool night. This happens to your car windows. Since glass is not a hygroscopic substance, how does this occur? The moisture forms because of condensation.

So, what do you do if you have some water in your fuel system? Do you stick a straw in and suck it out? No, you add a product like Dry Gas. Dry Gas is one or more type of alcohol, and ethanol is alcohol, meaning that you use ethanol to solve the problem of water in your gasoline tank. That's right, ethanol solves the problem! In the years since E10 has been in regular use, the overall incidence of car owners needing to use Dry Gas has declined dramatically. In April 2003, NPR Radio's Tom and Ray Magliozzi (*Car Talk - Click & Clack*) responded to a question on this issue:

> Dear Tom and Ray: Every time I fill my gas tank, I buy half a quart of dry gas. My father says it really helps with the fuel line and keeps things clean. Is this stuff legit, or am I wasting my cash? - Chris
>
> RAY: Well, we think you're wasting your cash, Chris. This stuff is mostly alcohol, which absorbs any moisture that is in your gas tank.
>
> TOM: The idea is that by getting rid of any water in the tank, you eliminate the chance of the water freezing in the fuel line and preventing the car from running.
>
> RAY: But to be honest, we haven't seen a frozen fuel line in our garage in a decade or more. Why? It could be because gasolines are better formulated now. All gasolines have alcohol already mixed into them.

Ethanol doesn't actually absorb the water; it breaks the water down so that ignition and combustion of the gasoline can take place. The water molecules are then expelled in the exhaust. In other words, ethanol aids combustion, it does not inhibit combustion. Gasoline, on the other hand, doesn't have the same ability to break down the water molecules; so, if it's cold enough, the water separates from the gasoline and freezes or gets sucked into the piston chambers where it cannot combust.

Does this mean that you could pour five gallons of water into your gasoline tank and the available ethanol will break it all down? Of course not. Don't be silly; too much of anything can be bad—even ice cream.

What does one of America's leading auto mechanics says about ethanol? Read this article from Bobby Likis: "Can't We Just Get Rid of Ethanol Ignorance?"

https://ethanolrfa.org/2015/03/cant-we-just-get-rid-of-ethanol-ignorance/

For those among you who are boat owners and have been swayed by the ethanol-hygroscopic lies related to boats, you should listen to the Mercury Marine Ethanol Webinar from August 2011. You will be shocked at their study findings, a position they continue to support. You can hear the webinar at: "Boating Industry Myths of Ethanol" (*https://www.youtube.com/watch?v=GbkmY7TMKMg*).

For those who are too impatient to listen to the entire webinar, this is their finding:

"There is no active transfer mechanism for ethanol molecules to reach out and grab water molecules out of the air. Under normal storage conditions, even in a vented fuel tank it just does not happen. ...The primary cause of water collecting in fuel tanks is condensation from humid air...."

From this point forward, let there be no more discussion about ethanol sucking the water right out of the air. What sucks is not the ethanol.

The Irrelevance of BTU Rating:
Big Oil's Gimmick to Hoodwink the Public
BTU is Big Oil Bunk. It's a marketing and labeling scam

Originally published October 19, 2015

The majority of passenger vehicles and light trucks on the road today are powered by gasoline internal combustion engines (ICE). The selection of gasoline as the primary engine fuel was not made because gasoline was the most effective or efficient fuel, but because it was the most economical to use.

Initially, gasoline's economical advantage was a result of high government taxation on alcohol (ethanol), the other most widely considered engine fuel.

After the Free Alcohol Act was passed in 1906, thc cost of denatured alcohol (aka ethanol fuel) became as inexpensive as gasoline. Because of ethanol's characteristics permitted higher compression engines to be used, ethanol was considered the better fuel. The First World War, as often happens during a war, spurred the rapid advancement of technology in general and the technology of the internal combustion engine in particular. Ethanol was becoming more important in allowing automobile manufacturers to build bigger, faster, more powerful vehicles.

This was a problem for the oil industry.

The advent of Prohibition presented gasoline with its next economic leg-up by virtually prohibiting the manufacture and sale of alcohol at any price, including denatured undrinkable alcohol. With the discovery by General Motors' scientists that adding tetra-ethyl lead to gasoline would mimic ethanol's anti-knock characteristics (thereby allowing gasoline to be used in higher compression engines), gasoline's dominance as the primary ICE fuel was assured. General Motors was able to patent the process and partner with Standard Oil and DuPont to make billions of dollars from leaded gasoline. There was no reason for the world's largest automobile manufacturer to build vehicles that could operate on any other fuel.

As engine design progressed, engines were devised to maximize the efficiency of the new leaded gasoline by refining piston stroke length, spark timing, and air flow to meet the characteristics of the fuel. In other words, the fuel didn't inherently make the mechanical components of the ICE function better; the mechanical components were adjusted or altered to fit the fuel. If ethanol was not ham-stringed by gasoline's economic advantages, engines would have been designed and built to optimize the characteristics of ethanol.

Although Prohibition was repealed about a dozen years later, the advantage given to gasoline during the critical post-war advances in motor vehicle technology and design remained an insurmountable obstacle for the re-emerging alcohol industry. If alcohol was not drinkable in its natural, non-denatured form, the alcohol industry would probably never have recovered in the United States.

Faced with the new threat from an old competitor, the oil industry set out to malign ethanol as a fuel. Among the lies used by the oil industry was the exploitation of the coincidental difference in BTU rating (British Thermal Units) between gasoline and ethanol. In the 1930s, this was easy for the oil industry to do because automobile engines were no longer being designed to allow driver controlled manual adjustments to carburetor air flow and spark timing. Since the engines were now automatically restricted to run on the characteristics of gasoline, any fuel with different characteristics would cause the engine to not operate as efficiently or not to operate at all. The oil industry explained this by claiming that the higher BTU rating of gasoline made gasoline a superior fuel. The truth was that there were simple mechanical explanations for the difference in operation efficiency, not reasons related to so-called energy content. It's like comparing the efficiency of a fork versus a spoon when eating soup by explaining the great difference on the raw material used to make the fork, instead of the obvious reason.

America, and much of the world, bought the oil industry lies to the point that there are otherwise reputable people still claiming that gasoline's dominance is divined by the immutable Laws of Physics.

Interestingly, the BTU argument had been raised during the earlier days of automobile development. Testimony presented at the Free Alcohol Hearings conducted by the House of Representatives' Ways and Means Committee in 1906, numerous studies (such as the U.S. Navy gasoline and alcohol tests on internal-combustion engines in 1907-08), and the well-known results of simple manual driver adjustments proved that the BTU variance between gasoline and ethanol was meaningless.

The oil industry could have relied on another line of reasoning to explain the difference in why a gasoline-powered ICE gets better performance from using gasoline than ethanol. They could have argued, "Gasoline has a gold color, while ethanol is clear and has no color; everyone knows that gold is more valuable than nothing." This argument makes as much sense as the BTU argument, but I presume they went with the BTU story because at least the BTU argument has the ring of scientific authenticity.

Ironically, there's a gaping hole in the oil industry's BTU argument that deflates the whole issue. The hole is caused by the oil industry's petroleum diesel fuel. Petroleum diesel fuel has an even higher BTU rating than gasoline. If the difference in so-called energy content explains why a gasoline-optimized engine gets better miles per gallon of gasoline than ethanol, then surely diesel fuel must produce even higher mileage if used in a gasoline ICE.

That's not what happens. If you tried to use higher BTU diesel fuel in a gasoline engine you get less miles; in fact, you'd get no miles if you put enough diesel fuel in the gasoline tank. How can this be? Gasoline and diesel fuel are both produced from petroleum oil; they're both produced during the same refining process; and they're both used as ICE engine fuels. So, if a lower BTU fuel like ethanol gets fewer miles when used in a gasoline engine because it has less energy content, then surely the Laws of Physics must divine that the higher BTU diesel fuel has to deliver greater miles per gallon when used in a gasoline engine. The results are due to engine optimization, not energy content of the fuel.

My guess is that if diesel engines had continued to be fueled with peanut or seed oil, the petroleum oil industry would have found a way to make peanuts illegal.

The BTU argument is a marketing and labeling strategy used by the oil industry. There is no Law of Physics involved. The oil industry uses British Thermal Units like the Wizard of Oz used smoke and loud noises to confuse the public; but behind the curtain, it's all due to mechanical contrivances.

An ethanol-optimized internal combustion engine running on ethanol will match or out-perform a comparable gasoline-optimized internal combustion engine running on gasoline. An ethanol-optimized ICE will power a vehicle to go faster and go farther per gallon of fuel than a comparable gasoline-powered vehicle. And gasoline used in an ethanol-optimized engine will get less MPG than if ethanol is used.

Here's another take on the BTU issue. In 1936, William J. Hale, Ph.D. published a book titled *Prosperity Beckons: Dawn of the Alcohol Era*. Hale was a leading chemical engineer of his day. He was Director of the Dow Chemical Organic Research Laboratory, Chairman of Division Chemistry and Chemical Technology, National Research Council, Washington, D.C., consultant to the Chemical Warfare Service, visiting Professor of Chemurgy at Connecticut and President of the National Agrol Company.

In his book, Dr. Hale writes that the inefficiency of burning gasoline in a gasoline-optimized internal combustion engine loses 25% of its energy value, whereas alcohol (ethanol) "enters into complete combustion" in the engine. Therefore, instead of comparing gasoline's 116,000 BTUs to ethanol's 76,000 BTUs (and the oil industry advancing the idea that ethanol has 33% less energy than gasoline, as the oil industry lays out their story), you should deduct one-fourth of the gasoline BTUs. This then makes the comparison 87,000 BTUs for gasoline versus 76,000 for ethanol, presumably giving ethanol 12% less BTUs, not 33% less. "Thus," Hale writes, "the custom of comparing fuels on their potentially available British Thermal Units becomes at once obsolete."

Coincidentally, this difference of about 12% fewer miles is much closer to real world MPG results when an ethanol-gasoline blend is used instead of ethanol-free gasoline.*

The significance to William Hale's position is that even if you want to discard my explanation for why ethanol might deliver fewer miles per gallon of fuel in a gasoline-optimized engine (that is, that the difference is related entirely to mechanical adjustments and not inherent energy content), you now have an expert-chemist's explanation of why the stated BTU ratings for the two substances is irrelevant when discussing their use as a fuel in an internal combustion engine.

(The adoption of terms British Thermal Units (BTUs) and energy content isn't the first time the petroleum oil industry co-opted a familiar technical term to add some special significance to gasoline and diesel fuel. In 1892, at the Geneva Congress on Organic Nomenclature, representatives from the oil industry (John D. Rockefeller's Standard Oil) promoted the use of the term to describe petroleum oil fuels to make them appear to be a scarce and finite resource, thereby giving it greater market value. Petroleum oil does not come from dead dinosaurs or the decomposed remnants of living micro-organisms and plant life. In the following video, Fletcher Prouty explains how gasoline and diesel fuel became known as fossil fuels).

https://www.youtube.com/watch?v=zSff0pwc1Xc

If you're a man of science, or an engine mechanic or designer, who is using the BTU argument to promote gasoline over ethanol, stop! Be ashamed of yourself.

Gasoline is poison. It's a rip-off. It's unpatriotic—responsible for killing hundreds of thousands of American servicemen and women, and wounding millions more.

*Independent testing has shown that the use of E30 to E50 ethanol-gasoline blends in flex fuel and non-flex fuel vehicles produces MPG results that are better than E0, E10, and E85. My own personal experience, as well as anecdotal evidence from others, also confirms this result.

For more information on the tests conducted that prove the irrelevance of energy content in comparing gasoline vs. ethanol read the following:

"Free Alcohol," Committee on Ways & Means - House of Representatives, 59th Congress, 1st Session. Feb-Mar 1906. *https://archive.org/details/cu31924094115254*

As a power producer, alcohol is about equal to gasoline...There is far more safety in the use of alcohol than in the use of gasoline...a fire produced by alcohol is readily put out by water...(whereas) water only spreads and increases the danger from a gasoline fire. Cleanliness is another attractive feature in the use of alcohol (ethanol). Cylinders and valves do not get clogged by the left-over products of (gasoline) combustion. Odors arising from (ethanol) are scarcely perceptible and not unpleasant." (page 60)

Three years ago, the increase in the cost of gasoline compelled us to look for a suitable substitute. I went over to Germany to visit our parent house and study the means which had been adopted to procure a cheap and effective power producer. I found that denatured alcohol (ethanol) was very largely used in internal combustion engines. I had occasion to take part in several shop tests made with these alcohol engines.... The results we obtained showed that out of an engine of a given size...we got on average 20% more power than out of the same size engine operated on gasoline." (page 96)

Other and more scientific tests were made at the same time. I remember one test...when the exhaust gases were analyzed...the object being to determine...the degree to which atmospheric air would be contaminated...The results showed that an alcohol (powered) engine produced about 20% less of obnoxious constituents than the exhaust gases from a gasoline (powered) engine." (page 97)

I have also examined the interior portions of alcohol (powered) engines that have been in continuous use for three years and have found them to be in good working order, except such wear as necessarily takes place in all internal combustion engines." (page 99)

Notwithstanding the fact that the heating value of alcohol, or the number of heat units contained (BTU value), is much less than in gasoline, it is found by actual experiment that a gallon of alcohol will develop substantially the same power in an internal combustion engine as a gallon of gasoline. This is owing to the superior efficiency of operation when alcohol is used. (page 107)

"Comparisons Of Gasoline And Alcohol Tests On Internal-Combustion Engines," Department of the Interior - United States Geological Survey 1909

http://pubs.usgs.gov/bul/0392/report.pdf

By using alcohol in an alcohol engine with a high degree of compression...the fuel-consumption rate in gallons per horsepower per hour can be reduced to practically the same as the rate of consumption of gasoline for a gasoline engine of the same size and speed." (page 6)

"An alcohol engine with the maximum degree of compression for alcohol will have an available horsepower 30 per cent greater than a gasoline engine of the same cylinder size, stroke, and speed." (page 7)

"Alcohol has many advantages over gasoline or kerosene as a fuel. The exhaust from an alcohol engine is never clouded with a black or grayish smoke, as is the exhaust of a gasoline or kerosene engine. ...The odors of denatured alcohol and the exhaust gases from an alcohol engine are also not likely to be as obnoxious as the odor of gasoline and its products of combustion." (page 36)

Notes

"Fuel Economy and Power Generation of 30% Ethanol (E30)"
http://www.sdfu.org/assets/docs/uploads/gle-e30-challenge-white-paper-1-19-17final.pdf

"Effects of High-Octane Ethanol Blends"
https://bioenergykdf.net/system/files/1/Thomas_Effects%20of%20High-Octane%20Ethanol%20Blends%20on%20Four%20Legacy%20Flex-Fuel%20Vehicles%20and%20a%20Turbocharged%20GDI%20Vehicle.pdf

"Study Finds Certain Ethanol Blends Can Provide Better Fuel Economy Than Gasoline"
http://www.greencarcongress.com/2007/12/study-finds-cer.html

"The Effect of Ethanol-Gasoline Blends - University of Nottingham"
https://eprints.nottingham.ac.uk/29749/1/546471.pdf

"The Effect of Ethanol-Gasoline Blends"
http://www.stonis-world.com/docs/SuTCAF.pdf

"High performance Wayne State ethanol car wins 1998 Ethanol Vehicle Challenge"
https://wayne.edu/newsroom/release/ 1998/06/04/high-performance-wayne-state-ethanol-car-wins-1998-ethanol-vehicle-challenge-1464

"Ethanol Vehicle Challenge"
http://avtcseries.org/competitions/ethanol-vehicle-challenge/?section=innovations-highlights

"Freedom of choice: E20 ethanol blends take the value crown from gasoline, E10"
http://www.biofuelsdigest.com/bdigest/2013/06/20/freedom-of-choice-e20-ethanol-blends-take-the-value-crown-from-gasoline-e10/3/

"ACE Optimal Ethanol Blend Level Study"
http://www.speedperf6rmanc3.com/content/ACE_Optimal_Ethanol_Blend_Level_Study.pdf

Detroit News Story Goes Overboard in Slamming Ethanol

Is this canard the latest effort by the gasoline lobby to derail alternative fuel initiatives?

Originally published September 30, 2009

On Monday, September 28, 2009, David Shepardson of the *Detroit News'* Washington Bureau published a story titled "Carmakers Fight Hike in Ethanol at Pump," that is as irresponsible a piece of journalism as you'd ever find.

The story's sub-headline and first paragraph insinuate that ethanol is too costly and that its use can damage engines. Unfortunately, this scurrilous report created significant urban myth damage of its own by hindering efforts to free America from gasoline tyranny. They presented a hodge-podge of erroneous factoids, disjointed unrelated issues and half-truths.

For example, Shepardson took comments made by Mike Stanton, CEO of the Association of International Automobile Manufacturers, and made it seem as if Stanton and the association he represents are against the use of ethanol. On the contrary, in a phone conversation with Mr. Stanton, The Auto Channel's Bob Gordon confirmed that Stanton supports the use of ethanol and that members of his group have implored Congress and the Administration to provide a definitive guideline on alternative fuel selection. Stanton feels "we" must get away from gasoline.

As if to prove my charge that Shepardson's preposterous urban myth has a lasting detrimental effect on efforts to replace gasoline with ethanol as the country's primary engine fuel, Stanton and Kim Custer (AIAM's Director, Communications) cited an incident in the Shepardson story about the Baltimore Police Department's patrol cars being disabled because the proportion of ethanol used in the fuel they used was too high (more than 10%). Coupled with Shepardson's repeated unsubstantiated remarks that "increased ethanol blends could corrode engines," Stanton and Custer, and casual readers, were left with the impression that ethanol (alcohol) does indeed damage engines.

I learned from phone conversations with spokespersons for the City of Baltimore and the city's fuel vendor, IsoBunker, that the fuel did not have a higher proportion of ethanol than previously used; there was NO damage done to the vehicles' engines; only a specific make and model year of the city's vehicles were affected (other vehicles were unaffected), and that after the fuel was pumped out and replaced with new fuel using the same formulation, the vehicles operated without problem. Moreover, the cause originally suspected by the city's mechanics was not inflated ethanol levels, but that diesel fuel had somehow gotten into the gasoline blend. According to IsoBunker, it seems the most likely reason for the malfunction was because the fuel was the first "winter-blend" used this year, and the vaporization level of the fuel was not properly recognized by the 2006-7 model year Chevrolet Impalas involved.

Ethanol played no part in the malfunction of the Baltimore patrol cars. Ethanol did not corrode their engines or cause any damage.

Shepardson goes on to report that "automakers" (although he doesn't identify which ones) and some politicians want more study, more testing, more millions of dollars to be spent on research—as if there has not already been far too much money spent, and way too much testing already done to prove what the world has known for more than one hundred years:

- Engines can run efficiently on ethanol (alcohol).
- Alcohol can be produced simply and cheaply and locally.

Ethanol (alcohol) production does not require the use of corn or any other plant that we use for food, so any concern about ethanol production affecting feed and food prices are ignorant or trumped up excuses. In addition, most modern passenger cars can use E85 (without being officially designated as flex-fuel vehicles), and virtually all vehicle gasoline engines can be inexpensively converted to use 100% alcohol (ethanol), so we needn't ever have a situation where we squander raw resources on vast never-to-be-used supplies of ethanol (as fearfully projected in Shepardson's story).

Incidentally, using alcohol (ethanol) actually causes LESS wear and tear on an engine and its parts.

Calls for additional testing and more research, in lieu of just turning the ethanol spigot open to immediate consumer availability, are nothing more than gasoline-lobby sponsored tactics designed to impede the general implementation of any non-gasoline fuel that would reduce their bloated profits.

Why Do Small Engines Suffer From Ethanol Problems?

Hint: They don't!

Originally published September 1, 2015

I recently responded to a knuckleheaded online radio broadcast featuring Lauren Fix, who bashed ethanol—something she has done in the past even though ethanol is something she seems unable to discuss with any accurate information. The broadcast was an episode of "The Jacki Daily Show," hosted by Jacki Pick. A link to the complete episode of the broadcast and my comments can be found in Section 3 of this book under the title "Lauren Fix Takes Ethanol Opposition to a New Level of Stupidity."

But this editorial is not really related to Ms. Pick's interview of Ms. Fix, it concerns a short comment made at the beginning of the show, when Jacki Pick read an email supposedly written by a male colleague of hers. This is what she read on-air: "I ruined a weed trimmer and a chainsaw on this stupid E15 ethanol gasoline."

The alleged writer of the email offered no further details of what he meant by "ruined," and Ms. Pick didn't elaborate on the issue, other than to use it as the springboard to introduce Lauren Fix to the show.

This comment is just one instance of negative claims made against ethanol in regard to small internal combustion engines (two-stroke engines). These are the engines typically found in lawn and garden equipment, motorcycles, and boats.

There are three complaints made about ethanol-gasoline fuel blends. The first is that the ethanol causes sludge, tar, gunk, goo, crap (or whatever you want to call it) to appear in the fuel tank, thereby clogging the filter, spark plugs, fuel injectors, carburetors, etc.

The second complaint is that the ethanol corrodes the rubber hoses and some metal parts.

The third is that the ethanol in the fuel-blend magically sucks water right out of the air and drops it in your fuel tank.

Of the three, I'd say that the first, "the sludge quagmire," is most often heard. And without knowing anything further about what Jacki Pick's colleague was referring to, I'd say that this is what he might have meant—assuming the email was real and not just a device to open the show.

To get to the bottom of this issue, basic questions must be asked:

Question #1 - Does sludge/goo/gunk/crap form in an internal combustion engine? The answer is yes.

Question #2 - Is the sludge/goo/gunk/crap because of ethanol? The answer is no; it is because of the gasoline and oil used in the fuel.

While ethanol is a fuel, it is also a very effective solvent. Solvents are used to clean and remove the residue and deposits left after a substance has been used, such as the residue and deposits left after gasoline and/or oil is burned.

Gasoline is also a solvent, but it's not as effective a solvent as ethanol. Gasoline and petroleum oil products are very effective in leaving residue. They are champs in leaving behind deposits.

In short, if you were choosing up sides to compete in a national residue-making competition, you would be better to select gasoline or petroleum oil for your team. But if you were choosing teammates for a national clean-up competition, you'd be better off picking ethanol.

How much more residue does gasoline or a gasoline-oil combination create as compared to ethanol? Watch this video from 2009 to find out:

https://www.youtube.com/watch?v=WOAyoCo3xXA

Now, imagine the glass jars shown in the video are the insides of an internal combustion engine. Imagine, if instead of burning just fuel-soaked wicks for a few seconds, you were to burn the fuel-soaked wicks for an hour or two every day for a week, a month, a year. Based on the results of the video, could you imagine that the residue and deposits from the gasoline and gasoline-oil fuels would be caked on? Would the residue be a half-inch deep? Would it be an inch deep? More than an inch?

Would the caked-on residue affect the functionality of an engine's pistons? Would the solid, caked-on bits break off and affect the functionality of other parts of the engine? Of course, they would. This is one of the reasons engines fail and why mechanics have to clean out an engine. To mitigate this problem, gasoline companies add "detergents" to the gasoline to help dissolve the buildup. If gasoline was a more effective solvent it might be able to remove the residue that used gasoline creates.

In an article published by GasBuddy.com on June 29, 2013 (*https://www.gasbuddy.com/go/blog/*), Gregg Laskoski wrote about automakers' efforts to encourage the development of more effective gasoline detergents. Mr. Laskoski is a Senior Petroleum Analyst at GasBuddy and had been Managing Director of Public Relations at AAA Auto Club South. Here are some excerpts from that article:

> The automakers formed a consortium, Top Tier Gasoline, that certifies retailers as meeting their standard for detergent. They say it's necessary because detergents prevent deposits of leftover material from building up in engines and exhaust systems. The deposits are like the ash that remains in a fireplace, and their presence in your car's engine can reduce fuel economy and performance...
>
> We strongly recommend Top Tier detergent gasoline to keep your engine clean," GM fuel specialist Bill Studzinski said. "Fuel economy, emissions and acceleration all suffer when there are deposits in an engine.

I hope you noticed that there isn't one single mention of ethanol in these two paragraphs attributing the deposits and residue problem to ethanol. And if you have the opportunity to read the entire article, you'll see the word 'ethanol' never appears anywhere in the text.

Imagine having a fuel that creates much less residue and at the same time is a very effective solvent. In other words, it's something that might make a small mess, and then cleans up after itself—along with cleaning up the mess left by the petroleum-based fuels. Isn't that better? How could it not be better?

But wait, some engines do have problems with sludge, tar, gunk, goo, crap (or whatever you want to call it) after the fuel has been switched to ethanol or an ethanol blend. There, I said it; I admitted it.

If you could fault ethanol fuels for anything, it's that they are too efficient at cleaning the residue and deposits. But that's like complaining that your toothpaste whitens too much; that your hand soap gets too much dirt off your hands; that your greasy pots and pans clean up too quickly.

If you have an engine that has used some type of gasoline fuel or gasoline-oil fuel, the engine will have substantial deposits throughout, and you will have to have it serviced. You will have to change filters, spark plugs, anything and everything that came in contact with the burned fuel deposits. Throughout the life of the engine and the device that the engine powers, you may have to do this multiple times.

The good news is that once you have had this service completed, ethanol-gasoline fuels will keep your engine clean longer and better than any non-ethanol gasoline with detergent. Mercury Marine, the world's largest manufacturer of boat engines, agrees with this. You can see their entire Ethanol Seminar webcast at:

https://www.youtube.com/watch?v=GbkmY7TMKMg.

So, if petroleum oil products, like gasoline and diesel, are the cause of the buildup of all the engine "ash," or engine "plaque," why do some people and organizations claim that ethanol is the cause of the problem?

Well, you could say it's a lie to blame ethanol. In fact, I will say it. It's a lie.

You saw the video demonstration above, you read the quotes from the *GasBuddy.com* article, and if you have just the most rudimentary knowledge of internal combustion engines, then you know what I've written is true. So why do the people, the supposed experts in automotive care, not know this very rudimentary information? Shouldn't they be embarrassed to show their faces in public?

Does Ethanol Damage Rubber Parts and Metal Components?

Automobile and engine parts break. They wear down. They corrode. It's what they do. It's what they've done since movable devices were invented. Human parts do the same, by the way. The auto parts and engine repair industries didn't start with Congress' passing of the Renewable Fuel Standard. There were parts stores all over America and all over the world for many dozens of years. Engine repairmen have been around seemingly forever. The question is, "Is ethanol more corrosive to engine parts more than gasoline?" The answer is that, at worst, it is no more corrosive than gasoline; and in some instances, it is less corrosive.

If you're prone to dislike ethanol you will now say, "Yeah, then why does my engine mechanic (or father or uncle or husband) say it is?" The easiest answer is that they either don't know, and they're just passing along bad information that they've heard; or that they needed an excuse, so they used ethanol as the villain. In short, they misdiagnosed the cause.

At this point, the disbelievers say, "B… S…!"

There are two ways to prove the point. One is by using a visual presentation of the comparative corrosive effects, and the other is by a scientific comparison. Let's look at the video tests conducted by Jacob Kaul. You'll see how after just a short period of time fuel line hoses react to being exposed to different fuels.

https://www.youtube.com/watch?v=0Zt6WzZaqro
https://www.youtube.com/watch?v=u6Ez6s2WEcE

Here are two videos of long-term tests conducted by Erron Spalsbury proving that ethanol fuels do not cause damage to metals or rubber hoses.

https://youtu.be/xkob6fBNrOg
https://youtu.be/HvQp3R_dNQ0

Here's a video about testing E85 in lawn mower engines conducted by Owensboro Technical College:

https://youtu.be/0oQ6CV550WQ

Let's look at the scientific comparisons. A company called Custom Advanced Connections has a website comparing a wide variety of chemicals with different types of rubber (https://www.customadvanced.com). They provide the results via both a dynamic database and a static list. Some of the direct links to information are:

ETHYL ALCOHOL AND RUBBER - *https://www.customadvanced.com/chemical-resistance-chart.html?chemical=Ethyl*

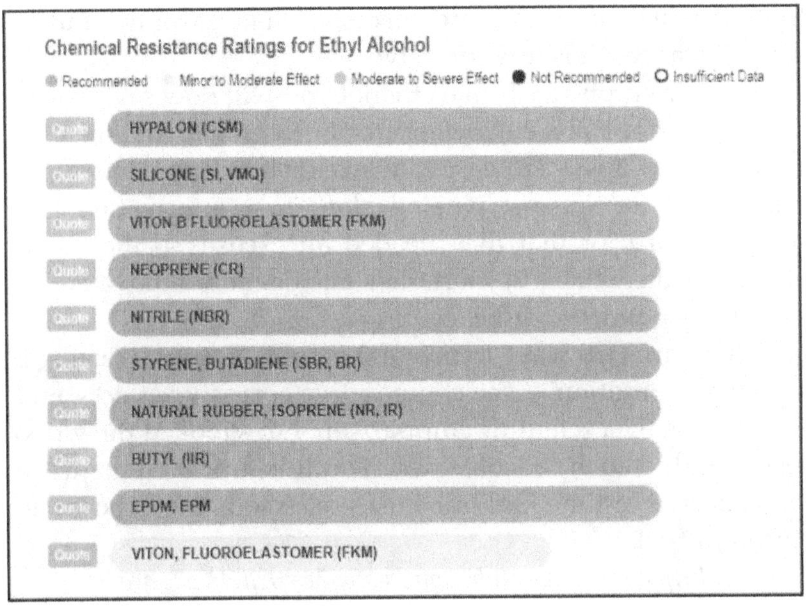

Denatured Alcohol and Rubber - *https://www.customadvanced.com/chemical-resistance-chart.html?chemical=Denatured+Alcohol&rubber=*

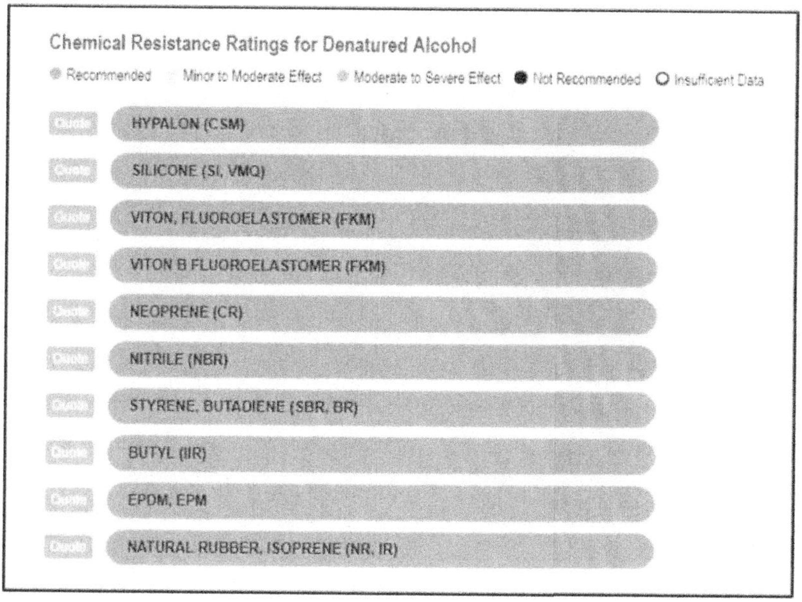

Gasoline and Rubber - *https://www.customadvanced.com/chemical-resistance-chart html?chemical=Gasoline&rubber=*

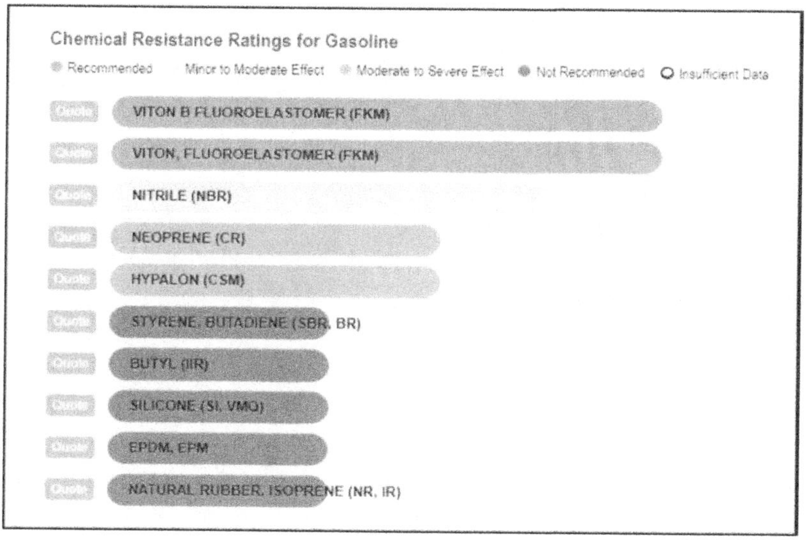

Benzene and Rubber - *https://www.customadvanced.com/chemical-resistance-chart.html?chemical=Benzene&rubber=*

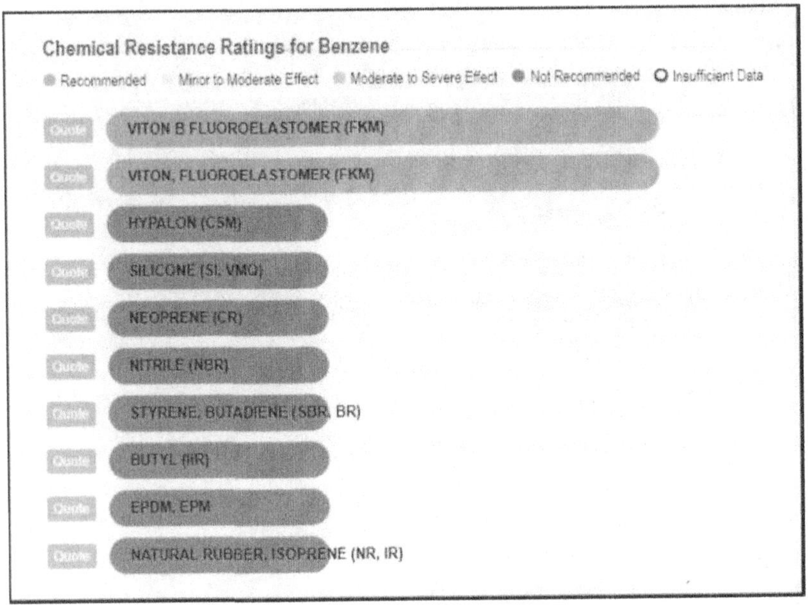

Toluene and Rubber - *https://www.customadvanced.com/chemical-resistance-chart.html?chemical=Toluene&rubber=*

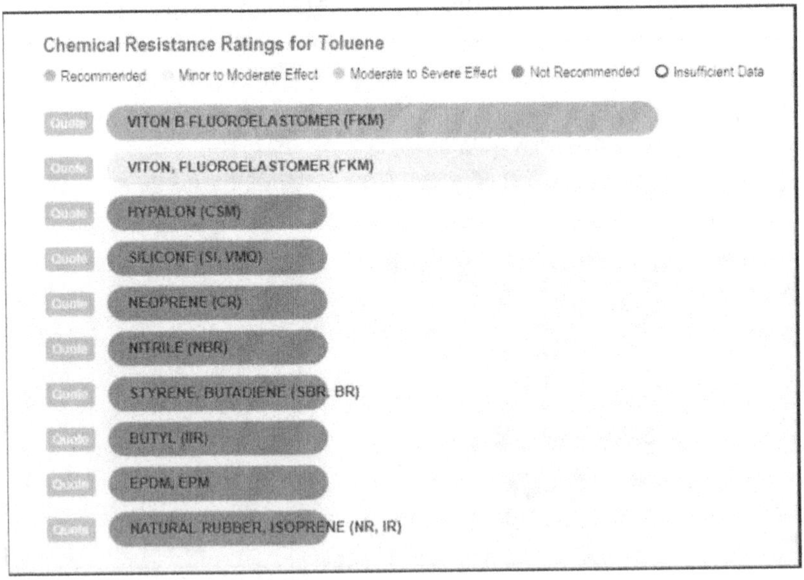

XYLENE AND RUBBER - *https://www.customadvanced.com/chemical-resistance-chart.html?chemical=Benzene&rubber=*

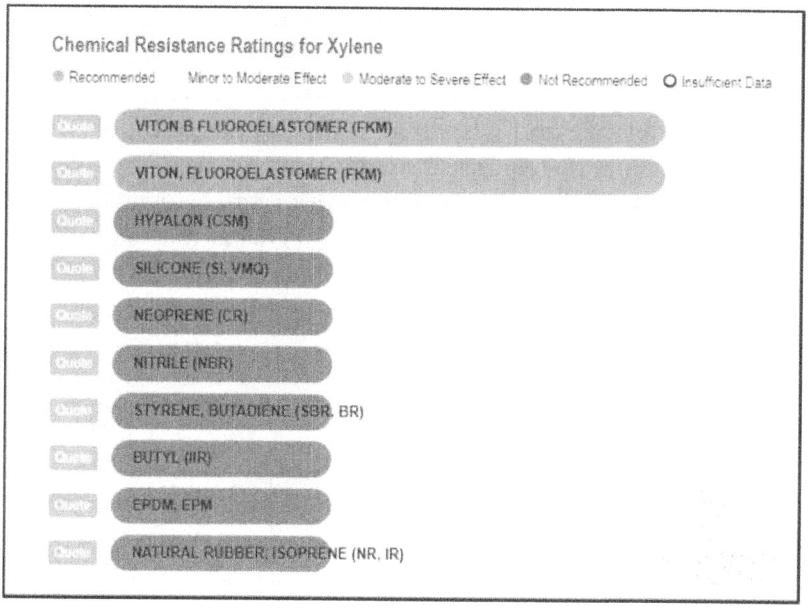

The results show ethanol is the least harmful fuel compared to gasoline and various "aromatic" substances. The one type of rubber that shows ethanol to be slightly less compatible, and at the same time shows gasoline and the aromatics to be better than ethanol, is Viton, Fluoroelastomer (FKM). In any event, the rating for ethanol is still in the highly acceptable category, and nowhere close to the "severe damage—not recommended" categories that gasoline and aromatics fall into with natural rubber, silicone, and neoprene, to name a few.

Viton, Fluoroelastomer is a rubber invented by Dupont in the 1950's and was first used in automobiles in the late 1950s. There are two important points about this:

- The first is that Dupont was one of the three partners in the leaded-gasoline patents (with General Motors and Standard Oil), and because of gasoline's—especially leaded gasoline's—corrosive characteristics, Dupont obviously had it in their best interest to develop a rubber that was very resistant to leaded gasoline.

• The second point is that the use of Viton in automobile engine components beginning only in the late 1950s would explain why ethanol and ethanol-gasoline blends in all the previous years never caused any problems, and why there is an absence of reports citing ethanol-related damage during the many decades that ethanol-gasoline blends were extensively used outside of the United States. (This is the salient point behind the editorial I published in December 2017: "The Hypocrisy of Big Oil and API," which you can find in this book.)

CHEMICAL COMPATIBILITY

CHEMICALS	METALS				
A: Excellent, B: Good, C: Fair to Poor, D: Not recommended - No Data	Aluminum	Carbon Steel	Cast/Ductile Iron	304 Stainless Steel	316 Stainless Steel
Ethanol (Ethyl Alcohol)	B	B	B	A	A
Denatured Alcohol	B	-	B	A	A
Gasoline (high-aromatic)	D	A	A	A	A
Benzene	B	A	B	B	B
Toluene	A	-	A	-	A
Xylene	B	B	B	B	B

The website of another company, Graco, Inc., provides a similar comparison, but they include metal substances. Ethanol is ranked slightly lower than some varieties of gasoline with metals—but is still in the "good to excellent range." It is about equal to benzene, toluene, and xylene. These results and comparisons put to rest any claim that ethanol is harmful (or more harmful) than other fuels to metals that might be used in an internal combustion engine and fuel system.

(http://www.graco.com/content/dam/graco/ipd/literature/misc/chemical-compatibility-guide/Graco_ChemCompGuideEN-B.pdf)

A third resource is the website for CPLabSafety. They specifically analyze and compare different chemicals to aluminum and show results similar to the Graco website metals comparison. One interesting result is that "high aromatic gasoline" ranks as not recommended for use with aluminum in both the Graco and CalPacLab comparisons.

(https://www.calpaclab.com/aluminum-chemical-compatibility-chart/)

Sometimes, it's claimed that ethanol damages plastics. However, you can store ethanol in ordinary plastic water bottles. Aromatics (benzene, toluene, and Xylene), the substances used to replace ethanol in E0, are far less compatible with different types of plastics.

See a comparison chart at:

http://www.plasticsintl.com/plastics_chemical_resistence_chart.html.

The CPLabSafety website confirms ethanol's compatibility with plastics with the charts at:

https://www.calpaclab.com/chemical-compatibility-charts/.

Does Ethanol Suck Water Out of the Air?

No, it absolutely does not. It is a ridiculous myth. I wrote and published an extensive paper on this issue, which you can find in this book under the title "Ethanol Does NOT Suck Water Out of the Air."

Special Update

The other day I was reminded of another category of small internal combustion engines that have been used for years by hundreds of thousands, perhaps millions of people around the world on a regular basis. These small 2-stroke and 4-stroke internal combustion engines prove once again that alcohol fuels are safe to use, and that the corrosion problems and debris problems blamed on alcohol fuels are either non-existent or overly exaggerated. These engines are the IC engines used in model cars, airplanes and boats. The engines are miniature versions of full-size engines, and they work in exactly the same manner as their larger counterparts. The fuel they use consists primarily alcohol (in this case, methanol). The volume of methanol typically runs from about 95% to 60%. This would be equivalent to M95 to M60 (think E95 to E60). The RC fuel usually also contains some nitro - nitromethane - for starting and a small amount of oil for lubrication. The need for the oil is like the use of 2-stroke oil in other small engines such as those in lawn equipment, go-karts, and some ATVs.

Radio controlled (RC) internal combustion engines, which are simply miniaturized versions of full size engines, are NOT susceptible to methanol corrosion. There is a difference between methanol and ethanol...ethanol is compatible with more materials than methanol...ethanol is even less corrosive than methanol.

RC internal combustion engines do experience problems, but the problems are exactly the same as the problems caused by petroleum oil and gasoline. If the engines are not regularly cleaned (many people say after each use), the oil gets gummy and it can harden into a varnish. The easiest and least expensive solvent to use to clean engines and parts is to use some type of alcohol. So, once again, this proves that ethanol used in ethanol-gasoline blends is not the cause of full-size engine problems; it is the solution to many internal combustion engine problems. The typical cause of the problems blamed on ethanol are actually gasoline and aromatics.

There one other thing to consider. Alcohol can be used to clean sensitive, expensive electronics, too. You can use isopropyl alcohol or ethyl alcohol. Among those items, the rubbing alcohol can be used on are video and audio heads on cameras and tape machines. Typically, you would use a swab or a q-tip dipped in alcohol to rub the alcohol over the item requiring cleaning. The alcohol dries quickly and cleanly, leaving no sticky residue.

Final Analysis

How do the bought-and-paid-for "experts" and media trolls who work for API get away with these blatant lies? How do the media bots hired by the oil industry get away with repeating ridiculous "fake news"? Small engine mechanics should understand this and should all be at the forefront to argue for—not against—ethanol. How can you trust a mechanic who can't or won't give a correct diagnosis of a simple problem?

In the broadest terms possible, it could be said that all fuels used in any type of engine will contribute to the engine breaking down. It could be a spark-induced internal combustion engine, a compression-ignited internal combustion engine, a steam engine, a water or wind turbine, or an electric motor. They all break down, and the fuel that supplies the power helps to cause it. Engines have moving parts; moving parts wear down. Internal combustion engines suffer from thousands of small explosions per minute.

When the anti-ethanol people talk about a fuel damaging an engine, they make it seem as if only ethanol causes the wear and tear. They make it seem as if the gasoline somehow cushions the explosions or coats the engine components to prevent deposits from forming. This is a lie. This lie is told to keep us subservient to the oil industry and the terrorist countries who are the key suppliers. There is no fuel used in an internal combustion engine that is worse for the engine than gasoline. Gasoline damages engines. Small engines suffer from gasoline problems. Ethanol helps to clean the engines.

If gasoline was not harmful to engines, America would not have had thousands of engine repair shops and mechanics spread throughout the country during the age of leaded gasoline.

Small engines and big engines don't suffer from ethanol problems, they suffer from gasoline problems.

Special thanks to Al Fletch for his contributions to the information contained in the materials comparison section.

Rolling Oil and Gasoline Bombs
Another Great Reason to Switch to Ethanol

Originally published May 17, 2015

One of the trendiest stories on Sunday TV news shows is about the increasing danger and damage caused by what people call "Rolling Bombs." The term describes what happens when railroad trains carrying petroleum oil products derail or slam into obstacles on the tracks. The incidents often result in human deaths and almost always cause significant damage to nearby homes or commercial structures. Even when there are no collateral damages caused by the explosions, oil spill damage affects human and animal life and usually costs millions or billions of dollars to clean up.

In the past year alone, there have been several such incidents in the U.S. and Canada. One incident in Quebec resulted in 47 deaths. Because of increased petroleum oil and gasoline shipments, the fear is that the regularity of these disasters will increase and threaten towns and cities along every railway used.

The oil industry contends that if new projects like the Keystone Pipeline were given the greenlight, these incidents would be far less. Of course, we all know that oil pipeline spills occur frequently and can be just as dangerous and costly as any oil train collision.

It's important to note that the oil industry opposes new federal oil train safety rules. This is not surprising since the oil industry always denies any risks associated with petroleum oil products and always opposes new rules designed to make their products safer. Remember, these are the same fine people who gave us tetraethyl-lead gasoline and reformulated gasoline with MTBE (methyl tert-butyl ether). It seems like the oil industry never met a poison they didn't like and want to force upon the public.

https://www.youtube.com/watch?v=WpXfQMFR_Qs

One sure way to diminish or entirely remove the threat of railroad oil and gasoline disasters is to replace petroleum oil products with a safer alternative. The best and most easily obtained substitute is ethanol and other products made from ethanol, such as biodiesel. There are four immediate safety advantages to using ethanol. First, ethanol is nowhere near as explosive and flammable as gasoline. A collision with an ethanol tank doesn't necessarily mean that the ethanol will ignite and explode. Ethanol's flash point is 55 degrees F, while gasoline's flash point is -45 degrees F. *

Crashing trains: Rolling gasoline bombs.

Second, unlike gasoline, in the event of an ethanol spill, the vapors from ethanol evaporate quickly, which mitigates the possibility of continuous secondary explosions. Gasoline vapors remain even after the liquid is cleaned, and gasoline vapors are more explosive than gasoline liquid.

Third, unlike gasoline and diesel fuel, ethanol fires can be extinguished with water.

Fourth, a mature and comprehensive ethanol industry would eliminate most need to transport ethanol via long distance rail or pipelines. This alone reduces the risks of using ethanol versus petroleum oil products. The optimal ethanol production scenario would be similar to the milk industry model. That is to say ethanol production would take place near retail distribution outlets. Instead of having to carry ethanol over hundreds or thousands of miles, the distance from production center to filling station could be little more than a couple of miles.

From an economic perspective, ethanol also provides great advantages. We wouldn't be dependent upon foreign enemy and terrorist regimes for our primary engine fuels; we wouldn't be sending hundreds of billions of dollars outside of America every year; we would create a tremendous number of new fulltime jobs—not just relatively few temporary jobs to build the pipeline, and we would be help support our domestic farming industry without government subsidies. (This would be true regardless of the crop used to produce the ethanol.)

And if America didn't have to defend foreign oil producers and ocean shipping, we would save billions more each year and not lose thousands of American military personnel in useless wars.

The Figurative Ethanol Blend Wall is a Fictional Ethanol Blend Wall

Originally published June 13, 2014...but as true as ever!

My business partner, Bob Gordon, called my attention to a story that recently appeared on the online publication, TheCuttingEdgeNews.com*, titled "Corn Based Ethanol Hits a Figurative Wall." The story was written by Steve Baragona, a science and health reporter for Voice Of America.

The story takes a swipe at the so-called ethanol "blend wall" and touches on a couple of negative myths about ethanol, which I can only assume were added because there was no real information about why or how corn-based ethanol has hit this figurative wall. (Read Steve Baragona's story: *https://thecuttingedgenews.com/index.php?article=84031&pageid=21&pagename=Energy)*

I'm puzzled as to why Mr. Baragona chose to bother writing this story, even though it's relatively short length must have taken no more than a few minutes to compose. If I was really suspicious, I would surmise he received a missive from someone in the oil lobby saying they would pay a few hundred dollars to anyone in the media who wanted to write something negative about ethanol. And with the advent of summer, he figured it would be nice to have some extra walking-around money. However, I hope he only tackled the topic because he or his editors felt it was newsworthy.

In any event, whatever his reasons, the headline and the negative comments about ethanol are incorrect; that's why my headline states the "figurative wall" is a fictional wall.

The essence of the ethanol blend wall argument posed by Big Oil is that with the decline in gasoline usage in America (due to improved engine MPG and less driving), government regulations calling for increased national ethanol usage can't be safely met. And the reason that the increased national ethanol usage can't be safely met—according to the oil lobby—is that the only way to comply with Federal regulations is to increase the blend level of ethanol in every gallon of gasoline, which they claim can damage a vehicle's engine.

The claim is, for all practical purposes, false. Using this claim to force the continued reliance on gasoline as our primary engine fuel is a lie. Every vehicle manufactured for American roads since the mid-1990s can safely use gasoline-ethanol blends higher than 10% (E10). In fact, they can use blends that are substantially higher than 10%; they can use blends as high as E85, without being flex-fuel certified. And since the overwhelming number of gasoline-powered passenger vehicles on the road today were manufactured after the mid-1990s, this means the blend wall issue is a non-issue.

Government tests have proven that E15 fuel is safe for all vehicles manufactured after 2001. Moreover, the government tested E20 and found that there is no difference in the results. Ricardo Laboratories, the world's most prestigious private fuel testing lab, concluded that E15 was safe in automobiles manufactured since 1994. Consequently, we can safely accept and use much, much more ethanol. To read more about the EPA and Ricardo tests, read the chapter titled "Big Oil's Anti-E15 Campaign Filled with Gross Exaggerations and Misinformation" in Section 3 of this book.

As I have written on many occasions, I own a 2002 non-flex Ford Taurus that I run on high ethanol-gasoline blends without any damage being done to the engine or fuel pump. I have run the car exclusively on E85, although I typically run it on a E30 to E40 level. The reason I run the car on this average blend level is because sometimes I am in areas where I can only purchase E10 regular gasoline so that when combined with the E85 already in the tank, it tends to average out to the E30 to E40 level. When I do fill up at a service station that offers E85, I "splash blend" it with E10 to achieve the E30 to E40 level.

I do this is because I have found from trial and error that this level will not cause my check engine light to illuminate. Like most people, I'd prefer to not have a red light on my dash, even though I know that it is simply illuminated because my engine's computer has detected that the fuel is different and not because there is any catastrophic problem. While my results are anecdotal, these results have been universally mirrored by other objective owners of late model non-flex fuel automobiles who have used E85.

The Oil Industry Has Been Poisoning Us for Decades, and They've Always Known It

The Oil industry has more ways to kill us than global terrorists...

but wait, they are them!

Originally published July 29, 2016

Highly explosive gasoline, tetraethyl lead, MTBE, toluene, benzene, and the deadliest of all—WAR, are some of the ways the petroleum oil industry has killed us. By "us" I mean Americans, Canadians, Mexicans, Uruguayans, French, British, Irish, Swedish, Polish, Russians, Italians, Chinese, Japanese, Vietnamese, and Sudanese, all killed by or for petroleum oil fuels. These numbers are not in the thousands or tens of thousands, but in the millions.

Tetraethyl lead's deadly characteristics have been known for years, even though the oil industry tried everything possible to hide the truth from the public, much in the same way that the tobacco industry tried to hide how dangerous tobacco smoking and chewing can be. But tetraethyl lead hasn't always been a component of gasoline, and in most instances in America, it hasn't been used very much during the past two decades, as health concerns were finally addressed.

Unfortunately, for the public, what is still used in gasoline is benzene, and the oil industry would like to increase its use by eliminating the use of ethanol (which was primarily adopted in recent years in place of tetraethyl lead and then MTBE).

So, what's wrong with benzene? Everything. In the late 1940s, the American Petroleum Institute sponsored a study of benzene. I assume they were hoping for a miracle finding to absolve them of any guilt or responsibility for killing people. What they did, instead, was shoot themselves in the foot—well, both feet.

The report, called the "API Toxicological Review," was published in September 1948. It "summarizes the best available information on the properties, characteristics, and toxicology of benzene. It offers suggestions and tentative recommendations pertaining to medical treatment, medical examinations, and precautionary measures for workers who are exposed to benzene. It was prepared at the Harvard School of Public Health, Boston, Mass., under the direction of Professor Philip Drinker."

The report was made available to me by the people at FixOurFuel.com and UrbanAirInitiative.com. The complete report can be found at:

http://fixourfuel.com/wp-content/uploads/2016/05/API-Benzene-Toxicology-Review-2.pdf

The report cites various symptoms of exposure to benzene, including the final ultimate symptom—death. It discusses possible treatment for exposure and sets forth the levels at which benzene can cause these symptoms and/or death to occur. The quantities range from very high exposure to relatively minute quantities, as permitted by government laws. It can go from 20,000 ppm (parts per million) in the air we breathe, to just 50 or 100 ppm. The upper quantities can result in immediate death, the lower quantities take a little longer.

The most critical part of the report, I think, is the part in which they give what they believe is a safe ppm amount of benzene in the air we breathe. The report states:

"Inasmuch as the body develops no tolerance to benzene, and as there is a wide variation in individual susceptibility, it is generally considered that the only absolutely safe concentration for benzene is zero."

Zero, as in NO safe level. As in there shouldn't be any concentration of benzene in the air we breathe. As TheAutoChannel.com's editor-in-chief, Mark Fulmer, says, "Benzene is what helps give gasoline it's unique odor...it's the smell of death."

There are some who claim that we (the editorial "we") owe a debt of thanks to petroleum oil fuels; that we wouldn't have enjoyed such advanced living conditions without petroleum oil fuels. Their underlying supposition is that the cost of achieving such advanced living conditions is the price of an insignificant number of humans (100 million dead, for example, is statistically meaningless compared to the 100 billion that have lived since the invention of petroleum oil fuels).

The problem with that kind of thinking is that we don't owe any gratitude to petroleum oil fuels or the people who make it. We owe our thanks to the people who invented the machines that use the fuel. The people who invented the original machines didn't invent them to be powered by petroleum oil fuels. Samuel Morey's internal combustion engine—perhaps the first internal combustion engine—was powered by a mixture of wood turpentine and alcohol. Rudolph Diesel didn't invent diesel fuel, he invented the diesel engine, which was powered by peanut oil. Nicholas Otto's more advanced internal combustion engine was also powered with an alcohol fuel.

It's only because petroleum oil fuels were given huge financial advantages that kerosene, gasoline, and petroleum diesel fuel became our primary fuels. They were not needed to advance the state of engine technology. In fact, gasoline could not and cannot be safely used in advanced engines (high compression engines) without the inclusion of ethanol or some other additive that emulates ethanol's anti-knock characteristics.

And what are those non-ethanol additives? Tetraethyl lead, MTBE, benzene and toluene—all poisons. The oil industry would like to turn back the clock on how and why ethanol was re-introduced into common use in America; they want to replace ethanol with more benzene and toluene. There are equally alarming studies about how dangerous toluene is, such as:

https://www.epa.gov/sites/production/files/2014-03/documents/toluene_toxicology_review_0118tr_3v.pdf

The bottom line is that petroleum oil fuels were never needed; alcohol fuels could have done the same job without all the loss of life attributed to the petroleum fuels and their additives. We don't need petroleum oil fuels now, but as long as they're there, for the foreseeable future, there's no reason why we have to suffer from the full consequences of that poison. The immediate acceptance of higher blend levels of ethanol-gasoline fuels will make it possible to meet all environmental/health goals almost immediately, not in 10 or 20 or 30 years. It will be a tremendous boost to our domestic financial condition, and it will diminish the threats from foreign terrorists.

We may still have had to fight a Great War, and then World War II and the two Gulf Wars, and the war in Afghanistan, but maybe they wouldn't have been so deadly and so prolonged.

How dangerous are benzene and toluene? Watch these videos to see:

https://www.youtube.com/watch?v=udJSDyZUolo
https://www.youtube.com/watch?v=sg6sZq8Sefk
https://www.youtube.com/watch?v=xoZIduAaKS0

What surprises me, is that with all the legal firms across America who are active in filing lawsuits against companies for asbestos poisoning, or faulty breast implants or any one of a dozen other class-action suits, the oil industry continues to sell their poisons with relative impunity.

Latest CARB Rules Screw Californians Yet Again

From the People Who Helped Bring Us Poison MTBE: We Get a New Set of Irrelevant and Dangerous Regulations

Recent Comments from California's Senate President Pro Tem, Darrell Steinberg, Brings This Story Back to the Front Burner

Originally published February 4, 2012

On January 27th, 2012, California's Air Resources Board (CARB) unanimously approved a package of new emissions rules they claim will save drivers money, create jobs, and cut smog and greenhouse gases under what is labeled The Advanced Clean Cars Program.

https://ww3.arb.ca.gov/msprog/zevprog/factsheets/advanced_clean_cars_eng.pdf

It seems the new rules will do almost exactly the opposite of what is intended or needed, except in the area of creating new jobs. Unfortunately, the problem with these new jobs, I think, is that the majority will only be opened by increasing new authoritarian government positions.

Allow me to explain.

We are either facing an imminent catastrophic environmental disaster, or we merely have a pollution problem. If we're facing catastrophic disaster, why is our government fooling around with half-hearted band-aid solutions that slowly kick in over the next several decades? We'll be dead before the problem is solved.

If we merely have a pollution problem, then it's time to own up to it and put into place -once and for all—measures that will eliminate pollution. Let me be perfectly clear. Buying carbon credits and changing CAFE (Corporate Average Fuel Economy) standards does nothing to eliminate polluting activities. It is a scam that shifts blame and allows exemptions—meaning that it permits on-going pollution as long as politicians and bureaucrats continue to get paid off.

In any event, both environmental scenarios require that we stop using gasoline and petroleum oil-based diesel as our primary engine fuels.

Basically, if you want to stop reading this story now, and say to yourself "OK, what is the take-away from this article?" it would be that governments should mandate the end of using gasoline and petroleum oil-based diesel.

For those who choose to leave this page now, thank you for your time. I hope you learned something.

For those who wish to learn more about why the latest CARB (California Air Resource Board) rules are just another dangerous manipulation of the public, or if you're hoping to find reasons to argue with me, please read on.

The new CARB rules (you can read all about them via the link in the first paragraph) rely upon electric-powered zero emission or plug-in hybrid passenger cars and light trucks to achieve the goals of The Advanced Clean Car Program.

The reliance on these two types of vehicles is where the problem lies.

First, zero emission electric vehicles are still in the realm of science fiction: The manufacturers can't profitably produce enough within the targeted time frame. Battery technology is still too premature; and even if the technology improves, the batteries themselves rely on components (elements) and/or manufacturing facilities that seem to be mostly in the hands of 'enemy' nations who could do to us what OPEC has done and is doing.

Studies from car makers, think tanks and governments indicate electric vehicles may not really come into use as we would like until as late as 2090. The Auto Channel has links to a number of these reports.

https://www.theautochannel.com/news/2009/12/07/457466.html

Second, plug-in hybrids, as we commonly know them, use gasoline-powered internal combustion engines to supplement or provide most of the power used to put the vehicles in motion. The key word here is "gasoline," the same gasoline that is at the heart of the pollution. There is no requirement or mandated provision for the engines to be fueled by either a non-fossil fuel or a domestically produced fossil fuel that causes less harmful emissions. Consequently, gasoline and diesel fuel pollution will continue by government mandate. This is like saying that the government has now mandated cigarette smoking.

Third, the number of zero-emission and plug-in hybrid vehicles mandated is insignificant to bring about the program's desired results. Even if there is some amount of reduction in the amount of fossil fuel-borne pollution that emanates from each zero emission and plug-in hybrid vehicle, the number of total vehicles on the road will continue to grow, thereby creating as much or more pollution then we have at present.

It's also important to understand that classifying a vehicle as a plug-in hybrid doesn't require it to conform to any better emissions standards, nor does it signify that it offers better fuel efficiency than full internal combustion engine vehicles. It simply defines the vehicle as being able to operate in two modes. A large plug-in hybrid vehicle operating in gasoline ICE mode can get decidedly worse fuel efficiency and cause more pollution than a smaller gasoline-powered vehicle. Indeed, some newer diesel-powered cars get equal or better fuel efficiency with less harmful emissions than some hybrids. CARB's use of the blanket-term "plug-in hybrid," without requiring anything special of the vehicle in terms of fuel efficiency, is an exaggerated category.

Fourth, the new rules provide no relief from the use of gasoline by the existing vehicles on the road in favor of less harmful fuels. Nor do the new rules provide an opportunity for the owners of the existing vehicles to convert them into lower polluting machines.

CARB continues to ban the conversion of existing gasoline-powered vehicles to CNG or propane. This ban not only grossly exceeds the EPA's onerous restrictions on conversions of existing vehicles, it flies in the face of all global experience and scientific testing that proves that CNG and propane—as well as the conversion process itself—are safe, economical and efficient.

Let me state further that CARB does nothing to expand the use of domestic ethanol or methanol or to encourage the purchase of newly manufactured CNG/propane powered passenger vehicles by mandating or even suggesting the installation of additional alternative fuel filing stations.

The Line in the Sand is Completely Imaginary

The worst part of the new CARB rules—regardless of whether the same rules are adopted by the EPA—is that the enforcement is toothless; and this is where the public really gets screwed.

Let's say battery/electric technology doesn't improve. Let's say that the car manufacturers just cannot make enough zero-emission electric or plug-in hybrid vehicles to satisfy the requirements, either because they can't afford to make that many vehicles and lose money or because public demand is not there for a high-priced or inferior performing vehicle. What happens then?

What happens is that CARB and/or EPA allows gasoline and diesel-powered vehicles to continue to be manufactured but impose penalty fees on consumers when they register the non-compliant vehicles or on the manufacturer for the right to sell the vehicles (which would of course be passed along to the consumer in a higher vehicle price). Therefore, what we have here is one great big dodge to increase government revenues without doing anything to alleviate the pollution problem or solve our general economic problems.

The New Jobs Fallacy of the New CARB Rules

The Advanced Clean Car Program claims it will create a significant number of new jobs. I don't think there's any wording that says the program will create a "significant number" of new jobs. After all, what's the benefit of an all-encompassing mandated program creating new jobs if it's not for a significant number?

In reading the new CARB rules information, I couldn't see where the new jobs will be. It looks as if they said it but left out the details.

The new jobs won't come in the manufacturing process. Vehicles are built in other states or countries. It's too expensive in California and the majority of California's politicians continually prove that they are disinterested in attracting new private-sector business.

The jobs won't come from increased oil drilling and refining. California doesn't want either because of environmental reasons.

Will the jobs come in the form of additional people needed to work at fueling stations? Probably not, since most stations are automated or self-service, and they will continue to be so as electric charging becomes as commonplace as filling a tank with gasoline.

Will there be more repair and maintenance jobs? There might be if the vehicles of the future are manufactured poorly and they require more service then is currently required. But this is unlikely because the move has been to improve vehicles so that they require less, not more service.

So Where Will the New Jobs Be?

As I wrote earlier, the only obvious place is in the creation of new government jobs to ensure the public dutifully observes the government's restrictive regulations (I guess they might be fascist or communist regulations depending on your political perspective), and to administer additional new revenues collected from the public in pursuit of doing nothing to improve the environment.

The Better Way—The Alternative Way—To Improve the Environment and the Economy

The development of a domestic alternative energy program is a "domestic jobs program." If the fuels to be used are domestically produced, they must be produced by domestic workers (hopefully U.S. citizens). The facilities to produce the alternative fuels must be located domestically (hopefully built by American contractors using U.S. citizens).

The raw materials used in the production of some of the alt-fuels must be grown domestically, on farms or plantations that employ domestic workers (hopefully U.S. citizens, or at least legal workers who are "paying their fair share into the system").

If CNG and propane conversions to existing vehicles are allowed, there are about two hundred million vehicles in the country that could be candidates. These conversions open the door to thousands of new mechanics, and for an increase in jobs at conversion kit manufacturing facilities. We already have excellent educational institutions in the U.S. with programs in place to train these new mechanics.

The nearly half-trillion dollars that leaves our country each year to pay for foreign oil stops. That money then gets spent domestically to purchase other domestic products or services, which winds up supporting and enlarging an entire range of additional domestic businesses and their employees. The Auto Channel refers to this as the "Ethanol Dividend." It could just as well be termed the "CNG Dividend" or the "Propane Dividend" or the "Methanol Dividend." You get the point. You can learn more about this in an earlier chapter of this book.

A Surprising Omission

In perusing California Air Resources Board's website and their description of this new mandated program, I found it quite surprising that the authors of the CARB report chose to liberally use information from the "Green Car Journal" website to support their position. I'm not surprised they chose this website as a supportive resource, since Ron Cogan, its creator, has done an excellent job in advancing alternative fuels and energy. In November 2011, the Green Car Journal's annual "Green Car Of The Year Award" went to Honda's Civic GX,* powered exclusively by CNG. This was the second time in the past few years that the Civic GX won this award. Minus changes or improvements to the design of the car, its engine and drive train is the same as it was.

https://www.theautochannel.com/news/2011/11/17/015046-honda-civic-natural-gas-named-2012-green-car-year-2011.html

So here we have an important technical award that has, in most recent years, gone to hybrid, diesel and electric cars, coming full circle back to a vehicle with a propulsion system that isn't even part of the new CARB program.

What Is Really Behind the New CARB Rules?

I think you have to follow the axiom, "Follow the Money." And since "money" usually equates to "power," the axiom could be stated as "Follow the Climb to Power." So, who benefits? Clearly the oil industry and OPEC. As my business partner, Bob Gordon, has opined ad nauseum, the promotion of electric vehicles as a solution to the energy and economic crisis is a diversion used by Big Oil to keep the world addicted to petroleum oil products. There is no information or argument that refutes this accusation. Even if it is not an official policy of the American Petroleum Institute or OPEC or any individual gasoline company, it is a matter of fact. As long as there is a carrot on the stick, we will plod along without change to the status quo.

Who else benefits on the money or power angle? The people who run CARB and the EPA, as well as those politicians responsible for creating or overseeing these agencies. Padding jobs, particularly in a time of high unemployment, give those in power more minions willing to follow them—which translates to retention of an office or position that provides princely rewards.

America needs a revolution in its thinking and approach to many issues. The policies that subjugate, whether they be from the private sector or the public sector, must be thrown off. It's not enough to "occupy" a position against one oppressor and let another get off scot-free, especially when the other is the worst of the lot. And the worst of the lot is Big Oil and the bureaucrats and media elites they control.

https://www.youtube.com/watch?v=17xh_VRrnMU

Note

*For 2012, GX has been dropped as the designation for the CNG-powered Honda Civic model. I use it for easier identification. Also, I am midway through a one-week test drive evaluation of a 2012 Honda Civic CNG vehicle.

Motorcycles and Ethanol (E10)
The Shocking Truth

Originally published April 24, 2017

When you read or watch anti-ethanol stories on the net, you ultimately find stories about how E10 destroyed someone's motorcycle engine. I've had any number of run-ins with bike owners who blame ethanol.

To hear them tell it, if you use E10 this is what happens:

"The ethanol starts going bad in one month."

"The ethanol attracts water that sinks to the bottom of the tank and stops the engine from starting."

"The ethanol caused my hoses and connectors to rot, and the ethanol-gasoline leaked out of the bike onto the garage floor."

"The ethanol hardened into a hard lacquer and froze the cylinders."

"The ethanol turned to a gummy gel, clogged the injectors, and ruined my engine."

Kawasaki Vulcan 900 Classic

I own a 2009 Kawasaki Vulcan 900 Classic. On April 23, 2017, I went into the garage to start my motorcycle and take it for a ride. The last time I started and rode the motorcycle was about five months earlier, at the end of October 2016. I hadn't planned to let the bike sit for that long, things just happened to get in the way. Normally, living where I do, there are usually warm spring-like days even in the dead of winter, and this gives me the chance to take the bike out to charge up the battery and get some "fresh" fuel. We did have a few nice days in January and February, but this year we also had lots and lots of rain. So, on those really nice days, instead of taking my motorcycle for a ride, I took my large German Shepherd to the dog park instead. (He's way too big to ride as a motorcycle passenger).

From past experience with a lawn mower in the 1990s, and with other vehicles I've owned, I know you are not supposed to let an internal combustion engine sit with fuel in it for an extended period of time without putting something like Sta-Bill in the tank; and five months is an extended period of time.

When I first tried to start the bike, nothing happened. There was no electric power to start the bike. The battery was dead. I swapped it for a fully charged battery. I turned the key, the electric light indicators and headlight illuminated. I flipped the ignition switch on and pushed the starter. It immediately started. I let it run for a few minutes to see what would happen. Nothing happened, it continued to run normally.

I turned the bike off, went to get my riding gear and boots, and came back out to the garage. The bike again roared to life instantly. I took the bike out for a half-hour ride. It was fine. I filled the tank with fresh E10, drove a short while longer, then headed home.

Before I left the garage, I made certain that there was no pool of ethanol-gasoline lying beneath my bike; there wasn't. There was no "dried" spot from where any fuel might have leaked. When I arrived back home and drove the bike into the garage, I let it run for another minute or so to see if any fuel would leak from the now fully fueled motorcycle. There was no leak.

The truth can sometimes be shocking if you're expecting something else to happen. But sometimes the truth is just calmly reassuring.

California Government Seeks New Ways to Screw the Public Using Energy Concerns

Originally published February 4, 2015

CALSTART is a California based organization of more than 140 firms, fleets and agencies supposedly dedicated to supporting a growing high-tech, clean transportation industry that cleans the air, creates jobs, cuts imported oil and reduces global warming emissions.

On February 3, 2015, the president and CEO of CALSTART and the president of the Advanced Biofuels Association issued a joint news release titled "California should stay on track with standard on cleaner fuels." (You can read the entire release at: *http://www.sacbee.com/opinion/op-ed/soapbox/article8839505.html)*

If I was to write a kindly response to this story I would say, "Looks like they missed the mark again."

The problem with that response is that I would be dishonest. The honest response is to say that once again the California government and its self-interested appointee organizations are about to level another scam to milk and bilk the working public of their hard-earned income.

If CARB or CALSTART really had any interest in low-carbon fuels, freeing us from foreign oil dependency and trying to make the environment healthier, they would have long ago mandated the use of the two fuels that are available to do the job. These two fuels are compressed natural gas (CNG) and ethanol. Instead, they continue to favor electric power, a solution that won't be affordable and in meaningful availability for two, three, four, five and maybe six decades from now.

CNG and ethanol are available right now and can be used by the overwhelming majority of passenger vehicles that are currently on the road. In the case of ethanol, these vehicles would not even require any engine conversion to use the fuel.

Starting immediately, we could make an enormous reduction in harmful emissions, while saving the public billions of dollars.

Nearly three years ago to the day, California's Air Resources Board unanimously approved a package of new emissions rules that they claimed would save drivers money, create jobs, and cut smog and greenhouse gases under what was labeled "The Advanced Clean Car Program."

This program only referred to electric. They never talked about the contribution that ethanol and CNG could make. What was particularly ironic was that the report from CARB stated they relied in part on Ron Cogan and his Green Car Journal to help them make the decision to support their electric car program. But there was a problem with this—they never consulted Ron Cogan, and Mr. Cogan was not in favor of CARB's program. I know this because my business partner and I discussed it with him at the Los Angeles Auto Show a few months later.

Even more interesting was that prior to CARB unveiling this program, Mr. Cogan had bestowed (once again) his company's Green Car of the Year Award on the Honda Civic GX, a car powered exclusively by CNG. Honda won this award for virtually the same vehicle just a few years earlier.

Although CNG has received great press in the past couple of years, California continues its refusal to allow existing gasoline-powered passenger vehicles to be converted to CNG. If you purchase a vehicle that was legally converted to CNG in another state, and bring it into California, you cannot register the vehicle. CNG conversions are performed everyday around the rest of the world. If there was a danger in doing so, we would be hearing terrible stories about this every day, rather than the terrible stories we hear every day caused by people supported by oil dictators.

Although the CALSTART article finally mentions CNG, they never said anything about the dearth of available new CNG powered cars. For the past several years Honda is the only company regularly producing CNG-powered vehicles—if you can describe annual production runs of about 2,000 vehicles as regular. The auto manufacturers make more electric vehicles per year than that, and that low number is meaningless. If California (and the EPA) removed their unjustified restrictions on CNG conversions, millions of cars could be converted, even if there is no increase in the production of new CNG vehicles.

The number of electric cars being produced annually is inconsequential, and they are too expensive for consumers and auto manufacturers. Consumers who purchase an electric car are paying a significant premium over the true value of the car, and auto manufacturers lose thousands of dollars on every electric car they sell. Sergio Marchione, CEO of Chrysler-Fiat said it best when he described his company's electric Fiat 500 as great, but asked people not to buy them because Fiat loses too much money on them.

Added to the cost and availability problems with electric cars is that battery production makes us dependent on China. At present, there's little to suggest that reliance on China is any better than reliance on foreign petroleum oil.

One hundred percent of our CNG and ethanol supply is produced domestically, and none of the ethanol producers are owned by foreign or terrorist-controlled regimes. No American serviceman or woman has ever been killed defending the domestic production and distribution of ethanol.

So when CARB, CALSTART, Jerry Brown and the California legislature say "Let's Keep Going," what they mean is let's keep milking and bilking the working public; let's keep padding our staffs with additional friends, relatives and campaign donors.

I recognize that some alternative fuel producers and organizations have signed on to this effort. I also am certain that they have done so only in the hope that California politicians and appointees will throw some crumbs their way. They know, as I know, that efforts to get California to act like the "green" concerned state that people assume it is are all exercises in futility.

Taking On EPA, CAL Air Resources Board, Obama, and Big Oil
An Open Letter to Sierra Club

Originally published March 16, 2017

On Wednesday, March 15th, 2017, the Trump Administration announced that it was recalling the fuel economy rule of 2012. This announcement was met with a flurry of frantic press releases from a whole host of "concerned" organizations. The first one I received was from the Sierra Club. Sierra Club Executive Director Michael Brune released the following statement:

> We need to put clean car standards in the fast lane to keep our air clean and our climate safe. Donald Trump and the automakers are endangering the health of our children and families by abolishing life-saving vehicle emissions protections that cut down on dangerous smog pollution and asthma attacks.
>
> The historic vehicle efficiency standards set by the Obama administration are a proven success with widespread support. Not only are our cars and trucks more efficient than ever before, but U.S. auto sales have reached record highs since these safeguards were put in place. It's easy to see why. These standards help clean up our climate and benefit our pocketbooks. These standards are improving the fuel efficiency of vehicles across all classes, spurring innovation of electric vehicles and other technologies, cutting oil use, reducing emissions, and saving drivers money in fueling costs.
>
> Donald Trump's executive order is reckless and puts the health and safety of American families in jeopardy. As we fight back against any efforts to

scrap these life-saving emissions reductions, we will be on strong legal footing.

The following was my response to Sierra Club and to the American Council for an Energy-Efficient Economy, The Natural Resources Defense Council, The American Chemistry Council, and other like-minded organizations and individuals:

Hi Michael, Lauren and all Sierra Club staff:

I received your press release titled "Trump Vehicle Executive Order Will Increase Gas Prices and Carbon Pollution While Putting Americans' Health in Danger."

As you may know, my business partner and I, and our entire editorial staff at The Auto Channel, have worked for many years to curtail the use of gasoline and petroleum diesel and make America energy independent. We are very strong advocates of clean air, water and the environment. We are among the foremost proponents of clean and affordable alternative fuels in the automotive media world. We have taken on the petroleum oil industry and their lobbyists in the strongest tones possible. If you don't already know this about us, you and your organization should make it a point to do so.

Having now reminded or acquainted you with our efforts, I must tell you that your press release presents shamefully wrong and ignorant information.

The Environmental Protection Agency has been a mixed bag when it comes to automotive-related environment and health issues. They have hindered as much or more than they've helped. They have blocked the acceptance and application of fuels and technology that could have dramatically reduced poisons that spew into the atmosphere and water; they've supported false studies conducted by the petroleum oil industry and readily disseminated its incorrect data; and they have helped establish deceptively irrelevant miles-per-gallon goals for gasoline-powered internal combustion engines.

As lukewarm as the EPA has been, the California Air Resources Board has been a total farce. If members of CARB are not on the oil industry payroll, they act like they are. Where the EPA only made the application of affordable alt-fuels like CNG outrageously expensive, CARB made it impossible. Where the EPA provided some assistance in helping advance ethanol acceptance in the American marketplace, CARB ignored it. I have written about CARB's malevolent blindness to these fuels on multiple occasions, but CARB continues to focus and waste money on electric solutions that may never materialize. Even if electric powered cars one day become the exclusive vehicles on the road, we are decades, if not a century or more, away from that time. Meanwhile, CARB will continue to allow poisonous emissions from gasoline and diesel fuel to pour into our air and water. (Read: "California Government Seeks New Ways to Screw the Public Using Energy Concerns")

We are well past the point at which the manufacture of all new gasoline-powered engines should have been discontinued. I wrote about this in 2008. (Read: "No New Gasoline-Powered Vehicles in the U.S. by 2014.")

We are past the point where benzene and toluene should have been completely eliminated from any gasoline blend. I wrote about this again and again over the years. (Read: "The Oil Industry Has Been Poisoning Us for Decades.")

Unfortunately, your press release doesn't stop with bestowing unearned praise on the EPA and CARB. Your chief executive makes a ridiculous statement commending Barack Obama for success in achieving "historic vehicle efficiency standards" and causing U.S. auto sales to reach record highs.

Barack Obama was not responsible for any successes, especially not when it comes to the automobile and transportation industries. While

Obama did deliver a very promising energy speech in August 2008, prior to winning the Democratic nomination and the presidency, he accomplished nothing once he was in office. Although new vehicle sales in 2015 and 2016 hit record highs, it was due to a pent-up consumer need for new vehicles to replace aging vehicles (traditional new vehicle buying cycles were disrupted by the long recession that began in 2000). The increase in sales was helped by a flood of dangerous subprime auto loans that will probably wind-up causing another economic crash similar to the housing bubble.

Moreover, 2016's record sales were due primarily to the increase in sales of SUVs and pick-up trucks, which deliver the worst energy efficiency, so there is absolutely no justification in crediting these sales to Obama's non-existent energy accomplishments.

In any event, worrying about increasing MPG in new vehicles is totally irrelevant; it is addressing the problem from the wrong end. We have hundreds of millions of gasoline-powered vehicles on the road. Between trying to extend gasoline MPG a few percentage points every few years and waiting for a time when electric cars become dominant (if ever), we are on track to poison all humans. And if man-made climate change is real, and greenhouse gases from vehicles are the primary cause, then we will all be dead long before electric cars ever get the chance to take over.

Right now, the vast majority of vehicles on the road can safely, economically, and healthfully use far higher ethanol blends than E10. And the overwhelming majority of the vehicles on the road can use E85 with little or no engine modifications. This would have an immediate and dramatic effect on all harmful emissions. It would also boost the horsepower in most cars and light trucks while saving us all money. It would also be a great boon to American farmers and the American ethanol industry. The EPA has known this for years. The EPA and the Obama Administration have stood in the way of making it happen. The Bush Administration was equally complicit.

Instead of wishing to make Donald Trump responsible for the evils of pollution from the internal combustion engine, fight for proven solutions that can save lives quickly.

Today We're Spotlighting Big Oil's Favorite Alternative Fuel: Electricity

Clean, Efficient, Cheap; What's Not to Love? No Wonder the Oil Industry and Their Star Spokespeople Support Electric-Powered Cars.

There's one other thing that makes electric the oil industry's favorite alternative non-fossil (non-abiotic) fuel: There's no chance it will ever replace gasoline and diesel for decades to come, if ever.

Originally published March 24, 2017

The reason The Auto Channel decided to spotlight electric-powered vehicles is because I received a press release on Thursday, March 23, 2017 from the NRDC (National Resources Defense Council). The press release alerted The Auto Channel and other media that California Air Resources Board (CARB) will discuss the status of the states' clean car standards following the results of their formal review.

The bottom line of CARB's formal review was that the State of California intends to stick with their existing standards for model years 2022-2025. They will begin considering post-2025 standards and will remain fixed on electric-powered vehicles as their solution to achieving those standards. The impact of the standards, according to the release, "...will be felt far beyond California because 12 other states have also adopted California's standards, collectively protecting 113 million U.S. residents from '**dangerous pollution**'." (The emphasis on the words dangerous pollution is mine.)

The press release makes it clear that the action is being taken specifically to spite President Donald Trump's decision to reopen Federal clean car standards, which CARB claims, is "the first step to weakening them."

The NRDC alert also connects CARB's actions to New York State's new 'Drive Clean' rebates program announced on

March 21, 2017. The New York program is also centered around battery-electric and plug-in hybrid electric vehicles in order to create the "clean" in Drive Clean.

All of this would be fine if enough electric vehicles could be put on the roads by the goal dates to protect us from the dangerous pollution caused by gasoline and petroleum diesel. It would be useful if the entire effort wasn't just a scam by state governments to suck penalty fees from auto manufacturers and consumers.

This is the bottom line of CARB's bottom line: Five years ago, CARB approved a package of new emissions rules they claimed would save drivers money, create jobs, and cut smog and greenhouse gases under what they labeled "The Advanced Clean Car Program."

At that time, we published a response to CARB that was sent to many members of their board. I pointed out that the numbers of electric and electric-hybrid vehicles they are counting on will not be enough to meet reduced pollution goals. Nothing has changed in the technology used or in consumer interest in buying these vehicles that would result in any meaningful reduction of dangerous pollution. In fact, because ownership of on-road gasoline and petroleum diesel vehicles will continue to increase between now and 2025, there will be more, not less, total dangerous pollution.

You can read how I laid this all out in "Latest CARB Rules Screw Californians Yet Again" at:

https://www.theautochannel.com/news/2012/02/04/023534-latest-carb-rules-twist-knife-californians-yet-again.html.

I also laid out the inevitable results of not reducing harmful emissions and getting enough new "approved" vehicles on the road. Those results are fines levied against auto manufacturers and increased consumer vehicle registration fees. The state profits, the citizens lose.

The other winner is the oil industry because, once again. Instead of government doing something to take necessary steps to protect us against dangerous pollution, they kick the can down the road to the next generation.

The oil industry is using electric vehicles as a diversion, and politicians are getting rich from it.

If California and New York really wanted to improve conditions and protect us, they would enact new rules to mandate the use of higher ethanol-gasoline blend levels in all gasoline-powered vehicles, including hybrid electric cars. They would also mandate increased use of biodiesel fuel made from ethanol. There would be an immediate and dramatic effect on reducing harmful emissions. Work could then continue to perfect electric vehicle technology so that in 30 or 40 or 100 years from now, a true transition to electric vehicles might take place.

https://www.youtube.com/watch?v=17xh_VRrnMU

• SEE ALSO: Electric Cars - Solution or Diversion
https://www.theautochannel.com/news/2009/12/07/457466.html

The Irrelevance of BTU Rating Revisited

Originally published July 14, 2018

Ever since I started writing seriously about the irrelevance in the difference of energy content between gasoline and ethanol for internal combustion engines, I've had some very funny and spirited exchanges with individuals who are quick to defend petroleum-based engine fuels (gasoline and diesel fuel). Often, the exchanges are with persons claiming to be mechanical engineers or chemical engineers.

Early on, I would be challenged to prove the existence of any reliable sources supporting my supposition. As time went by, I would supplement my own deductions with references to others' statements and findings that confirmed my position. In some cases, people were convinced and went on to argue about other things. Sometimes the references I used were attacked. Whenever I added more supportive sources of information, the attacks would become more hostile.

After I wrote and published "The Irrelevance of BTU Rating: Big Oil's Gimmick to Hoodwink The Public" with references that date back to more than a hundred years ago, the attacks suggested that I was relying on people or information that is too old. I always found this comment ironic because the idea that gasoline and diesel fuel will produce the highest miles per gallon based on energy content is based upon old information relevant to powering steam engines (not internal combustion engines).

The references I cited aren't too old, they simply prove that the irrelevance of energy content in internal combustion engines has been known for a very, very long time and it didn't originate with me. The knowledge has simply been shunted aside by the overwhelming voices that the big wallet and political power of the petroleum oil industry can generate.

In preparation for this book, I thought I would write an addendum to "The Irrelevance of BTU Rating: Big Oil's

Gimmick to Hoodwink the Public" with statements I often use when discussing this issue face-to-face with naysayers:

The issue comes down to this. If all internal combustion engines (ICE), regardless of how the fuel is ignited, and regardless of how the engine is tuned and adjusted, achieved peak efficiency (miles per gallon of fuel) when using the highest energy-content fuel, then energy density of the fuel is a significant and a primary factor. But this is not the case. A spark-ignited ICE optimized to run on ethanol (tuned/adjusted/outfitted with the appropriate fuel injector) will achieve equal or better MPG than the same engine optimized to run on gasoline. Furthermore, in the case of compression-ignited internal combustion engines (diesel engines), biodiesel that contains about 10% less energy content than petrol-diesel will deliver the same MPG without any adjustments to the engine.

Look at this a different way. Let's say the development of internal combustion engines followed the original course set by Samuel Morey, Nicholas Otto, and Rudolph Diesel and that ICE machines continued to be powered by alcohol, alcohol/wood turpentine blends, or nut oils. Engine technology would have developed just as it has, except the engines would have been designed and built to be ethanol-optimized (ethanol can be used to produce diesel fuel).

A hundred years goes by and some enterprising person says:

> Hey guys, instead of using ethanol to power our vehicle engines, let's try this new concoction made from the same gooey stuff we use to make asphalt. Let's give it a quirky name...we'll call it gazoline. And because the gazoline has a higher energy density than ethanol, we'll probably get much better MPG.

So if they fill their fuel tanks with gazoline and do a road test with assorted vehicles to determine MPG, they would return to calculate the miles traveled and the fuel used, and discover the vehicles using gazoline got fewer miles per gallon than those fueled with ethanol. They scratch their heads and consult a mechanical engineer and possibly a chemical engineer. The engineer(s) ponders the information. The results defy everything they were taught in school. He (or she) would be confused. He/she would say:

> I don't get it. Gazoline has more energy content per gallon than ethanol. When I use gazoline as fuel to heat water it boils the water faster than the same amount of ethanol. It must produce more miles per gallon, but it doesn't.

Eventually, the engineer(s) would make alterations to the vehicle engines to fit the characteristics of how gazoline is ignited and burned. At that point, the engineer(s) would say:

> Wow, when it comes to internal combustion engines, energy content...that is, energy density...in other words, BTU rating...doesn't matter. What matters is how the engine is optimized.

I'm not saying that gasoline and diesel fuel don't have a higher energy content; I'm not saying that the BTU measuring system is faulty; I'm saying that BTU rating (energy content) is irrelevant when it comes to internal combustion engines, and the oil industry's entire story of greater energy content is just a marketing scam to sell the poison they call gasoline and diesel fuel. It's said that elephants never forget. Neither should the public. The oil industry is only interested in money and their stockholders, not in the general public, not in their specific customers, and certainly not in patriotic national ideals.

ethanol™ CHRONICLES

Ethanol Chronicles: An Almost Daily Blog
July through December 2015
Lively, Spirited and Sometimes Humorous Repartee Concerning Energy Issues

On a fairly regular basis, Bob Gordon and I answer questions and respond to comments that are either emailed to us or we find on other outlets. The following are some of the best of the banter sessions.

July 2, 2015

Posted by LanceM:

I've seen cars that sat only six months to a year completely disabled by this garbage ethanol. it destroys everything it touches. Gas tanks look like they've been coated with a thick nasty coat of black tar and crusty, super-rust on the inside. The odor of it will make you vomit it's so nauseating. All metal lines, carbs and pumps are the same. Rubber rots from the inside on it. I've replaced more gas tanks and the parts above than I care to mention. Fact is I've lost count. When the old unleaded or non-ethanol fuels sat, they simply evaporated sooner or later. If they left anything, they left a bit of powdery yellow substance that just washed out and was burned with the new gas running through. Ethanol destroys EVERYTHING it touches.

More ethanol, more destruction. More vehicle repairs, parts sold, the whole beeswax. So, it's not just farmers who have a stake in ethanol being increased in our fuel supplies. I wish it was as simple as egging our representatives, but it's not. They do what's best for THEM, not US.

Reply from Marc:

Lance - When ethanol blends (of any percentage) are used in vehicles that have hardly ever or never used an ethanol blend there is a great likelihood that there will be what you describe as "a thick nasty coat of black tar and crusty." This is because ethanol is a solvent. In other words, what is happening is that the ethanol is cleaning the engine. The residue is actually caused by the inefficient burning of gasoline.

In time, the engines would need to be cleaned anyway, and the chances are that the product used to clean the engines would be either ethanol-based or a product that was formulated to replicate the cleaning characteristics of ethanol (alcohol). As I'm sure you know, in the several decades that the only gasoline available in America was ethanol-free gasoline, it was common for internal combustion engines to experience all the problems, and more, that you attribute to ethanol-gasoline blends.

There's no question that if a person goes directly from long term ethanol-free gasoline use to using high ethanol-gasoline blends, the cleaning process can exacerbate the need to clean the engine and fuel line parts. So, the answer is to slowly, over several fill-ups, increase the ethanol-gasoline blends you use. In a sense, it's like changing the food you feed a dog: You should never go cold turkey from one food to the other. Even though the new food might be the best for your dog, the sudden change will cause digestive problems that can also result in something that looks like a thick nasty coat of tar.

Ethanol is a corrosive substance, but so is gasoline; and gasoline can also be used as a solvent, but its flammable, explosive qualities make it much more dangerous than alcohol. Air is also corrosive; and of course, water is one of the most corrosive liquids on Earth. The not-so-secret secret is to use materials that are resistant to the substance you are dealing with.

Take water for instance. We can't live without it, and we consume it as if it has no corrosive characteristics whatsoever. We swim and bathe in water, including the most corrosive of all water, saltwater. It's not that the water we consume and bathe in is not corrosive; it's that our bodies are not very susceptible to water corrosion.

It's the same with engine fuels. You can't just put gasoline in any old container; it must be a container that is not vulnerable to the fuel. The same is true for alcohol (ethanol). Everyone knows you can leave scotch, rum, vodka, whiskey and brandy in a glass bottle or metal flask in your home for years and there will be no degradation of the alcohol nor the container during that time. There will be no so-called "phase separation," even if you leave the top off the bottle; although eventually, you will lose some or all of the liquid because of evaporation.

The same is true of rubbing alcohol, regardless of what the rubbing alcohol is made from. Some rubbing alcohols are just regular grain alcohols with a denaturizing ingredient to render it non-drinkable. And if you have rubbing alcohol in your home, chances are it is in a plastic bottle. The question to ask yourself is why the alcohol doesn't eat away at the plastic bottle. It's because the bottles were manufactured to be resistant to ethanol's solvent characteristics.

During Prohibition, when alcohol was supposedly not available in America, and in the decades subsequent to the end of Prohibition, automobile parts makers used materials that were not highly susceptible to gasoline corrosion. The parts did not fail to corrode because the gasoline was not corrosive, but because of the materials used. If ethanol fuel or blends had been America's primary engine fuels, then auto parts manufacturers would have used parts resistant to ethanol.

In parts of the world (like England, where ethanol-gasoline blends were always available), they didn't experience any greater problems using the ethanol-gasoline blends, because the automobiles built for use in those countries used parts that were resistant to gasoline and to ethanol. Moreover, you have your rubber facts backwards. Ethanol is compatible with many more types of rubber than gasoline and any of the aromatic substances (benzene, toluene, and xylene) that the oil industry would like to use in place of ethanol.

Historically, gasoline was highly corrosive to rubber when DuPont invented a new rubber called Viton, that was not susceptible to gasoline or the aromatics.

There are another couple of points I'd like to address. You wrote that the odor of the thick nasty coat of black tar will make you vomit, it's so nauseating. The odor comes from the gasoline. Ethanol often has a very sweet, pleasing odor, even after extended periods of time.

You also wrote that the old unleaded or non-ethanol fuels left in an idle engine will simply (and safely) evaporate sooner or later. What you left out is that the gasoline will turn to a varnish-like substance that can destroy the engine and fuel system. In the decades in which ethanol was not used in fuel, the problem of leaving gasoline in an idle engine for protracted periods of time always existed. This is not a new problem caused by ethanol, nor is it a problem made worse by ethanol.

So, when you write that you've "replaced more gas tanks and the parts above than I care to mention," and imply that it is only because of the ethanol, you are being terribly disingenuous. An American engine mechanic in the 1920s, 1930s, 1940s, 1950s, 1960s, 1970s and 1980s could have made the exact same declaration, and it would have pertained only to non-ethanol gasolines.

It is true American farmers are among the top beneficiaries of ethanol production; and this is true whether the ethanol is made from corn, sugar, beets, or any other crop. But the thing I always say is that I would rather have my fuel money go to support American farmers than to foreign regimes and terrorist countries. If we're talking about doing what is best for the U.S., the best is to keep as much money as possible here and to employ and many Americans as possible.

Also, remember that no American service men and women have ever died defending domestic ethanol production and distribution. Depending upon which wars you can describe to be oil-related, it could be said that more than a million Americans have died defending Arab oil countries and Russia.

I welcome any and all replies you would like to make, Lance. I only hope you will carefully read what I just wrote and avoid inventing facts or taking my words out of context. If need be, please read the sentences two or three times to get the context correct.

Follow up from LanceM:

I HAVE. Read my post above! I collect older '60s-'70s police vehicles. If a car ever sits for an extended length of time with ethanol in it, you can kiss its fuel system goodbye. You WILL be replacing every component.

I've got a stack of at least six fuel tanks and the carbs and fuel pumps that went with them in my scrap metal pile. The stuff is HORRIBLE for vehicles and actually costs more petroleum to create it than the simple straight up leaded and unleaded fuels we used prior to the Gov. needing to find a use for all that tons of extra corn they paid farmers to grow every year.

Reply from Marc:

Hi Lance - Good to hear from you again so soon. Speaking purely anecdotally, my own personal experience is different. My experience with older cars and different gasoline fuels concerns a 1956 Bentley S1 that I owned and drove regularly for more than 20 years. I purchased the vehicle when leaded gasoline was still the only gasoline fuel commonly available. My ownership lasted through the unleaded MTBE era and into the E10 era.

Bentley Owners Club members were very concerned when unleaded gasoline was proposed and mandated. The frantic claims were very similar to the frantic claims later made against E10. I never experienced any problems switching to unleaded gasoline with MTBE. When MTBE was finally ordered off the market, I heard all the same complaints about the coming of E10. Fortunately, I personally never experienced any problems with using E10.

There was a time when I did have to replace the fuel pump and some of the hoses, but this occurred less than one year after I purchased the vehicle in England and had it shipped to the U.S., during the leaded-gasoline era. It was my understanding that this was the first time the car needed to have its fuel pump replaced.

What you fail to address is that many older cars were built during a time when ethanol-gasoline blends were used around the world with great success. By great success, I mean that you cannot find any reports of vehicle owners in the 1920s, '30s, '40s, '50s, or '60s complaining about corrosion problems related to using the ethanol-gasoline blends. The complaints are modern complaints, probably invented by the oil industry, and wrongly blaming ethanol for gasoline's corrosive characteristics.

The point you may be making is that you and other owners of older cars have to make unnecessary changes, or changes that are not original (or original replicas) to the vehicle. I know this is something that collectors of anything hate to do. First, the parts will need to be replaced at some point anyway, as I discovered with my Bentley. Second, the reasons for mandating a switch away from leaded gasoline were to benefit all of society, not just a relatively minute number of enthusiasts. The argument that was made to Jay Leno after his AutoWeek article was: "Why should we have to be concerned with your collection. You should be concerned with us...the very vast majority."

Lance, I will take exception to your comment that "The stuff (meaning ethanol) is HORRIBLE for vehicles and actually costs more petroleum to create it than the simple straight up leaded and unleaded fuels we used prior to the Gov." This is untrue.

The basis for the claim that ethanol is EROEI (energy returned on energy invested) negative comes from the study conducted by Ted Pimentel and Tad Patzek. The study was incorrect and has been discredited in other university and government studies. In fact, subsequent studies showed that gasoline is far more energy negative than ethanol. David Blume has written about this for years, and we at The Auto Channel have written extensively about this. For more information about the EROEI studies please read the following page:

https://www.theautochannel.com/news/2013/06/11/082057-gusher-lies-book-review-and-reply-to-robert-bryce-pt.html

July 10, 2015

Posted By Babalugatz:

Ethanol is a failure. Costs MORE to produce than it's worth, highly subsidized & never mind old cars...it's junk within 6 months, NO MATTER the age of the car. Turns to varnish rather quickly.

Reply from Marc

Hey, Babalugatz: I'm glad you had the opportunity to join the conversation. The "ugatz" part of your screen name made me laugh; it reminds me of my youth growing up in Brooklyn and my Italian friends.

In any event, you couldn't be further from the truth, from a factual standpoint and from a metaphoric perspective.

What I mean by "metaphoric" is that you can't say it costs more than it's worth, because it is worth what people will pay. The point I think you are trying to make, and it's a typical incorrect criticism of ethanol, is that it requires more energy to produce a gallon of ethanol than the energy it produces. This is often referred to as EROEI (energy returned on energy invested). The notion that ethanol is negative EROEI is incorrect. The study conducted by Ted Pimentel and Tad Patzek is pretty much the originator of the negative EROEI theory. That study has been discredited several times over by subsequent studies conducted by the U.S. government's labs and by universities, including the university that hosted Patzek (UC Berkeley).

The subsequent studies showed that gasoline and other petroleum oil products are EROEI negative. If you have a real interest in learning more about this, you can find an hour-long video online of a C-SPAN debate between Pimentel/Patzek and Bruce Dale/John Sheehan.

The question in my mind is whether you know you are wrong and just posting the information for some unknown purpose, or do you not know you are incorrect?

Because I'm not sure what motivated your post, I'll respond as if I'm talking to other readers who are seeking truthful information, and you can just follow along.

All sources of energy have been highly subsidized by the government. This is true whether we are discussing nuclear, hydroelectric, thermal, solar, bio, or so-called fossil fuel. But the single largest recipient of governmental subsidies has been for petroleum oil-related products. There are subsidies that were granted to the petroleum industry a century ago and have never been retired. Remember that the petroleum industry is an extremely profitable industry. If all subsidies and allowances granted to the oil industry were removed today, and the oil industry had to pay for U.S. military protection for defense of the oil fields and worldwide shipping, gasoline could cost $15 to $20 per gallon. To single out ethanol is an unfair and wrong criticism.

Moreover, while there had been an at-the-pump government subsidy of ethanol, that subsidy was discontinued in 2013. Yet, E85 without the subsidy is still less expensive at the pump than E10 or non-ethanol gasoline.

To say that ethanol left in the fuel system of an idle engine is junk within six months may be true, but it's also true of gasoline, perhaps more so. It's not helpful to debate which is worse. The point is that you should not leave fuel in the fuel system of an engine that sits idle for six months. If you do, you can expect to experience some big problems. This is true no matter the age of the car.

The "varnish" that you are referring to is caused by the gasoline component, not the alcohol component. If you have ever worked with finished wood items, you would know that varnish is often removed or softened by using alcohol. Ethanol/ alcohol is a solvent. It cleans engines.

I look forward to any reply you would like to make.

Follow Up From Babalugatz:

Well, I'm off on the subsidy thing but you can't eat oil or gas. Ethanol is/has been driving up food prices across the board. Cattle feed, thousands of food products [are] made from corn.

From Marc:

I'm glad to see that I was helpful in straightening out the subsidy issue. You're correct, you can't eat or drink petroleum oil or gasoline. But you are incorrect about ethanol production causing food prices to rise.

The issue of ethanol causing food prices to rise came to a head in 2008 when the World Bank published a report that blamed rising food prices on increased ethanol production. The oil industry was quick to jump on this and spent tons of money to get media outlets and on-air media personalities to hawk the story. It caused quite a sensation.

However, the report was wrong. In 2010, the World Bank published another report on food prices, in which they reevaluated the data on the earlier report blaming ethanol. They now correctly blamed the skyrocketing cost of petroleum oil products for the food price increases. The petroleum products that caused the skyrocketing crude oil prices included gasoline, diesel fuel, plastic (for food packaging), and printing inks (also for food packaging).

In 2012, the World Bank published another report on the subject and reiterated it was rising petroleum oil prices that were the cause of the food price increases.

In 2010, a $4 box of cornflakes contained about 6 cents worth of corn. Even if corn prices rose 50%, that would mean that there are only 9 cents worth of corn in a $4 box of cornflakes. There's a lot of space between 6 or 9 cents and $4. So, clearly, it's not a rise in corn prices that is the problem.

As a matter of interest, corn prices as of June 2015 (per bushel) are about the same or slightly higher than they were in the 1970s, 1980s, and much of the 1990s. You can check this out for yourself at:

http://www.tradingeconomics.com/commodity/corn

As we all know, oil and petroleum oil fuel prices have also been lower in recent months. Yet food prices keep rising. If oil prices are down, and corn prices are down, why are food prices rising and why does the blame continue to fall on ethanol production? Food prices are rising because there are many other factors to consider, such as the significant increases in minimum wages paid to employees at food processing plants and retail stores, higher costs in real estate, higher taxes of all types, and higher advertising/marketing costs.

The oil industry never corrected the false stories; and without the appropriate financial incentive to do so, the media outlets and media personalities had no financial motivation to revisit and recant their previous anti-ethanol broadcasts. Consequently, there far too many uncorrected media reports that are all too easily available. People like you read or hear these reports and assume they are (still) accurate.

One other thing, Babalu, most corn used for ethanol production is not grown for human consumption. It's not a question of corn-laden trucks suddenly being diverted from the mouths of obese junk-food eaters or starving peoples in far-flung lands. Corn grown for ethanol provides work and profits for one of the most important sectors of the American economy and society: domestic farmers.

July 12, 2015

Posted by TS:

Ethanol is subsidized, but you have to remember that it has less energy per gallon. I don't have the specifics for pricing in your area but let me throw out a few numbers to illustrate. Say E85 is $2.70 and regular gasoline (E10 unfortunately since it's virtually impossible to find gas without ethanol added, even in Texas) is $3.00.

- E100 has a specific energy of 26.4, while unadulterated gasoline has a specific energy of 44.4.
- E85 has (.85*26.4+.15*44.4) = 29.1
- E10 has (.10*26.4+.90*44.4) = 42.6
- Energy per $
- E85 = 29.1/$2.70 = 10.78
- E10 = 42.6/$3.00 = 14.20

Even though in the example E85 is $2.70 and regular gasoline is $3.00, the regular gasoline gives you 31% more energy per dollar spent.

Reply from Marc:

Hi TS - Please allow me to jump in here. Internal combustion engines are inorganic, non-sentient machines. They do not react to the "food" they consume in the same way that a live animal does. There are also no emotional highs or lows felt by the machine to affect output. The energy output of an internal combustion engine is entirely related to engine optimization (mechanical adjustments, parts and computer software).

The comparative measurements of fuel BTUs is an irrelevant intellectual exercise.

A comparably sized engine that is optimized to run on ethanol will produce the same or better MPG than an engine optimized to run on gasoline. When arguing energy balance issues as it pertains to automobile fuel, it is inaccurate to discount ethanol's lower BTU rating into the final net energy equation. At the least, one gallon of ethanol is equal to one gallon of gasoline.

It doesn't matter that gasoline is rated at about 114,000 BTUs per gallon, and ethanol is rated at 76,100 BTUs per gallon. MPG efficiency makes them 1:1 for all practical purposes.

Moreover, in actual use, a vehicle using E85 instead of E10 or ethanol-free gasoline does not experience the same loss in MPG as the BTU differences on paper indicate. Therefore, if you lose 8-15% in MPG but save 15-25% in cost per gallon, then using E85 is more economically efficient.

FOLLOW UP FROM TS:

You are chemically incorrect, and it has nothing to do with how the engine "feels." The reality is that gasoline simply has a lot more power density than ethanol so you will go a lot further on a tank. You cannot change the laws of chemistry.

Now, granted, ethanol can be burned at higher compression. So in an optimized set-up can actually produce more horsepower than gasoline, but that still requires more fuel for the energy produced.

REPLY FROM MARC:

Hi TS - Thanks for your latest reply. I think you missed the sarcasm in my point about emotion.

It's not a question of being "chemically" correct or incorrect. It's a question of mechanics, that is, engine optimization. If we were discussing heating water to create steam to power a steam engine, then the BTU difference between ethanol and gasoline would be important. But there's no water heating involved in an internal combustion engine. The factors that come to play are the mechanics of the ICE.

To test the irrelevancy of BTUs, as I'm sure you know, diesel has a higher BTU rating than gasoline. Yet, if you put diesel fuel in a gasoline-powered engine you would get far less miles, perhaps none. And if you put gasoline in a diesel-powered engine you wouldn't get 10% fewer miles, you would probably not get any miles as the engine might not start.

An internal combustion engine optimized to run on ethanol will deliver MPG that is equal to, or greater than a comparable internal combustion engine optimized to run on gasoline.

Follow up from TS:

Thanks, but you're still wrong. Yes, engines need to be designed for a specific fuel in order to run well, but a diesel-optimized engine will still get more MPG on diesel than a gasoline-optimized engine on gasoline than an ethanol-optimized engine on ethanol and the reason is simply that when using the proper fuel their relative efficiencies compared to the energy contained in the fuels are very similar so the energy density of the fuel dominates.

Reply from Marc:

Hi again, TS - Thank you for your recent reply.

I'm not incorrect, and I think you are confusing things a bit.

If you can match a diesel-powered engine to a comparable gasoline or ethanol powered engine, and the diesel engine gets more MPGs, it's because a diesel engine is a more efficient ICE. This has been known for more than 125 years. The reason why diesel engines didn't become the primary engines used in automobiles was because they cost more to manufacture, and they were much heavier.

On the other hand, gasoline and ethanol powered engines were never selected for use as train engines or heavy trucks because they didn't have the power of a diesel, and the weight and cost of manufacture was much less relevant. In modern times, cost and weight of diesel engines have been mitigated by new materials and manufacturing techniques/volume. It's why there are so many proponents of diesel-powered passenger vehicles. Unfortunately, other oil industry issues have kept diesel fuel from being readily available, and available at a lower price. When I was young, diesel fuel always cost less than gasoline, and very recently diesel has been selling in many areas for less than gasoline—but for 20 years (give or take) diesel was more expensive than gasoline.

Keep in mind that diesel fuel doesn't need to be produced from petroleum oil. Rudolf Diesel, inventor of the diesel engine, used peanut oil to make the fuel, and renewable diesel or biodiesel (made cheaper from ethanol) will deliver exactly the same mileage in a diesel-powered engine as petroleum diesel fuel.

The reason for my bringing diesel into the discussion that you and I are having is to show that BTUs (so-called 'energy content') is not relative in the argument between gasoline and ethanol. The issue is simply engine optimization. In fact, the thing that I really find amazing is that a substance other than petroleum-based gasoline can be used in a gasoline-optimized engine at all. It's sort of like discovering that you can use a music CD on an old record player. And what's more amazing is that even if you lose some MPG when you use ethanol in a gasoline-optimized engine, the cheaper price for the ethanol more than makes up for the loss in MPG.

Again, the whole creation of BTU rating and understanding was to determine what it takes to heat water one degree. This was important (and still is important) when dealing with steam engines or water heaters or cooking using fire. BTUs have no importance in internal combustion engines. Engine optimization is the key.

If there was a significant higher cost in building an ethanol-optimized engine versus a gasoline engine, then that information could be used against the ethanol engine. However, the engines are virtually the same. The differences are merely mechanical and software adjustments that require no additional practical cost.

July 14, 2015

Posted by Erocker

Corn ethanol does not lower CO2 compared to gas.

Corn ethanol causes a larger dead zone in the Gulf of Mexico.

Corn ethanol leads to Nitrogen fertilizer polluted ground water.

Corn ethanol leads to pollution from pesticides.

Corn ethanol leads to plowing of grass lands to add corn fields.

Corn ethanol leads to destruction of forest lands to add corn fields.

Corn ethanol is increasing the Ogallala Aquifer Depletion.

Corn ethanol pollutes the air with formaldehydes and Acetaldehyde.

Corn ethanol use leads to higher levels of ozone pollution.

Corn ethanol leads to higher food prices.

Corn ethanol uses as much energy to make as it delivers.

Corn ethanol separates out of gas over time (shelf life 3 months).

Corn ethanol has only 2/3 the BTU power of gas.

Corn ethanol costs more per BTU than gasoline.

Corn ethanol can be very corrosive.

Corn ethanol is often distilled using coal as a heat source.

Corn ethanol distillers exhaust high levels of VOCs (volatile organic compounds) pollution.

Reply from Marc:

Hi Erocker: All of your points are either outright lies, gross exaggerations, or irrelevant. I presume you found this list somewhere and have merely re-posted it.

If you created the list on your own and posted it to make a personal statement, then you are simply trying to deceive other readers.

To another person named 'Carney' you wrote, "You are so into your corn ethanol lobby lies you actually cannot see the truth when it is presented to you." Actually, Erocker, you are so into your anti-ethanol lies that you don't know the truth.

If you're interested, a very good place to start is my 60+ page report, "Truth About Ethanol: Book Review and Reply to Robert Bryce's Gusher Of Lies." The entire report is contained in this book.

The only individual point I don't address in the 60+ page report is the issue of ethanol being corrosive. I didn't address it because it was not discussed in the Robert Bryce book I was reviewing. In any event, of course ethanol is corrosive; it's a solvent. Among other benefits, ethanol is used to clean engines.

But everything is corrosive: Gasoline is corrosive, the wind is corrosive, and water may be one of the most corrosive substances of all. To state an obvious characteristic and include it in a list to denigrate ethanol is meaningless. All corrosive substances and forces (wind) can be mitigated or eliminated entirely by using materials resistant to the corrosive substance.

I'd be happy to continue this conversation with you once you've read my report.

But if all you're going to do is just reply with the same hollow generic statements, then you should save your energy.

Follow Up From Erocker:

All are true and I can get you current news articles on any of them. Which one would you like to see? They are easy for you or any other person to find with the use of Google. I can see you are probably a farmer and don't want people to know the truth about this product because you are probably on the welfare program known as the ethanol mandate, and this would expose this program and all the lies that the ethanol lobby is putting out.

Reply from Marc:

Hi again E -

No, I'm not a farmer. There were no farms in Brooklyn when I grew up.

If you would read the report that I suggested you read, you would see that I am the co-owner and co-publisher of the Internet's largest automotive information resource.

I know that the misinformation you provided is easily available on the Internet, and that's why I wrote the 60+ page report plus dozens of others that respond to the lies.

The reason I asked you to read my report is because if you are simply reposting information that you have not verified yourself, then you shouldn't be doing it unless you work for the oil industry.

By the way, I do now grow some vegetables in my backyard, and I have a plum tree and a couple of orange trees. But I still think that this does not make me a farmer.

I look forward to your next reply—after you have read my report.

Follow Up From Erocker:

I have noticed a lot of bad information coming from you (sic) writings. If you write it in your book that doesn't make it true. It appears you are a book writer and need to sell your books and that is what you are about. Not interested in you (sic) book of lies.

Reply from Marc:

Hi E-

First, let me get a little housekeeping out of the way. You wrote "you" twice when what you really meant was to write "your." I'm only bringing it up because it makes you sound ignorant, and I know that's not the image you would wish to portray.

Let me turn my attention to your comments. You write "I have noticed a lot of bad information coming from you (sic) writings." Okay, could you point out what that information is, and could you provide references that set the bad information right?

Then you write, "It appears you are a book writer and need to sell your books and that is what you are about." Yes, I am an author in addition to being a filmmaker. But I'm not trying to sell you a book. The report that I referenced is completely free. You don't even have to register to get access to it—it's just there and FREE!

As a matter of interest, the book I wrote about was not free, I had to buy it. The author of that book, Robert Bryce, knew I was in the media but still did not offer me a complimentary copy.

Although you do not wish to avail yourself of the information I have provided, and you wish to be very combative, I still thank you for responding. At some point someone else will come across your original post, then they will read my reply. Then they'll read your responses where you refuse to look at what I have written. This will pique their curiosity, and they will read what I wrote (after all, it's free, so why not). When they read what I wrote and compare it to the lies you presented, they will laugh at you. And I will have accomplished what I set out to do, which is to inform people of the truth.

I hope you have a great rest of the week and a fun summer.

Follow Up From Erocker:

Good luck with your book of outright lies, gross exaggerations or irrelevant. Got to quit wasting my time with you (sic) nonsense.

Reply from Marc:

Thank you, and I hope your grammar and spelling lessons go well.

July 16, 2015

Posted By Liberty And Justice:

If energy content is not relevant, let's consider a "zero BTU" fuel, like water. Can you make an optimized ICE that runs on water and gets the same miles per $ or miles per gallon or whatever metric you choose? What about something in between a hydrocarbon and ethanol, maybe a carbohydrate. Can you make an equally efficient optimized ICE that runs directly on corn waste or pyrolysis oil?

Useful energy out = energy content * efficiency of converting energy content to useful energy.

If you have less energy content, you would have to have a greater efficiency in order to get the same useful energy out (which is proportional to things like "miles"). Is there a reason that ethanol would make it easier to develop an engine *more* efficient than (equal cost, to be fair) gasoline or diesel engines?

Reply from Marc:

Hi LJ - You're asking some questions that I can't answer for two reasons. First, I'm not a chemical or mechanical engineer. Second, I don't know if what you're asking even makes sense to ask.

However, I think you're trying to compare apples and hammers (I use hammers instead of the proverbial oranges because the comparison is far greater than just comparing two different fruits).

Water, in its normal liquid state, cannot, by definition, be used as a fuel for an internal combustion engine; it is not combustible. On the other hand, if you use the water to create hydrogen, the hydrogen could be used as a fuel in an ICE. Unfortunately, other than BMW, no other automakers seemed to be interested in building hydrogen-powered ICE vehicles, so this possibility seems dead.

The hydrogen could be used to power fuel-cell vehicles, and when and if issues related to batteries, etc. can be ironed out, fuel-cell electric vehicles will probably become the dominant form of passenger vehicles (in 40 or 50 or 80 years from now).

Follow Up From Liberty And Justice:

Over the short term, the amount of available U.S. farmland is relatively fixed. Very little of it is unused. And it is decreasing with time due to development. An acre of ground used to grow corn for ethanol is an acre not used to grow corn for food or used to grow soybeans or wheat or even hay crops which also feed into the food pipeline. If "ethanol corn" increases, food crops decrease. Generally, less food means higher prices for food. Why wouldn't this be the case?

And "profitable demand for corn for ethanol production" is profitable because of subsidies and mandates, isn't it?

Reply from Marc:

Hi again LJ - Thanks for your comments and question. America has 440 million acres of 'cropland,' and 939 million acres of 'farmland.' According to land definitions, farmland is not quite as good as cropland, but suitable for growing things such as specialized energy crops.

According to David Blume, who's one of the most experienced and knowledgeable guys in the world on ethanol and ethanol production:

> Of its nearly half a billion acres of prime cropland, the U.S. uses only 72.1 million acres for corn in an

> average year. The land used for corn takes up only 16.6% of our prime cropland, and only 7.45% of our total agricultural land. Even if, for alcohol production, we used only what the USDA considers prime flat cropland, we would still have to produce only 368.5 gallons of alcohol per acre to meet 100% of the demand for transportation fuel at today's (2007) levels. Corn could easily produce this level—and a wide variety of standard crops yield up to triple this.

According to the U.S. Energy Information Administration we have been using less fuel since 2007, so while this quote from David Blume may seem outdated, it is accurate today. Also, corn yields per acre have increased since 2007.

Corn crops have not impacted other crops. You'll notice that we've had no food riots in America and there is no shortage of corn-on-the-cob, canned corn, corn chips, or corn flakes.

Regarding the subsidy issue, no energy sector is more subsidized than the oil industry, and that industry is highly profitable. There was a fairly large subsidy for ethanol at the pump, but it was cancelled more than two years ago. E85 is still usually 50-70 cents cheaper per gallon than E10. E10 and E85 are far cheaper than ethanol-free gasoline.

Farmers grow corn because they can make a profit. For America it's better than growing crops that can't make a profit, which we then have to subsidize or even pay farmers not to grow them.

As far as mandates go, would you rather go back to leaded gasoline or gasoline with MTBE? Without tetraethyl-lead or MTBE or ethanol what would we use as the anti-knock agent in gasoline? What would we use to help reduce harmful engine emissions?

The concern about mandates is over-blown. For example, during Prohibition there was a mandate to use only gasoline, a fuel that is inferior to ethanol. And from the time of Prohibition repeal up to the time that ethanol was mandated as the additive to gasoline, we were virtually mandated to not use ethanol blends because the oil industry didn't want to use it. There was no freedom of choice until the government mandated ethanol.

I would like to see gasoline and petroleum diesel be discontinued to the greatest extent possible, but I'm still okay with having a choice as long as the petroleum oil choices don't include ingredients that are poisonous. If a consumer wants to buy ethanol-free gasoline, and they are okay with spending up to twice as much for it (as compared to an ethanol blend), fine. But I would like the oil industry to stop the ridiculous lies about ethanol. If their products are better let them prove it.

July 19, 2015

Posted By Charles W:

If ethanol in gasoline is a benefit, then why mandates and subsidies? How can it make sense to turn good food into poor fuel and consume non-renewable resources to do it?

Reply from Marc:

Hi Charlie - Most corn grown for ethanol in the U.S. is not intended for human consumption; it's grown specifically to be used for ethanol and/or animal consumption. Therefore, it's not a case of turning good food into bad food. You could ask, "Wouldn't it be better to grow it so that it is suitable for human consumption?" To which I would reply, "Why, have you had any trouble finding corn in your local grocery stores?"

The mandates have to do with the environment. Would you prefer to go back to leaded gasoline or gasoline with MTBE? If you don't use lead, MTBE or ethanol as anti-knock additives what would you use? Ethanol-free gasoline (which contains other ingredients to mitigate knock) often costs twice as much as E10. Is that your idea of a good solution?

Regarding subsidies: no energy sector is more subsidized than the oil industry. If you hate subsidies, then you should hate oil subsidies. In any event, one of the largest subsidies available to the ethanol industry was eliminated in 2013. E85 is still usually cheaper than E10, and in most instances far less expensive than ethanol-free gasoline.

I hope the foregoing helps you to understand the issues.

Posted By Scott R:

The real problem is that ethanol is not the answer to the problem. It reduces mileage by 20% (but based on ethanol percentage ~ 2% overall). It has driven the price of meat in the US almost to unobtainable levels because of the loss of corn for feed. Its use is a "feel good" patch so that Congress who would have a problem spelling ethanol, can feel like they have done something for the environment. If they really were serious, they would look at the technology of the power plants and focus on maximizing them. While they have made strides, they are far from going as far as they can. America has lost the will and ability to do the development needed and have left that to other countries. That is a national shame.

Reply from Marc::

Hi Scotty - Although it's already late Saturday, I hope the rest of your weekend will be lots of fun. I think that your desire to improve the technology at power plants is admirable. If you have or know of any workable solutions, I'd love to hear more.

Regarding mileage reduction due to ethanol added to gasoline, you are incorrect. Also, the food vs. fuel issue is not a real issue. It was only invented by the oil industry to denigrate America's potential to rely on ethanol.

I bring these points up because if you're like me, a patriotic American who wants what is best for our country, then I know you'll appreciate learning the facts.

Corn use for ethanol has no effect on food prices. The sole reason American farmers grow as much corn as they do is because there is the profitable demand for the corn for ethanol production. If this demand was not there, the farmers would not be growing the corn. In fact, they might not be growing anything.

I would prefer my fuel money to reward American farmers rather than foreign dictators and terrorist regimes.

In my opinion, Exxon/Mobil, like all petroleum oil companies, is a predatory organization whose only desire is to keep American enslaved by its poisonous fuels and destructive economic practices.

You'll probably have some questions, and you may even strike out at me as if I have been the one misleading you over this issue. However, I can assure you that I can provide point-by-point indisputable evidence to back up my position. All you have to do is be willing to read the material (or watch the videos, in some cases).

July 24, 2015

Posted By Granddad1:

If I am paying for gasoline, I should be getting 100% gasoline and not a watered-down substitute. Once again, Government intervening where they have no business to be. Article 1 Section 8 US Constitution outline Government responsibilities.

Reply from Marc::

Hi Granddad - You make an interesting point. It could be nice if you were able to buy 100% gasoline and have it work in your automobile. After all, when you hand over your dollar bills to pay for the fuel you are paying 100 cents per $1 dollar bill. It's not like you're paying with some bills that you printed that only reflect 90 cents per $1 dollar bill. Isn't that right?

The problem is that there is no such thing as 100% gasoline, and there's a couple of reasons for this. First, gasoline is not one single substance that is refined from petroleum; it's actually a composite of multiple ingredients.

Second, way back, about 100 years ago, as automobile technology and road building were improving, consumers wanted vehicles that would perform better and go faster. This required that vehicles need high compression engines. But gasoline couldn't be used in high compression engines because gasoline caused a hideous "knock" that would rip the engines apart. Some manufacturers, like Ford, built cars that could be manually adjusted to accept gasoline or an alternative such as ethanol. Ethanol (alcohol) doesn't produce the knocking problem that gasoline does.

Some fuel refiners solved the knocking problem by adding ethanol to the gasoline. But when Prohibition was enacted (largely because of Rockefeller providing money to pay off politicians to vote for Prohibition), alcohol was almost legally impossible to come by. This meant that a new fuel had to be created. Luckily for General Motors they had three very good scientists leading the charge. They added tetra-ethyl lead to gasoline, and leaded gasoline was born. The tetra-ethyl lead stopped the knocking and allowed vehicles to use engines that had exclusively high compression engines. And then, because tetra-ethyl lead is so corrosive, ethylene bromide had to be added to mitigate the corrosion.

From your prospective, Granddad, this would have left you in the same situation of not getting 100% gasoline for your 100 cent dollar bills. You were buying gasoline plus tetraethyl lead and ethylene bromide.

There was a worrisome issue with tetra-ethyl lead: it's poison. Handling it or just inhaling it could kill humans—and it did. GM and their fuel partner (Standard Oil, aka Exxon/Mobil) tried very hard to hide this fact from the public. They were successful for decades. It was similar to the tobacco industry paying off politicians and doctors to hide the health dangers of smoking and chewing tobacco.

Eventually the public and some politicians woke up and began to make changes. Tetra-ethyl lead had to be eliminated from gasoline, but then the knock problem would re-emerge. The oil industry could have chosen to use ethanol in place of the lead since Prohibition had been overturned long ago. The oil industry didn't like this idea because they were worried that people would remember that ethanol is inherently a better fuel. Also, the oil industry doesn't control alcohol production, so they would have to share some of the profits.

Consequently, the oil industry introduced their idea of a good solution: gasoline with MTBE (Methyl tert-butyl ether). MTBE is basically a substance that mimics ethanol's anti-knock characteristics. The oil industry used their financial muscle to get politicians and auto manufacturers to accept this new fuel. MTBE proved to be almost as bad as tetraethyl-lead, so it had to go.

By the way Granddad, this meant that you were again not getting 100% gasoline for your 100 cent dollar bills.

We finally came full circle and oil producers only had one alternative that was safe to humans and could be used safely in modern high-compression engines. That solution is ethanol.

For some reason, Granddad, no one explained all this to you. You were deceived by the oil industry. To make matters worse, they continued to deceive you by feeding you lies about ethanol. You think that you should be getting 100% of something that you wouldn't want to use if you could get it. Moreover, ethanol doesn't "water down" gasoline; that's not what happens. Ethanol also doesn't dilute the effectiveness of gasoline (if that's your meaning), because the addition of ethanol to gasoline raises the overall octane, unless the oil refiners purposely choose to dumb-down the gasoline's octane. They do that, hoping you will have an inferior experience using the ethanol-gasoline blend.

In your comments, you cited Article 1, Section 8 of the U.S. Constitution as a way to complain about ethanol-gasoline blends. I'm not a Constitutional scholar, so I looked up the source material. Article 1, Section 8 has nothing to do with public health and safety issues; it concerns taxes. And on that score, Article 1, Section 8 clearly allows the government to levy taxes, which we know the government does. I would say that if you wish to use the Constitution as a tool against ethanol, you should re-research the document.

I hope I was able to help clear up your misunderstanding of ethanol's use, and that you have a great rest of the weekend.

POSTED BY JEFF A:

Thanks for your historical analysis of ethanol. Now let's talk about the damage ethanol does, not to the engine's metallic parts, but the rubber ones. The cars of today are made with rubber specially formulated to handle alcohol. I believe Jay Leno's intent in writing his AutoWeek magazine story was that people should be able to buy unadulterated gasoline for older cars if they want it, so they don't have to replace the rubber seals in carburetors, fuel tanks, install new fuel pumps (the kind that bolt to the side of your engine), and let's not forget all the fuel lines. In Brazil they still went through the same painful transition we are going through today with our cars 15 years old and older. In Missouri you can get pure gasoline, but you have to shop around and its premium only. If I had an antique car, I would need this option to keep it all original.

Reply from Marc:

Hi Jeff - I'm glad you found my historical analysis helpful.

If Jay Leno was only referring to older cars, I wouldn't have published a lengthy rebuke of his article in February. I would have either not responded, or I would have said that I agree with the general advice of not using high level ethanol-gasoline blends in vehicles manufactured prior to the early 1990s.

Unfortunately, Jay didn't limit his criticism to older vehicles; and because Jay has never responded to the volume of criticism he's received by acknowledging that he only meant older vehicles, then I must assume he was making it as a blanket statement.

Quoting from the article I published on February 23, 2015 here, in part, is what I wrote:

> ...[T]he problem to me is that Jay didn't say, "Write to your legislators to demand more freedom of fuel choice to give us old car owners easier access to ethanol-free gasoline." Instead, he is calling for less freedom of fuel choice. More importantly, as much as I hate to say it, Jay is using information that is untrue and misleading to sway the argument. Since I think that Jay should know, and does know better, I believe he doesn't know what he is talking about in the *AutoWeek* story.
>
> For example, in the new *AutoWeek* story, Jay states that "ethanol will absorb water from ambient air, causing corrosion and inhibiting combustion."
>
> Ethanol doesn't absorb water from the ambient air. This lie is one of the oldest and most malicious one of the lies created by the oil industry to denigrate ethanol. The only thing new in how Jay used this lie is that he used the word *ambient*. I've not seen that before. I've

seen quotes that use the word 'thin' to denote ordinary air that we normally breathe, but not "ambient." This is not what occurs.

Many years ago, some clever oil industry person must have learned that ethanol (alcohol) is a hygroscopic substance, and that the general dictionary definition for a hygroscopic substance is that it can attract moisture from its environment. What the oil industry wag then did was to substitute the word *attract* with *absorb*, and *air* for *environment*. Thus, ethanol attracting moisture from its environment magically became absorbing water out of thin air.

What I'm getting at is just because you can play semantic word games with the definition of hygroscopic, that doesn't mean the result is relevant and correct.

To prove that alcohol will not absorb water right out of thin or ambient air, I always offer this simple at-home experiment: Fill any open container halfway with alcohol and place it on your kitchen counter. Allow it to sit for one or more days. If alcohol absorbs water right out of the air, then when you check the level of liquid in the ensuing days you would find it has risen. If you find that the level of the liquid in the container has risen (without any manipulation, change to the environment of your indoor kitchen, or interference to the natural process) and you can document it, I will pay you $1,000.

Incidentally, cotton is also a hygroscopic substance. So just as additional proof that being a hygroscopic substance doesn't mean that it absorbs water right out of the air, place a ball of cotton on the other side of your kitchen counter and see if it gets saturated with water from just sitting out in the open.

Moving on to one of Jay's other points, if you were to pour a gallon of water in your gasoline tank, your vehicle will probably have great difficulty starting. But that's not how water gets in your gasoline tank,

> unless you're very, very drunk when you go to the filling station. You can get water in your fuel system because of condensation. What do you do if you have some water in your fuel system? Do you stick a straw in and suck it out? No, you add a product like Dry Gas. Dry Gas is ethanol, meaning that you use ethanol to solve the problem of water in your gasoline tank. That's right, you add ethanol to solve the problem!
>
> Ethanol doesn't actually absorb water; it breaks water molecules down so that ignition and combustion of the gasoline can take place. The water molecules are then expelled in the exhaust. In other words, ethanol aids combustion, and not, as Jay stated, inhibits combustion.

Jeff, you write about the possibility of needing to replace and/or repair some of the components that might get damaged by the ethanol. There are two points I would like to make on this issue:

Speaking anecdotally, my personal experience is different and concerns older cars and different gasoline fuels. I owned a 1956 Bentley S1 that I drove regularly for more than 20 years. I purchased the vehicle when leaded gasoline was still the only gasoline fuel commonly available. My ownership lasted through the unleaded MTBE era and into the E10 era.

Friends belonging to the Bentley Owners Club members were very concerned when unleaded gasoline was proposed and mandated. The frantic claims were very similar to the frantic claims later made against E10. I never experienced any problems switching to unleaded gasoline with MTBE.

When MTBE was finally ordered off the market, I heard all the same complaints about the coming of E10. Fortunately, I personally never experienced any problems with using E10.

There was a time when I did have to replace the fuel pump and some of the hoses, but this occurred less than one year after I purchased the vehicle in England and had it shipped to the U.S. It was during the leaded-gasoline era. It was my understanding that this was the first time the car needed to have its fuel pump replaced. I never again had to replace those parts.

The second point is that all parts will need to be replaced at some point anyway, as I discovered with my Bentley. If it's necessary to switch out a part, switch to a part that is not susceptible to ethanol. I know that car collectors would rather not use non-original parts when possible, but that's the price that you pay for progress. The reason for mandating a switch away from leaded gasoline was to benefit all of society, not just a relatively minute number of car enthusiasts. The argument made to Jay Leno after his AutoWeek article was: Why should we have to be concerned with your collection? You should be concerned with us, the vast majority.

I'm sure you would agree that Jay can afford to change parts, and he can definitely afford to have non-ethanol gasoline shipped to his garage, if he doesn't already do so. It is my opinion that his beef and attempt to sway national policy is exaggerated, greedy and self-centered. I've briefly met Jay Leno on a couple of occasions at automotive events. That does not qualify me to know what kind of guy he is. However, I do have a very good friend who was helped by Jay; Jay went out of his way when he clearly did not have to. Based on this experience I would say that Jay was not a greedy and self-centered person. Combining that with Jay's other public articles and videos in which he has expressed great favor of ethanol and other alternative fuels, I was very surprised at Jay's about-face on ethanol. I believe it happened because Jay was looking for sponsorship from oil companies for his new automotive TV show.

July 27, 2015

Posted By Brian W

Mr. Rauch,

While you made some good points about ethanol, it looks like there are other automobile writers who disagree with you. I found this article from Edmunds.com (https://www.edmunds.com/fuel-economy/e85-vs-gasoline-comparison-test.html) while searching the Internet. They say that ethanol delivers poorer performance and a decrease in MPG. I was wondering if you would comment on it?

Reply from Marc:

Hi Brian - Thanks for your email and for taking the time to contact us. I hope you're having a very enjoyable summer.

The 8-year old story from Edmunds that you submitted is interesting but merely one of any number of anecdotal stories that could show one fuel is better than the other. For those not familiar with the Edmunds article, and who don't wish to read that article before continuing, I'll give a brief summary: Two of Edmunds' writers compared gasoline against E85 by driving a 2007 Chevrolet Tahoe LT roundtrip from Las Vegas to San Diego on two successive days. The first day they used gasoline, the 2nd day they used E85. The authors did not specify if the gasoline was E10 or some type of ethanol-free gasoline.

The result of their test was:

Gas Result: From San Diego to Las Vegas and back, we used 36.5 gallons of regular gasoline and achieved an average fuel economy of 18.3 mpg.

Gas Cost: We spent $124.66 for gasoline for the trip. The average pump price was $3.42 per gallon.

E85 Result: From San Diego to Las Vegas and back we used 50 gallons of E85 and achieved an average fuel economy of 13.5 mpg.

E85 Cost: We spent $154.29 on E85 for the trip. The average pump price was $3.09 per gallon.

Gas/E85 difference: The fuel economy of our Tahoe on E85, under these conditions, was 26.5 percent worse than it was when running on gas.

The Edmunds article closes with a comment about ethanol being energy negative (takes more energy to produce than it delivers).

Right off the bat, environmental conditions, such as the wind the authors talked about, could have easily accounted for any difference. Unless the wind blew in precisely the same directions, for the same length of time, and exactly the same speed (not just close to the same speeds), you can't make an accurate comparison between the two roundtrips.

In addition, as the writer pointed out, there were different traveling speeds in the outbound and inbound drive. While you can say "X will make up for Y" in a general sense, you don't really know if X plus a wind speed of 10 MPH can be directly offset by Y plus the same wind speed. And this doesn't even account for how the vehicle burns gas at the different speeds. It's possible that if the vehicle's optimum MPG efficiency is attained at 60 MPH, 65 MPH could produce significantly different MPG results than 55 MPH, even though both speeds are 5 MPH different than the optimum. Over a protracted distance traveled, even the slightest difference can add up to an alarming difference. Unless that distance is traveled on a regular basis by an individual, the difference could be irrelevant in their consideration of which fuel to use.

In my estimation, a more correct test would be to have two identical vehicles, using different respective fuels, with equal payloads traveling alongside each other for the entire distance, whatever the distance might be. Either vehicle would have to avoid any "drafting" assistance that might occur if either vehicle followed another vehicle too closely. On the other hand, a pure engine bench test with equally optimized engines - one optimized for 100% alcohol and one optimized for 100% unleaded gasoline is probably the only way to do it. In my understanding, any bench tests conducted of optimized engines have shown that the MPG efficiency is equal.

I think the important point to learn from the Edmunds test is that the results were quite similar, and that the writer did not report that the vehicle suffered any damage to its engine due to using alcohol. As you may know, one of the pernicious "manufactured" complaints against the use of alcohol in a "gasoline" engine is that alcohol will damage it.

One more point on the Edmunds' test: If you go to a filling station that sells E85, you may notice a little sign that reads that the E85 may only contain 70% alcohol, not 85%. I don't know enough to be able to say if the difference in the 15% could produce better or worse results, but it seems likely that there would be some difference. If the vehicle was adjusted to run as an E85 flex-fuel vehicle but it was only being fed E70, it is conceivable the fuel economy would be affected.

The Edmunds story was written shortly after David Pimentel and Tad Patzek issued a very derogatory report on ethanol.

They claimed ethanol was energy negative. This report was commissioned by the oil industry, and the oil industry had their PR people working overtime to get the results out to the media. I would say that the two Edmunds writers were relying on the Pimentel-Patzek report, or a third-party report of the report.

As it turns out, the Pimentel-Patzek report was severely criticized by several other studies that showed the Pimentel-Patzek methodology and findings to be very wrong. I discuss this in greater detail in the lengthy report I wrote in June 2013. You can find this discussion in Part 5. (https://www.theautochannel.com/news/2013/06/11/082057-gusher-lies-book-review-and-reply-to-robert-bryce-pt.html)

I hope I have answered your request and addressed your issues.

August 9, 2015

Posted By Competition Accessories:

What I want to talk about is ethanol fuels and what it's doing to motorcycles. First off, pure ethanol is hygroscopic; it attracts water, to the point that it will pull it out of the air. Ethanol and gasoline will mix, but ethanol, gasoline and water will not; the ethanol-water mixture will come out of solution and settle on the bottom of your tank or carburetor bowl. Add a little oxygen to the mix, and you get rust. The water can also sink to the bottom due to its heavier density causing the carbs to become gummed or have a solidified gel/varnish that will also damage the carbs.

Reply from Marc:

Hi CA - You are correct, ethanol is a hygroscopic substance. However, what you've done, as others have and you've followed them, is to misuse the meaning of the word "hygroscopic" to attribute qualities to ethanol (alcohol) that do not exist. And this is where the problem lies.

Different substances are hygroscopic. In addition to ethanol, salt, cotton, diesel fuel, and even gasoline are hygroscopic. And the different hygroscopic substances react differently. If you leave salt in an open container it will clump up because of moisture in the air, but water does not form at the bottom of the saltshaker. In a general full definition of hygroscopic, a hygroscopic substance can attract moisture from its environment. You have wrongfully translated that to "ethanol sucks water right out of the air." The word "environment" means "near" or "adjacent," it doesn't mean "air." The word "attract" doesn't mean "absorbs" or "sucks."

There are two ways to prove this: Fill a drinking glass halfway with alcohol and leave it on your kitchen counter overnight (or a couple of days). If alcohol could attract water right out the air, the drinking glass would have more liquid in it as the days pass. But it won't have more liquid (it'll actually have less liquid). Or place a dry cotton ball on your kitchen counter and let it sit overnight. The cotton ball is a hygroscopic substance. If it could suck moisture right out of the air, the cotton ball will get wet. But it won't. However, if you place the dry cotton ball "adjacent" to a small puddle of any liquid, it will absorb the liquid in a wicking process.

For some reason you have forgotten, or never learned, that if your gasoline tank developed some water (usually through the natural process of condensation), you would use a product such as Dry Gas to solve the problem. Dry Gas is ethanol. If ethanol caused water to form, Dry Gas wouldn't be used to get rid of the water. Ethanol gets rid of the water problem by breaking up the water into molecules and allows ignition to take place. When the engine runs, it expels the water molecules in the exhaust. The gumming up and "varnish" comes as a result of the burnt gasoline residue. The ethanol works as a solvent to get rid of the gasoline residue and varnish. The result may be to turn the gasoline residue/varnish into a gummy substance, but over the course of time any internal combustion engine will experience this problem and need to be cleaned. You would then use an alcohol-based product, or a product that emulates alcohol's solvent characteristics to get rid of it. This has been true since the birth of internal combustion engines. The whole point of having a gasoline that has a "detergent" is to do what ethanol does...but ethanol does it better.

If you have only ever used ethanol-free gasoline, then your engine has a lot of residue. If you suddenly switch to an ethanol blend you may find that there is an overwhelming amount of gunk that is just too much for the usual running engine to handle on its own. But blaming the problem on ethanol would be like blaming a clogged drain in your bathroom sink on the soap for doing too good of a job. In any event, once the engine is cleaned out, the ethanol blend will help keep the engine clean. To better explain the "varnish" effect, you will notice that if you ever worked with a piece of finished wood, and wanted to remove the varnish, you would use an alcohol-based solvent to remove it. If alcohol caused the varnish to occur, you wouldn't use alcohol to remove it.

I've owned and ridden motorcycles for a long time. I have never, ever, experienced the separation problems that you and other ethanol opponents claim will happen. If you are experiencing problems, it is because the problems are caused by the gasoline and/or are just normal problems that will occur in any internal combustion engine.

Follow Up By Competition Accessories:

You may be the one rider that has not experienced any issues with ethanol fuel. The American Motorcyclist Association (AMA) is currently fighting to stop the mandate for E15 fuel blends because it is such an issue for small carbureted motors. With over 25,000 signatures already, ethanol fuel is clearly an issue. We can discuss the semantics of fuel separation and ethanol properties all day, but let's focus on the real problem instead. Ethanol fuel causes carburetor failure.

Reply from Marc:

There is no E15 mandate to fight; any claim by you or others is simply a scare tactic. The EPA has only ever approved E15 for use, they haven't mandated it.

The AMA may be receiving some financial inducements from the oil industry to go against E15.

Additionally, the only semantics issue is your misuse of words and definitions. The facts are the facts. Do the two tests

I suggested. I've owned more than one bike. Moreover, every day I see plenty of motorcycles whizzing by and they have no problems with E10, and where I live/work ethanol-free gasoline is not easy or cheap to come by. You have exaggerated the problems and blamed the wrong fuel for those problems. And it's clear that I am not the only motorcycle owner/rider to not experience problems with ethanol. All you have to do is read some of the other comments on your YouTube message board.

FOLLOW UP BY COMPETITION ACCESSORIES:

You are entitled to your opinion and in the auto industry the consensus may be different, but this is a moto channel. I'll stick by the 25,000 American Motorcyclist Association members that also have an issue with ethanol fuel and continue to warn of the shortcomings.

REPLY FROM MARC:

CA - Calling facts "just an opinion" is a goofy way out of the discussion.

There are more than 5 million licensed motorcycle riders in America (and more than 8 million registered motorcycles). Having 25,000 signatures represents less than 1/2 of 1% of licensed motorcycle riders. If the problems exist in the manner in which you claim, there would be signatures of far more than 1/2 of 1% of all licensed riders. Statistically speaking, you have virtually disproven your claim that ethanol routinely causes problems in motorcycles.

FOLLOW UP BY COMPETITION ACCESSORIES:

Here is a link to an article that the EPA admits ethanol fuel causes damage to engines not designed for blended fuels. It is true that many cars can and do produce higher power output with ethanol fuels. In the case of motors not designed to operate on blended fuels it can cause major issues. Our service department regularly services motorcycles and ATVs that have failed fuel systems as a direct result of blended fuels. Renewable fuel sources are important for our future, but an increase to E15 is not the answer.

http://www.americanmotorcyclist.com/News/14-04-21/EPA_acknowledges_that_ethanol_damages_engines.aspx

Reply from Marc:

And here is the 2010 press conference where the EPA announced that E15 and E20 was safe to use in all modern gasoline powered cars and trucks:

https://www.youtube.com/watch?v=6E0uZvtfFrg If your service department is intentionally describing the wrong causes of customers' problems and exaggerating the repairs necessary, it may be fraud. If your service department is simply misdiagnosing and treating the problem, you wouldn't be the first service department to not know what they are doing.

August 16, 2015

Posted By Pywaket 1:

This guy, Marc Rauch is a known ethanol industry apologist. Read his history (before he hides it) and you'll see that it's practically the only thing he ever posts about.

Reply from Marc:

Hey Py,

I hope you're having a great weekend.

I'm not an apologist. What would I have to apologize for?

I'm an investigative journalist, a TV-film producer, and co-owner and co-publisher of The Auto Channel (the Internet's largest automotive information resource). Unlike you and many of those who post wrong information about ethanol (and lots of other things), I don't hide behind little child screen names. My articles about ethanol have been widely published; I've been a guest many times on radio and TV shows; I've given presentations at all kinds of seminars; and I appeared before a Congressional committee to talk about ethanol.

You can Google my name or just visit my website and do a search for my editorials. I write about all kinds of things, including wine, cooking, and baseball. The best piece I've done on ethanol is a 60+ page report on ethanol that destroys all anti-ethanol myths.

I'm happy to take on anyone at any time, and I'll even publish whatever comments you care to make along with my corrections on The Auto Channel website. If you know anyone working at a petroleum oil company or any of the PR firms that work for the oil industry, please see if they have the courage to debate me on the issues.

Follow Up By Pywaket 1:

1) My name is Pywaket, not "Py", thank you very much. And yes, that is my real name. Shall I refer to you as Ma, from now on?

2) I would think that an investigative journalist would know the difference between "your" and "you're."

3) Any idiot can appear before a Congressional committee if they're friendly to powerful people with an agenda (Archer Daniels Midland/ADM, anyone?). If there's any group that has nearly as much influence in Congress as Big Oil, it's Big Ag, and you are definitely a shill for them.

4) Your online history largely shows you arguing with people about ethanol. People who are clearly unhappy about being forced to use a product they don't like, don't want, and that causes them problems. I trust them more than some self-important "journalist" who goes around telling them that they don't know what they are talking about. Especially when I've experienced the same problems they have.

4a) I fly airplanes. Many of the engines I've flown behind can be approved to run on Mogas, which is automotive gasoline. Every single one of those, though, says "no ethanol in the fuel," right in the STC that was approved by the FAA. I'm supposed to trust my life to your inane ramblings over the knowledge of the engineers and designers at Lycoming, Continental and Bombardier/Rotax? I don't think so.

5) "Widely published" means nothing, as does "being a guest". Most newspapers and television stations are desperate for cheap filler material, and you fit the bill.

Reply from Marc:

Hey Py, pretty good response.

1) Yes, you can refer to me as Ma if you want.

2) Good catch on my your/you're spelling error; thanks for the heads up. But if spelling was important to you, then you should have picked up on your misspelling of the word "practically" in your first post.

Also, as long as you're going to be so persnickety about common typos, then perhaps you should have been more careful in using the word "shill" to describe me in your second post. A shill is someone who pretends not to have any relationship with the person/people being touted. I never hide what I do. I would have thought that someone with sufficient intelligence to fly airplanes would own a dictionary.

3) There are a lot of idiots who do appear before Congressional committees. Those who testify against ethanol are great examples. I laugh at your comparison of ADM versus the oil companies. It's like comparing a little league baseball player to a major league player. They both play baseball but there's a huge difference between them. The amount of influence peddled by the oil companies greatly overshadows the influence ADM wields.

For example, in the same year that ExxonMobil reported their fiscal fourth quarter profit as $40 billion, Archer Daniels Midland reported their fiscal fourth quarter profit of $372 million. Although $372 million is nothing to sneeze at, it's less than 1% of ExxonMobil's profit. Corporate revenue and profits are often tied directly to political influence. Can you guess who has the greater amount of influence with Washington? (Rhetorical question, no need to answer)

And since any idiot can appear before a Congressional committee, when were you last there?

4) Yes, I do argue a lot with people who are spreading lies about ethanol. Some are true shills for the oil industry, and some are just people who have no idea what they are writing about. I believe that fuel independence is of primary concern to America and that ethanol can be a major player in ending dependence on foreign oil and restoring the health of our economy. I believe that we have lost too many military men and women fighting to support foreign terrorist regimes. Excuse me for caring. Excuse me for being vigorous in how much I care.

By the way, I hope you noticed that NO ONE is able to provide information that refutes the facts I present.

4a) I don't get your reference to airplane engines not using ethanol. You must know that planes can and have used ethanol. The reason the FAA has restricted its use is because the airplane engine industry is so small that they were not affected by the regulations regarding ethanol resistant parts. The government just didn't feel that planes were significant enough to worry about. It's simply a matter of swapping out parts. The Auto Channel has a great video of stunt pilot Greg Poe flying his ethanol-powered plane at the California Capital Air Show.

(You can watch it at: *https://www.youtube.com/watch?v=ierFrs5FkAU. And if you search Google you'll find other examples.*)

I don't think you'll find any comments from me suggesting that pilots should use ethanol in their planes. So, I don't know why you felt it necessary to exaggerate and pretend that anything I've written pertains to your safety while flying.

5) This point is sort of like your Point 3. Yes, media outlets are hungry for cheap filler material. That's one reason why there is so much junk published in favor of gasoline and the oil industry, even though both have proven to be poison to humans and other animal life.

But you shouldn't be too harsh on your criticism about the content that media outlets accept. After all, if it wasn't for their low standards, you would never get your empty comments published.

I am looking forward to your next try, because I think this could prove to be the most entertaining exchange yet on my Ethanol Chronicles blog.

August 23, 2015

Posted By Wrencher:

Corn based ethanol is indeed a waste of time and resources. Even the studies that show that you have a net gain of energy only have it at a razor-thin margin. But there's a lot of cash behind this whole thing and nothing speaks louder than $$$$. It's actually very strange since now you have environmental groups teaming up with oil companies to combat the ethanol lobbyists. Shared goals make strange bedfellows.

That said, and since the sugar industry is arguably more influential than the corn industry, why aren't we making biofuels out of cane sugar like they do in Brazil? The energy density of the final product is much higher and sugar cane produces more than double the amount of ethanol per acre of land. I know there are challenges to growing a tropical crop in a largely temperate region but if you've driven past a cornfield lately you'll know that we're pretty good at getting plants to do what we want.

On a personal note, I run my V-Strom (Suzuki motorcycle) on whatever's cheapest. It hasn't complained yet. E10 seems to have a largely negligible effect on performance and fuel economy.

Reply from Marc:

Hi Wrencher - I read through your comments and found them to be an interesting mix of positive and negative. On the one hand you have found that the use of E10 has no ill effects on your motorcycle, which mirrors my own E10 motorcycle experience, but on the other hand you make negative statements about EROEI (energy returned on energy invested).

I'm sorry to say that you're wrong about the corn's net energy gain or loss. The information that you're using to base the opinion on was incorrect in some instances and outdated in others.

The real point to make is that no "energy" resource comes without the expenditure of energy, and gasoline is terrible. In fact, gasoline takes more energy to produce than ethanol. The studies done to counter the Pimentel-Patzek hatchet job show this. Google Bruce Dale and you'll find this information. Dale participated in a televised debate against Pimentel and Patzek and you can watch it in its entirety here: https://www.c-span.org/video/?188549-1/ethanol-energy-policy

It is true that some environmental groups oppose ethanol, and it's primarily because they are still relying on the wrong EROEI information.

In addition, the "energy density" between ethanol and gasoline, or ethanol produced from sugar vs. corn is irrelevant. MPG is a result of energy optimization, not fuel BTU content.

Reply From Simon John (To The Comments Made By Wrencher):

Brazil's ethanol fuels are based on sugar cane and are quite different. Mechanics all agree ethanol causes problems particularly in older vehicles and should never be used in boats.

Reply from Marc:

SJ - Fuel ethanol is fuel ethanol. It doesn't matter if it's made from corn or sugar or beets or sorghum. To suggest that an engine and fuel system parts react differently is ludicrous. The engine does not say "Yech, this tastes like bourbon, I want rum."

Regarding your other incorrect admonishment of using ethanol-gasoline blends in boats, you might want to read/listen to the Mercury Marine Ethanol Webinar. The full webinar can be found at:

https://www.youtube.com/watch?v=GbkmY7TMKMg.

An abbreviated version of the Mercury webinar below:
http://www.boatus.com/magazine/2011/december/ethanol.asp

August 26, 2015

Posted By Dennis:

There's no ethanol-free gasoline in Houston. I have to ride 50 miles to the coast to find it. It is a ridiculous law perpetuated by graft from large agriculture lobbyists. Given the current level of corruption, I am afraid it will be here for a while.

Reply from Marc:

How about the political graft and corruption from the oil industry that forced us to use only poison gasoline for eight decades?

Follow Up From Dennis:

Are you talking about leaded gasoline?

Reply from Marc:

Leaded gasoline and then gasoline with MTBE.

Follow Up From Dennis:

They were both used until the unintended consequences caused them to be banned in this country. Still widely used in Asia. Next.

Reply from Marc:

Hey Dennis, thanks for the continued exchange.

I'll respond to your last point first. They also eat dogs in Asia, and I assume you've seen photos of the air in China.

By unintended consequences, do you mean the acknowledgement that lead and MTBE are poisonous, or do you mean the rigged OPEC supply problems and our dependence on foreign oil, or do you mean both?

NEXT, is right!

Follow Up From Dennis:

I would agree with all of your points other than the poisonous nature of MTBE, which is still under investigation. I am still not a lover of ethanol due to it causing food prices to rise.

I would love to use something besides gasoline, but nothing is available yet. Unless your electric vehicle gets its electricity from wind, solar, or geothermal, you are just transferring the CO_2 from your tailpipe to a smokestack. The consumers are not the enemy.

Reply from Marc:

Dennis, I was happy to receive this response from you. I was thinking you were going to tell me that dog tastes good and that we should eat them here, too.

I'm also glad to see that you didn't try to argue tetra-ethyl lead's dangerous qualities.

Anyway, the inhalation of MTBE is dangerous and causes cancer. There is no dispute on this. The uncertainty concerns how dangerous it is in drinking water. Now I acknowledge that drinking water contains traces of arsenic, and yet it doesn't routinely kill humans; and you can have minute quantities of chlorine bleach in drinking water and it won't kill humans, but why take the chance with MTBE?

You can argue that ethanol might also get into the drinking water supply, but ethanol spills evaporate before it can seep down into aquifers and drains. And humans can safely consume rather large quantities of ethanol (alcohol). Drinking fuel grade ethanol will make you sick and probably kill you, but that's because of the denaturing ingredient added to ethanol to stop people from drinking it.

Your only remaining argument is the food prices canard. When the World Bank first issued the report in 2008 blaming ethanol production for food scarcity and price hikes, it caused a great sensation, and the oil industry jumped all over it.

Since that time the World Bank retracted those statements on multiple occasions and placed the blame of rising food prices squarely on the shoulders of the oil industry. The myth continues to be spread because it's always much harder to retract a story after it goes viral. The oil industry paid many lobbyists to influence media personalities and publications to carry the story. When the World Bank retracted the blame, the oil industry didn't pony up the money to retract the story; they've continued to ride the wave ever since.

As for electric vehicles, no, that's not the solution and it probably won't be for 30, 40, 50 or more years. Ethanol is the solution. Methanol is a good runner up, and CNG is another good solution.

You are correct about American consumers not being the enemy. I just wish the American Petroleum Institute and its members didn't act as if we were.

September 17, 2015

Posted By Scott B:

Hi Marc - I'm a fairly new convert to E85. I started using it when my employer gave me a nice Ford F150 to drive for my work duties. Being a flex-fuel vehicle, and my being a curious person, I started using E85. I had all the same misconceptions that everyone else has. But I had to know for myself. Needless to say, I'm all on board now. When the truck kept running and the sky didn't fall down around me, I started reading everything

I could get my hands on about ethanol, which is when I came across your blog on the Auto Channel web page.

I enjoy what you have written. It's a learning experience for me and I'm enjoying it. And once you start paying attention, it's amazing what you start hearing others spew with nothing but hearsay to back it up. Big oil has brainwashed us well.

Now I know I'm not supposed to, but I have been using it in my 2011 BMW GS motorcycle, as well as my wife's 2013 Honda CRV. The CRV did drop MPG from about 32 to 22. And the check engine light came on after about 50 miles. But I knew this would probably happen. Of course, it went out after going back to E10. The bike LOVES the stuff. It wants 91octane anyway. Much cheaper this way.

Here in Minnesota, E85 is usually $.70 a gallon less than regular. You would think it would be less since we make so much of it here. Still, I paid $1.65 per gallon just this afternoon when I filled the tank.

Keep up the good fight. I know I will. Thanks for your writings. I'll keep reading.

Reply from Marc:

Scott, thanks for reading the blog and for writing in. Your experience is pretty standard; and you're right when you learn that the sky doesn't fall and your teeth don't fall out simply from using E85, it's a rather amazing feeling.

Using E85 in non-flex fuel vehicles does produce different MPG results, based upon the vehicle. The on-board computer isn't sure what to do. You may find that by "splash blending" your own mix to achieve something around E30 or E40, it will work better and give you a positive Cost vs MPG loss result. Splash blending is when you pump in X amount of E85 and then move to an E10 pump and fill up the rest of the tank.

At some point, you may wish to install an Alcohol Fuel Conversion Kit on your wife's car. The kit will help you to achieve better results and it can be used with E100. The kit can be purchased from David Blume's company at

http://www.permaculture.com/store/conversion_kits.

David's book, *Alcohol Can Be A Gas,* is fantastic. You can read about it and order it from his website.

In general, the retail cost for E85 should be less, but it's pretty much controlled by the oil company that's delivering the fuel. They don't want to hurt their regular gasoline sales. You may be able to find an ethanol producer in Minnesota that is selling ethanol directly to consumers at their plants, and that price might be more advantageous.

SEPTEMBER 26, 2016

POSTED BY GREG W:

AAA states that average gas price in my area is $2.34, with E85 offered at $1.99. Nationwide it's $2.47 and $1.97. One gallon of gasoline yields 116,000 BTUs, (British Thermal Unit). One gallon of corn-based ethanol yields only 76,000 BTU's. Let's take the lesser of the ethanol prices, $1.97, and multiply that by 1.53 = $3.01 is the over-under or break-even point of ethanol.

Average car drives about 1,000 miles per month (250 miles per week). If the car holds 15 gallons of gasoline and gets 25 MPG, the range per tank would be 375 miles.

If the same car uses E85 and gets only 65% of that typical mileage, then the range per tank is 225 miles. That's over four fillings per month. However, over a protracted period, let's say one year, you would have to fill your tank ONLY 22 more times with E85 than otherwise, (54 fill-ups vs 32 fill-ups). That's one more fill-up every 2 weeks.

Nationwide now:
with gasoline
12,000 miles per year / 25 mpg = 480 gallons
with ethanol
12,000 miles per year / 15 mpg = 800 gallons
800 x $1.97 per gallon = $1,576.00 per year with E85
$390.40 difference.

REPLY FROM MARC:

Greg - BTU ratings are irrelevant when considering fuel in an internal combustion engine; we're not boiling water. You should lose any references to BTUs in attempting to calculate fuel economy in passenger cars and trucks.

To simplify things even more, throw out any reference to AAA, unless you want to discuss where to get route maps. With today's GPS devices, AAA isn't even needed for that.

Your calculation of E85 getting only 65% of the mileage of E10 or E0 is based on the nonsense BTU issue, which has nothing to do with an internal combustion engine.

In practical, actual driving experience, E85 will only lose between 5% and 15% of MPG compared to E10 or E0, depending upon the vehicle. If you lose 10% in miles, but the price of E85 is 20% less than E10 or E0, then it's a net gain. If you lost 20% in MPG but the E85 is 20% cheaper then you break even.

I won't go through all the math on various prices and mileage losses, but it's clear that the scenario you painted is incorrect.

Posted By Hope-Full-Lee:

I can just see God saying, "Why are you people burning FOOD in your cars, I GAVE YOU OIL!!"

I have three beautiful children that I don't want to see go hungry at some point in their lives.

Reply from Marc:

What I think God might be really saying is "Why are you people allowing your young men and women to be killed fighting wars over oil that is controlled by countries that hate you? I gave you the ability to grow things that can be made into useful products. Why are you using corn to make unhealthful sweeteners, when corn can be used to produce fuel to power your cars and make the air clean?"

Maybe this is what you should be concerned about, considering that you have three beautiful children to protect and take care of?

Posted By Rick:

ADM (Archer Daniels Midland) lobbies and spends millions to sway (bribe) Congress to vote in ways that benefit

ADM—such as [using] corn for ethanol, which ADM grows. You are correct. The average citizen is no longer represented by Congress [and] hasn't been for just about a century. Congress is for sale and the buyers are shills for big corporations.

REPLY FROM MARC:

Rick - The oil industry spends far more money on lobbies than ADM; and the oil lobby has owned Congress for more than 100 years. What ADM spends is just a fraction of what the oil industry spends. Corn used to produce ethanol in America is grown by American farmers. The oil industry is controlled by foreign oil producers and terrorist countries. No American serviceman or woman was ever killed defending the production and distribution of ethanol.

Why is there a disconnect with you and others on this? Are you patriots or not? Do you know how to add and subtract? Don't you care about U.S. military personnel being killed?

By the way, Rick, even if you're a shill for the oil industry (it's not relevant if you are or not), answering these questions, if you are an American patriot, should elicit only one response.

SEPTEMBER 29, 2016

POSTED BY LMAC N:

Ethanol has FAR less energy-density per gallon than gasoline has. A perfect, stoichiometric ratio of air-to-fuel for gasoline is 14.7:1. For ethanol it's 9:1. In other words, it takes about 40% MORE ethanol to generate the same amount of energy—you need 1.4 gallons of ethanol to go as far as one gallon of gasoline will take you. So, which is a more efficient use of resources?

And speaking of efficient use of resources—corn is one of the WORST ways, in terms of gallons-of-ethanol-produced-per-acre, to make ethanol. It takes a LOT more corn to make a gallon of ethanol than it does, say, switchgrass or sugarcane (about twice as much, actually). Do you not think that corn-growers can be just as greedy as the oil companies? Do you not think that they have lobbyists in D.C. that are pressuring the government to require corn-based ethanol so that THEY can get rich? Is it only the oil companies who are screwing the consumer, or are corn farmers capable of doing it, too?

IF we were able to produce ethanol from any cellulosic plant material, THEN we would achieve an efficiency that would rival gasoline in terms of the cost-per-unit-of-energy-obtained. But we cannot do that yet. Technology will get us there eventually, I'm sure. But until then, ethanol is NOT the panacea that you make it out to be.

For the average car on the road, ethanol DOES do damage in multiple ways. It's EXTREMELY corrosive when compared to gasoline and eats away at aluminum fuel tanks and rubber fuel-hoses. Most cars on the road today do not have the stainless-steel fuel-tanks and fuel-lines needed to handle ethanol.

Reply from Marc:

Hi LMAC - As far as your "stoichiometric ratio" is concerned, you should stow your metric-ratio. It serves no purpose in this discussion.

The BTU rating of gasoline over ethanol has nothing to do with how engines perform. An engine optimized for ethanol delivers more horsepower per gallon of fuel than an engine optimized for gasoline. This issue has been erroneously raised numerous times on this blog. Repeating the erroneous information doesn't make it true, and my having to repeat the correct information multiple times doesn't make it false.

While corn is not near the top of the list in raw materials that can be used for ethanol, there are several good reasons why it is being used. The amount of acreage being used to grow corn is irrelevant, since we have enough land at this time.

In the future, other crops with better ethanol yields will undoubtedly come into higher use in the U.S.

Older cars (pre-early 1990s) may have components that are susceptible to alcohol deterioration. However, any pre-early 1990s cars on the road are susceptible to all kinds of problems anyway, regardless of the fuel used. When repairs are made, the new parts used should be alcohol resistant parts.

Rubber and plastic are generally not susceptible to alcohol, but the same can't be said of gasoline, which is one of the reasons why fuel system repair didn't just start when E10 fuel was mandated. One of the most corrosive liquids is water. It can be rendered rather harmless in engines and other places by using materials that are not subject to easy water corrosion.

Follow Up By Lmac N:

A stoichiometric mixture is determined by the laws of physics. Engine "optimization" has nothing to do with it. It requires 40% more ethanol than gasoline to make a stoichiometric mixture. Period. End of story. You cannot get around the laws of physics. Miles per gallon will ALWAYS be worse on ethanol than it will on pure gasoline. Period. That is part of the laws of physics. There is simply less energy in a gallon of ethanol than there is in a gallon of gasoline, no matter how much you may wish it to be different.

Reply from Marc:

Fuel optimization has everything to do with the issue. BTU ranking is irrelevant. This has been known for more than 100 years and was testified to before Congress in 1906.

If BTUs and not engine optimization were important, you would be able to achieve better mileage by using diesel fuel in a gasoline engine, which has a higher BTU ranking. As it relates to internal combustion engines, you can take your understanding of physics and use it as a physic.

Follow Up By Lmac N:

It takes X amount of work to move a car from point A to point B. Period. Doesn't matter what fuel you use. Doesn't matter how the engine is optimized. It takes X amount of work to get a car from point A to point B.

Ethanol has LESS energy available per gallon. This is a KNOWN FACT!! It is a LAW of PHYSICS. You cannot escape this fact! Gasoline has 114,000 BTU per gallon. Ethanol has 76,100 BTU per gallon. THIS IS A FACT!!!!!!!!!! LOOK IT UP!!!!!!!!!

Therefore, if ethanol has less energy available per gallon than gasoline does, it WILL take MORE ethanol to produce the same amount of work!!!! You CANNOT deny that! It is a mathematical FACT!

Using more ethanol to go from point A to point B means FEWER MILES PER GALLON OF ETHANOL!!!!! ALWAYS!!!!!!!! PERIOD!!! END OF STORY!!!!!

It doesn't matter how the engine is "optimized." The engine cannot change the amount of available energy in the fuel!!!! There is only so much energy available in a gallon of fuel—and the amount of energy available in that gallon of fuel is dependent upon WHAT KIND OF FUEL IT IS. It is NOT dependent upon how well the engine is optimized!!!!!

Pouring fuel into an engine does NOT change the characteristics of the FUEL!!!! No matter HOW the engine is "optimized"!!!!!!!

Reply from Marc:

It's not a question of the Laws of Physics, it's an issue of engine optimization.

In any event, the so-called Laws of Physics can be mitigated or rendered irrelevant by external forces. The use of pulleys to lift great weights is one example. You can make up an equation that expresses the energy required to lift a 10-ton cement block, but the use of a pulley makes the task nearly effortless.

A gasoline-powered internal combustion engine is optimized to run on gasoline. A diesel-powered internal combustion engine is optimized to run on diesel. Diesel's BTU rating is higher than gasoline, yet if you attempted to use diesel in a gasoline-optimized ICE, it won't run. Conversely, if you attempted to use gasoline in a diesel-optimized ICE, you don't get reduced performance, you get NO performance.

If we were discussing a steam engine and we had to calculate which fuel to use to heat the water to create the steam, then BTUs are important. Internal combustion engines do not work on steam.

Pouring fuel into an engine does change the characteristics of the fuel. It transforms from a liquid state. I understand your point on this and agree. However, an engine's characteristics can be changed to fit the fuel, and this is precisely what happens.

The fact that ethanol or methanol can be used at all in a gasoline-powered engine is incredible. It's normally perfectly correct to assume that anything used in a device that is not optimized for that thing will get reduced or zero results. But it has nothing to do with the Laws of Physics or with one substance being inferior to another. It's OPTIMIZATION. Any conversation about the difference in BTUs is a waste of time.

A Beatles song on CD, DVD, digital, 8-track cartridge, or vinyl will be the same. The song is the same; the lyrics are the same; the music is the same. But just because the song and its components are the same doesn't mean you can expect to use a CD in an 8-track player.

Two facts remain: 1. An internal combustion engine optimized to run on ethanol will outperform an internal combustion engine optimized to run on gasoline. The ethanol engine will provide greater MPG and more horsepower. This is a fact, and you can look it up.

2. If you use E85 in a gasoline-optimized engine and lose 10% MPG, and the E85 costs 15% less than the E10 or E0 gasoline, then you have a net gain. This is a fact, ask a math teacher to explain it to you.

By the way, the information that I present is nothing new, it has been known since the advent of the internal combustion engine, but the oil industry has done a very good job on confusing people like you. As I've mentioned previously, these facts were also testified to in Congressional hearing in 1906 and the oil industry was never able to refute the information.

Follow Up By Lmac N:

THE AMOUNT OF WORK IT TAKES TO LIFT A ONE-TON BLOCK IS THE SAME WITH OR WITHOUT A PULLEY!!!! If you lift a one-ton block 1 meter off the ground, you have applied the same amount of energy to that block with or without a pulley!!!! With a pulley, it took twice the amount of TIME to do it, therefore your energy PER SECOND decreased, but the TOTAL AMOUNT of work done IS THE SAME!!!!!

A 1-ton block that is 1 meter off the ground has the same amount of potential energy NO MATTER HOW IT GOT TO BE ONE METER OFF THE GROUND!!! If you lifted it directly, or if you lifted it with a pulley, IT HAS THE SAME AMOUNT OF POTENTIAL ENERGY!!!!!!!!

Reply from Marc:

You made a good point separating the difference between energy required and the force used in the example of lifting a block of cement. However, in doing so you help prove the point as to how and why BTUs are irrelevant for internal combustion engines.

One of the three characteristics of an ethanol-optimized engine is that it has a longer piston stoke. The longer piston stroke provides more force. Gasoline, even with its higher BTUs, can't effectively push the piston in an ethanol-optimized engine. With this extra force, an ethanol-optimized engine/car can travel faster and further on the same volume of liquid fuel (ethanol vs. gasoline).

Conversely, ethanol doesn't effectively use the shorter piston stroke and spark timing of a gasoline-optimized engine, so it uses more fuel to accomplish the same task.

Therefore, as I've maintained, engine optimization and not BTUs is the key. And the example that higher BTU diesel fuel can't be used in a gasoline engine further confirms it. It's why gasoline in a diesel engine doesn't just get 10 or 15% fewer MPG than diesel, it might get no MPG.

The following excerpt is from a paper to the American Society for Environmental History, Annual Conference March 26-30, 2003 by William Kovarik, Ph.D.:

> Studies of alcohol as an internal combustion engine fuel began in the U.S. with the Edison Electric Testing Laboratory and Columbia University in 1906. Elihu Thomson reported that despite a smaller heat or B.T.U. value, 'a gallon of alcohol will develop substantially the same power in an internal combustion engine as a gallon of gasoline. This is owing to the superior efficiency of operation.' (New York Times Aug. 5, 1906) Other researchers confirmed the same phenomena around the same time.

> USDA tests in 1906 also demonstrated the efficiency of alcohol in engines and described how gasoline engines could be modified for higher power with pure alcohol fuel or for equivalent fuel consumption, depending on the need. The U.S. Geological Service (USGS) and the U.S. Navy performed 2000 tests on alcohol and gasoline engines in 1907 and 1908 in Norfolk, Va. and St. Louis, Mo. They found that much higher engine compression ratios could be achieved with alcohol than with gasoline. When the compression ratios were adjusted for each fuel, fuel economy was virtually equal despite the greater B.T.U. value of gasoline.
>
> "In regard to general cleanliness, such as absence of smoke and disagreeable odors, alcohol has many advantages over gasoline or kerosene as a fuel,' the report said. "The exhaust from an alcohol engine is never clouded with a black or grayish smoke." USGS continued the comparative tests and later noted that alcohol was "a more ideal fuel than gasoline" with better efficiency despite the high cost.

In the "Ethanol Vehicle Challenge 1998" tests were conducted on ethanol-optimized Chevrolet Malibus versus stock Chevy Malibus. The tests showed that most ethanol-optimized vehicles tested on the dynamometer exceeded the fuel efficiency of the stock Malibu, with the best schools showing efficiency improvements of 13 to 15%. In the city portion of the dynamometer testing, all the vehicles demonstrated a higher fuel efficiency than the stock Malibu. See Ethanol Vehicle Challenge 1998.

(https://avtcseries.org/competitions/ethanol-vehicle-challenge/)

Additionally, tests and studies conducted by Matthew Brusstar of the U. S. Environmental Protection Agency show that ethanol-optimized engines deliver greater MPG. See:

"Economical, High-Efficiency Engine Technologies for Alcohol Fuels"

https://archive.epa.gov/otaq/technology/web/pdf/epa-fev-isaf-no55.pdf

and

"High Efficiency and Low Emissions from a Port-Injected Engine with Neat Alcohol Fuels"

https://www.sae.org/publications/technical-papers/content/2002-01-2743/

PERIOD, this is the end of the story!!!!

October 3, 2015

Posted By Robert P:

Marc, unlike you and The Auto Channel that keep defending ethanol, I have spent 31 years of my professional life working on automotive, marine, recreational, and power equipment. I have been a card-carrying member of the antique outboard association, and besides owning my own Marina, I am a A.S.E. certified master mechanic and have been for 27 years. I make thousands of dollars a year repairing damage to boat engines and fuel systems caused directly by the corrosive and hygroscopic nature of this solvent. It degrades over time, destroys seals, steel fuel lines, steel fuel tanks, hoses, carburetor floats, needle valves, metal screws inside the fuel system and metal sending units. It also forms all kinds of solids when left inside carb fuel bowls for extended periods and becomes almost inflammable after a few years. IT ALSO IS NOT EFFICIENT AND CREATES LESS POWER.

I have burned fuel from 1975 out of a 25ft. Sea Ray salvage boat that was 100% gasoline. It provided more power and required nothing more than a little stabilizer. I used to be a GM dealership tech for over ten years and when the flex-fuel trucks first came out, we had to recall them all due to the sending units flaking rust in the first six months of operation so bad they would clog the fuel system. Ethanol at this point has more minuses, and if it were not for government subsidies it would be unsustainable and drive up the cost of fuel.

Reply from Marc:

Robert, your comments provide a great example of how little some engine mechanics know about engines, even when they are so-called certified mechanics. Basically, everything you wrote is wrong or a gross over-exaggeration. And your experience as a dealership tech is rather meaningless because dealerships often have the worst mechanics.

Engines break down. Engines break down regardless of the fuel they use. Everything you blame on ethanol is also true of gasoline and diesel fuel, and it has been true since the invention of the internal combustion engine. If someone owns an older boat or car, they are always advised to not use ethanol blends. Always. I don't know any proponent of ethanol who tells owners of older vehicles to use ethanol—unless the engines and fuel systems have been fitted with modern components. If you have customers with antique boats and you let them use E10 at your marina, then you should have been responsible for any problems and costs.

In general, if you've experienced situations in which your customers' boats were subject to extensive clogging, it's because they should have had their engines cleaned and serviced, or because you did a lousy job in cleaning and servicing the engines. If you question my comments, argue with Mercury Marine. They'll tell you that you are wrong.

You use the term "hygroscopic nature," and I guarantee you don't know what you're talking about, because what you think it means isn't what it means.

Your comments about leaving "all kinds of solids...for extended periods and (it) becomes almost inflammable after a few years" is a joke. What mechanic would leave any type of fuel or solids in an engine for extended periods or even years?

To top it off, your claim that ethanol creates "less power," is proof positive you don't know what you're talking about.

Oh, in case you didn't get the memo, government subsidies on retail sales of ethanol-gasoline blends ended a few years ago, and ethanol blends are still cheaper than ethanol-free gasoline.

FOLLOW UP BY ROBERT P:

It's funny how you separate yourself from mechanics, showing that you are clearly not one and that you lack any high-volume, real-world perspective.

As for your reference to Mercury Marine's research, the boating industry does not concur. They have done they're [sic] own research on ethanol and other alternatives. They have found a better, less damaging alternative.

REPLY FROM MARC:

I'm not a mechanic. While that's not a problem for me, it should be for you. If I, as a layman, know more than you do about how internal combustion engines work, your customers should be running for the hills.

Mercury conducted significant research that completely overshadows anything you've done. They presented the information in a public online webinar in 2011, and stand by the results today. You say the boating industry does not concur. That's incorrect; Mercury Marine is a big part of the industry and their results have been published in multiple boating publications, which no other serious player in the industry has challenged.

My personal knowledge comes from more than four decades of working on my own vehicles with internal combustion engines. I've owned classic and antique cars, new cars, and an antique ELCO boat. Then I combine my hands-on experience with extensive research and mix it together with truth and common sense.

FOLLOW UP BY ROBERT P:

Please enlighten me on how you test for ethanol content in fuels using a test tube, water and fuel. Then explain how it is not hygroscopic. While you're at it, explain why gasoline free of ethanol can sit in a glass vessel for hours and not have water form across the bottom while a vessel filled with 10% ethanol/

gasoline quickly forms a water layer. Then explain why metal parts corrode in ethanol-based fuels and not in 100% gasoline. Then explain how efficient it is while at the same time, it produces less power per volume.

Reply from Marc:

Robert, what does testing fuel in a test tube have to do with something being hygroscopic?

If you place any fuel in an open container and water appears after a few hours it is the result of condensation, not hygroscopic characteristics. There is a difference. If you don't know this, it's proof you don't know what you are talking about.

Ethanol breaks water down. If water formed in an open container that has E10 in it, the water would mix with the ethanol and you would not be able to see the water unless the water content became too great to mix with the ethanol. Gasoline does not have the same miscible characteristics of ethanol, so if any condensation forms in a container with ethanol-free gasoline it will separate from the gasoline over time.

Your understanding of the word hygroscopic is the classic misunderstanding of what the word means. A hygroscopic substance does not absorb water from thin air. A hygroscopic substance attracts water from its environment. The word "environment" does not mean "thin air." It means adjacent. It works like a wick in an oil lamp. If you don't know what wicking means, look it up.

You make it sound like only ethanol is corrosive. All liquids are corrosive—especially water. Air, wind and solar rays are also corrosive. Gasoline is corrosive, which is why certain materials cannot be legally used to hold gasoline. The solution to the problem is to use materials that are resistant to the liquid being used. Modern vehicles use metal, rubber and plastic resistant to ethanol corrosion. Vehicles produced prior to the modern ethanol-fuel generation used materials that were not susceptible to gasoline or diesel.

These are extremely basic principles. If you don't know these things, it proves you don't know what you are talking about and means you should have your certification revoked.

It's funny how some mechanics use their ASE certifications as if they were medical degrees.

Ethanol has more octane than gasoline. This delivers more power, not less power. It's why race engines favor ethanol or methanol fuels. It delivers more power because you can use higher compression engines with longer piston strokes. Even in an engine that is not optimized for longer piston strokes, ethanol will provide additional horsepower. You should be ashamed for bringing this point up.

Consider yourself enlightened. I hope you use the enlightenment wisely.

Follow Up By Robert P:

In a gasoline engine, ethanol produces less power. Period. Also, water has a higher octane reading than gas.

Reply from Marc:

Robert, water has a higher octane reading than gas? Are you sniffing the fumes while typing?

Follow Up By Robert P:

You're the worst kind of moron. Dumb and overconfident. Closed minded. Always right and unwilling to look at facts. Read this:

https://web.archive.org/web/20120303172107/http://www.alcoholinjectionsystems.com/a11/116+-Octane-With-Water-Injection!/article_info.html

Reply from Marc:

I'm only right about the things I'm right about. On things where I am wrong, I'm wrong. What has you puzzled is that we haven't been discussing anything I'm wrong about.

It seems you didn't take the time to read the article connected to the link you sent me. It has nothing to do with "water having a higher octane rating than gasoline." You really don't know anything about engines, do you?

Follow Up By Robert P:

Yeesh, I don't have time for this, and you clearly require mental treatment. I will go back to fixing imaginary fuel issues.

Reply from Marc:

For the first time, I think you inadvertently revealed the truth, you just want to go back to "fixing imaginary fuel issues" (these are all your words).

October 13, 2015

Posted By Tom K:

Ethanol does have higher octane but has a lower BTU count, that's why you get such poor MPG with it. I think corn-based ethanol has been a failure in this country.

Reply from Marc:

Tom - It's nice that you decided to try this message board, but BTUs have nothing to do with MPG.

And ethanol has not been a failure, it's been a great success. All you have to do is go to any large city and look up. In fact, if you live in the L.A. area and if you can see the Hollywood Hills from LAX, thank ethanol. The skies over America are the cleanest they've been in decades.

Good try, though.

Posted By Bill B (Coming To Tom K's Rescue):

The clear skies in L.A. are not a result of using ethanol. Unleaded gasoline, better emission controls, more efficient automobile's, government standards and many other factors have done that. All ethanol does is keep the price of corn higher to pacify the National Farmers Organization and their lobbyists. It's been known for decades that ethanol is bad for a manifold of reasons.

Reply from Marc:

Bill, I love your comments. You made my day. It shows just how ignorant ethanol opponents are. In just a few words you made the case for ethanol.

There would be no unleaded gasoline if it wasn't for ethanol. Tetra-ethyl lead was added to gasoline in order to eliminate engine knock. The previous way to eliminate knock was to put ethanol in gasoline. Leaded gasoline was the invention of GM and the oil industry to duplicate the results of ethanol-gasoline blends. Leaded gasoline was preferable to GM and the oil industry because they could patent the formula and make billions of dollars.

When the oil industry and automobile industry were faced with the requirements to eliminate harmful emissions, they tried using MTBE in place of ethanol. MTBE did eliminate knock, but it did little to clean the air (MTBE is a byproduct of petroleum oil). In addition, MTBE is nearly as poisonous as Tetraethyl-Lead, so it was discontinued.

The mandate to use ethanol in blends came after the advent of catalytic converters and some improvements in engine design. However, the effect on air quality during the MTBE was slight. The big jump in cleaning the air came after using ethanol.

You also wrote that improvements in the air were a result of better emission controls and government standards. You're making a silly semantic argument. The controls brought to play were to eliminate, as much as possible, fuels that created too much airborne residue. Government standards, per se, didn't clean the air. The air didn't say "Oh we have to clean up because the government says we have to." Enhanced emission control is due to using ethanol. Government standards were met because of ethanol use.

It is possible to buy ethanol-free gasoline (E0). And modern ethanol-free gasoline does eliminate knock by using other additives called aromatics. These primary aromatics are benzene and toluene and they are petroleum oil by-products like MTBE. They literally stink. They are more expensive than ethanol, which is why any ethanol-free gasoline costs more than E10...sometimes lots more. Additionally, benzine and toluene are hazardous to breathe and are not as clean burning as ethanol.

If you would like to see some videos that demonstrate how harmful aromatics are go to YouTube and search for any of the following video titles:

1. Fuel Emissions - Gas, Ethanol, E85, Kerosene
2. E0 and E10 MVI 2090

3. Aromatics Styrofoam Balls
4. Swelling Plastic Video 120413
5. Aromatics Testing with Styrofoam Part Two
6. Aromatics Testing with Styrofoam

October 16, 2015

Posted By Tom K:

The amount of energy in any given fuel doesn't have a bearing on its MPG??? Is that what you actually said? Marc, Go back to school.... and don't waste my time.....

Reply from Marc:

Tom, you're back so soon - it was just a couple of days ago that you tried to make the same point and I soundly thrashed you. Well, okay, I'm happy to do it again.

The BTU rating of gasoline is irrelevant. A gasoline-optimized internal combustion engine delivers the MPG it does because it is optimized to run on gasoline. The same engine optimized to run on ethanol, and using ethanol, will deliver comparable or better MPG than the gasoline-optimized engine.

If you tried to use gasoline in an ethanol-optimized engine it will deliver far less MPG. If BTU was the important factor, then gasoline would achieve better results—even in an ethanol optimized engine.

If what you learned in school—or think you learned in school—is what you posted above, you wasted your own time.

A quick and easy way to prove engine-optimization and not BTU rating is the key, think about this: Diesel fuel has a higher BTU rating than gasoline. However, if you put diesel fuel in a gasoline-optimized engine you do not get better MPG, you would get less MPG.

I look forward to your next attempt.

Posted By David N:

MPG is per gallon. A gallon of ethanol doesn't not have the same amount of energy as a gallon of gas. The efficiency of a typical gas engine is pretty low, 20-30%, so even if you optimize the engine to run ethanol (ethanol has a higher octane ~110, so you could theroy run it has a higher compression, higher compression in carnot cycle leads to greater effiecney) but that doesn't mean you will get get better mpg's due to fact that if you had a 100% ethonal engine you still have significantly lower about of BTU's per gallon than gas. BTU's are the basic measurement of energy and the ability to do work.

REPLY FROM MARC:

Dave - Should you choose to post another message, you might want to use spell and grammar check before you hit the submit button.

Let me try to understand what you wrote and respond appropriately.

If you take two comparable spark-ignited internal combustion engines and optimize one for ethanol and the other for gasoline, the ethanol engine will deliver more MPG and more horsepower. This has been proven by studies and tests for more than 100 years.

The following excerpt is from a paper to the American Society for Environmental History, Annual Conference March 26-30, 2003 by William Kovarik, Ph.D.:

> Studies of alcohol as an internal combustion engine fuel began in the U.S. with the Edison Electric Testing Laboratory and Columbia University in 1906. Elihu Thomson reported that despite a smaller heat or B.T.U. value, 'a gallon of alcohol will develop substantially the same power in an internal combustion engine as a gallon of gasoline. This is owing to the superior efficiency of operation.' (New York Times Aug. 5, 1906) Other researchers confirmed the same phenomena around the same time.
>
> USDA tests in 1906 also demonstrated the efficiency of alcohol in engines and described how gasoline engines could be modified for higher power with pure alcohol fuel or for equivalent fuel consumption, depending on the need. The U.S. Geological Service (USGS) and the U.S. Navy performed 2000 tests on alcohol and gasoline engines in 1907 and 1908

in Norfolk, Va. and St. Louis, Mo. They found that much higher engine compression ratios could be achieved with alcohol than with gasoline. When the compression ratios were adjusted for each fuel, fuel economy was virtually equal despite the greater B.T.U. value of gasoline.

"In regard to general cleanliness, such as absence of smoke and disagreeable odors, alcohol has many advantages over gasoline or kerosene as a fuel," the report said. "The exhaust from an alcohol engine is never clouded with a black or grayish smoke." USGS continued the comparative tests and later noted that alcohol was "a more ideal fuel than gasoline" with better efficiency despite the high cost.

See Ethanol Vehicle Challenge 1998.
https://avtcseries.org/competitions/ethanol-vehicle-challenge/

Tests and studies conducted by Matthew Brusstar of the U.S. Environmental Protection Agency show ethanol-optimized engines deliver greater MPG. See:

"Economical, High-Efficiency Engine Technologies for Alcohol Fuels"

https://archive.epa.gov/otaq/technology/web/pdf/epa-fev-isaf-no55.pdf

and

"High Efficiency and Low Emissions from a Port-Injected Engine with Neat Alcohol Fuels"

https://www.sae.org/publications/technical-papers/content/2002-01-2743/

Follow Up By David N:

So, your best argument is a paper in 1906? Before direct injection, fuel injectors, etc. that have greatly increased the MPG's of the modern engine. Quote for the ethanol vehicle challenge "competition's opening ceremony, when a small engine fire broke out. The Wayne State vehicle, which attained 29.2 miles per gallon highway driving and a top speed of 81 mph from a standing start in a little more than 16 seconds." My 2015 Subaru Legacy averages 42 mpg highway, 0-60 in 8.4 seconds with AWD. Not sure what the top speed is but it still had more in it when I pushed to 95 mph.

Reply from Marc:

No, my best argument isn't a paper written in 1906, it was written in 2003. It quotes studies done 100 years ago and more. You didn't finish reading what I wrote, and you obviously didn't read the tests and studies I suggested. These studies were done fairly recently.

What I was showing you is that knowledge of BTU irrelevancy has existed since the invention of the internal combustion engine. There are guys like you running around spouting wrong information because of oil industry propaganda. You heard something about BTUs and you never stopped to consider that BTU measurement relates to heating water, not powering internal combustion engines.

To your point of modern engines and direct injection, etc., these improvements also improve the performance of ethanol fuels. In any event, these improvements prove that engine-optimization, not BTU rating, is what is relevant. Didn't you realize this?

Follow Up By David N:

BTU's (British Thermal Unit) are a measure of the energy output. As most all of engines run a carnot cycle (exception be[ing] rotary engines), you are talking about how much heat energy you produce in a given cycle of rotation in an engine. Few BTU's the less energy availability for the thermodynamic cycle of your car's engine which creates less MPG's.

Reply from Marc:

David, on this point you're exactly correct. BTU is a measure of energy output, specifically, the amount of energy required to raise the temperature of water by 1 degree.

However, internal combustion engines don't run on water or steam. If we were discussing steam-powered vehicles or in-home water heaters, then BTUs would be significant. Therefore, the rest of your comment is wrong.

Internal combustion engines are optimized to run on specific fuels. Optimization is achieved by mechanical adjustments. A gasoline-powered ICE is optimized (adjusted) to run on gasoline. Performance of that engine is dependent upon the optimization. The proof of this is that if you tried to use another liquid fuel that has a higher BTU rating (such as diesel) in a gasoline-optimized engine you do not get better MPG, you get worse MPG - in fact the engine would not run on 100% diesel.

A gasoline-optimized engine might operate on a blend of 80% gasoline and 20% diesel (if you accidentally put some diesel in the tank), but it doesn't get better MPG to correspond with the higher BTU diesel, it gets worse MPG.

Follow Up By David N:

BTU as is a measurement of Joules, the basic measurement for all energy. 1 BTU = 1055.05585 Joules. Energy is a measurement of how much work is done. An engine's primary purpose is to do work. Diesel is an ICE engine; it just will not work in a normal gas engine. 87 pump gas has a max compression ratio of about 10x while diesel is in the range of 18 to 22x, so your gas engine simply doesn't compress the fuel enough to produce the explosion to do work. Diesel engines don't use spark plugs to create the explosion but rather compression to create the explosion. No engine will run properly on incorrect fuel, e.g., if you put 2-stroke fuel in a 4-stroke engine, it might run, just not very well. Your response makes no sense.

Reply from Marc:

Exactly, it's all about engine optimization, not BTUs. Your continuous attempts to make BTU measurements important are what doesn't make sense.

Follow Up By David N:

You missed my point. A car will always go further on a gallon gas than a gallon of ethanol. Look at Brazil that uses

E100, in vehicles made for E100 as compared to E10, there was an increase in HP and Torque of around 8% but there was a drop in fuel efficiency of 20-25%. This was on the same car the VW Polo. So, you are burning more fuel to go the same distance which is only partly offset by the lower cost per gallon of E100.

Reply from Marc:

No, David, you have no point because you are incorrect. An engine (car) optimized to run on ethanol will deliver more MPG than a comparable engine (car) optimized to run on gasoline.

Go back and read the resource information I posted above.

October 17, 2015

Posted By Paul Schlueter:

Marc, I enjoy reading the conversations you have with people who do not understand the concept of how fuel BTU content is mostly irrelevant to car mpg. Maybe I can give you some additional material to use in these conversations. I have had this conversation with many people over the years and it amazes me how difficult it is for some people to understand. What I find most people understand more easily is explaining that an internal combustion engine is a device that converts fuel BTU's into mechanical power AKA Horsepower, and unfortunately not very efficiently, because most of the BTU's are wasted as excess heat. However, almost anyone who knows how engines work, knows that long stroke, high compression engines have mechanical advantage.

When I explain that gasoline-optimized engines can only convert about 25% of the gasoline BTU's into mechanical power, but high compression ethanol optimized engines can convert about 35% of the ethanol BTU's into mechanical power, this seems to help clear up the misunderstanding. The reason most current model flex-fuel cars have such a mileage penalty when running E85 is because no manufacturer is presently using an ethanol-optimized engine. Hopefully that will change in the not too distant future.

Ford has been working on some new engines to take advantage of ethanol. See this link, "Ethanol Optimized Engine."

https://view.publitas.com/the-auto-channel/robert-stein-avl-3-29-11/page/1

And, Cummins has developed an engine that is in testing right now. See this link, "A Cummins Gasoline Engine?"

https://www.dieselnet.com/news/2014/07cummins.php

And this link: "Updates On 5 Cool Projects At Cummins"

https://www.cummins.com/news/2015/12/09/updates-5-cool-projects-cummins

Reply from Marc:

Hi Paul - Thanks for your comments and links, all of which are greatly informative.

Let me address your last point first: You are absolutely correct that vehicles designated as flex-fuel vehicles are not optimized to use ethanol; they are merely more friendly to ethanol blends. They do not maximize the potential of the ethanol fuel. A flex-fuel vehicle typically has an adjusted software program that understands that a fuel other than non-ethanol fuel is being used. The flex-fuel vehicles still lack the ability to adjust to all three necessary changes to permit them to optimize the ethanol. Most people don't know this and when they argue with me or others about how flex-fuel vehicles get less MPG, they think it's because ethanol is inferior to gasoline.

The oil industry has done a very good job hoodwinking the public. They do a very bad job at protecting the public from poison, but they are very good liars and know how to finagle money from the public and public coffers.

A recent webinar presented by FuelsInstitute.org, "Electric and High-Octane Vehicles: Developments And Implications," addressed this issue and talked about how the only way for auto manufacturers to meet the coming emissions standards is to build new engines that are optimized to run specifically on higher octane fuel blends. They state that there are only two ways to achieve the higher-octane blends: one with toluene, and the other with ethanol. But, they point out, toluene is more expensive than ethanol and is poison. Consequently, the only safe and inexpensive alternative is ethanol. The webinar talks about all new "gasoline" engines being designed specifically to optimize blends like E30, or higher.

Let me address the BTU portion of your comment; and let me again thank you for your comments on this point, along with a big thank you to fblee for his insightful and informative comments. Any discussion about BTUs with respect to internal combustion engines is completely irrelevant. The term BTU (British Thermal Unit) is misused to explain something that has no relationship to how and why it was originated. It has become a convenient marketing tool, but that doesn't make its use correct.

As I'm sure you know, and as I've explained on many occasions. BTU was coined as a measurement of how much energy is required to heat water by one degree. The specific need and reason for this was for steam-powered engines (stationary and mobile engines). People wanted to know which source of fuel (such as wood, coal, kerosene, or alcohol) was more efficient to heat water. Boiling water has nothing to do with an internal combustion engine. In fact, if a person's internal combustion engine is boiling water (the water in the radiator or windshield washer container), they are in serious trouble.

The term originated around 1850, before the first commercially successful internal combustion engine was invented, and before the first modern internal combustion engine was built by Siegfried Marcus in 1864.

Ranking an internal combustion engine fuel by BTU is nothing more than a marketing or labeling tool. Always the best example of BTU irrelevance is comparing diesel fuel to gasoline. If BTUs were the significant factor, then you should be able to use diesel fuel in a gasoline engine and achieve better MPG.

If internal combustion engines could only be built to gasoline specifications and characteristics, it might be appropriate to use gasoline fuel as the benchmark to compare fuels. Because internal combustion fuels can be built and adjusted to the specifications and characteristics of various liquid and gas fuels, and the performance results of those engines/fuels are equal to or better than the gasoline ICE, then the BTU comparison between the fuels is simply window dressing.

I agree mathematical equations can be used to show the difference in the supposed energy content of the various fuels, but this is reverse engineering to make a point. I would compare these formulas to the following:

What would you say if I asked you "Which would you rather have, two apples plus two more apples; or 690 apples divided by 3 minus 227 apples? If you like apples you would rather have the four apples than the three apples, even though the second equation is so much sexier.

If engine "A" is optimized to run on ethanol (with 76k BTUs) and gets 30 miles per gallon of ethanol fuel, and you can buy ethanol for $2.10 per gallon. If engine "B" is optimized to run on gasoline (with 116k BTUs) and gets 27 miles per gallon of gasoline fuel that costs $2.59 per gallon, how important is it that gasoline has a higher BTU rating than ethanol? Answer: It's not important at all, and there's no scientific or mathematical equation that can make BTUs important in an internal combustion engine.

For the oil industry to pretend BTU ratings mean something special they might as well market gasoline this way: "Gasoline has a gold color, while ethanol is clear and has no color. Wouldn't you rather have gold than nothing?"

Gasoline is poison; it's a rip-off; it's unpatriotic; and it's responsible for killing hundreds of thousands of American servicemen and women.

November 8, 2015

Posted By Mickey Shay:

Hi Auto Channel! Recently there have been posts by groups like Smarter Fuel Future saying Iowan corn farmers are depleting Iowa's topsoil. Can you refute this in anyway or is it true that limited crop rotation is damaging the soil?

Reply from Marc

Hi Mickey - Thanks for reading our Ethanol Chronicles and for your comments and questions over the past several weeks.

You bring up a great topic for two reasons: There's the question of soil depletion, which is an important issue; and then you mention groups like Smarter Fuel Foundation, which definitely deserve to be exposed.

I'll address the soil issue first: The oil industry and their apologists and lobby have focused on a problematic point to make it seem as if ethanol (and other alternative fuels) are the sole cause of the problem. The issue of corrosion is a great example. Big Oil spreads exaggerated information about ethanol being corrosive and causing engine damage. As I've stated on a number of occasions, all liquids are corrosive, including and especially water. Air, wind, solar rays are also corrosive. Gasoline is certainly corrosive, which is why certain materials cannot be legally used to hold gasoline. But the oil industry doesn't tell you this. They make it seem as if only ethanol causes corrosion.

Big Oil's issue with soil depletion follows the same fallacious line of attack. But everything causes soil depletion. Growing crops depletes the soil; not growing crops depletes the soil; rain depletes soil; no rain depletes soil; wind depletes the soil; solar rays deplete the soil; walking on soil depletes the soil; driving motor vehicles on soil depletes the soil; animals walking on soil depletes the soil. But ethanol opponents won't acknowledge this. They make it seem as if only corn crops deplete the soil.

Let's look at the most extreme case scenario. Let's pretend that only corn crops damage the soil and that nothing can be done to stop the damage caused by growing corn. Let's also pretend the oil industry and groups like Smarter Fuel Future are filled with good people truly concerned about this problem. Why are they not calling for the elimination of all products that are made from corn? The soil doesn't know if the corn being grown is for ethanol or corn chips or corn syrup. If corn growing is so detrimental, why do the oil industry and groups like Smarter Fuel Future partner with the junk food industry? Why do we need corn chips and corn flakes and corn syrup sweeteners? Why are we feeding animals corn or corn by-products like distillers' grains? Can't chickens, ducks and cattle find something else to eat? Don't these animals know that they are helping to destroy all the good soil?

If growing corn is so bad, then why is corn grown anywhere in the world for use in any product?

The answer is that the oil industry's depiction of ethanol causing a soil problem is as grossly exaggerated and untrue as other lies they spread about alternative fuels.

Mickey, this brings me to your other question: lobbying groups, like the Smarter Fuel Foundation. The oil industry and their paid apologists set up these organizations in their attempts to legitimize lies and disinformation to denigrate all competition. These groups create names to make them sound socially concerned or scientific, or both. They care only for themselves and their agenda. People forget that petroleum oil and the industry are responsible for the death of millions of human beings in numerous wars fought to protect their assets and those who died from the effects of carbon emissions and pollution.

November 12, 2015

Posted By Raymond B:

Gasohol takes more energy to produce using fossil fuels than it produces. It is not a solution to the CO2 emissions from fossil fuels at all. It has caused world food prices to rise that has hurt least developed countries the most. In Brazil they have deforested large parts of their rain forests to grow sugar cane to make ethanol for gasohol. Is that a success story?

Reply from Marc

Hi Ray - Good to hear from you. As I count it you make three points against ethanol. All three are incorrect; They're based on outright lies or gross exaggerations.

I'm not criticizing you for making these incorrect statements because they are three of the most common incorrect statements made when arguing against ethanol. But what surprises me is that you have allowed yourself to be taken in by these misstatements. I say this because you seem to be a person who has done some investigation into the subject.

I read some of the other comments you've made in online forums and it's clear you know all energy sources and fuels require energy to find and produce. So, when you complain about ethanol requiring more energy from fossil fuels to produce than the amount of energy the ethanol creates, I'm disappointed in you. Gasoline is more energy negative to produce than ethanol. You should know this.

Regarding the claim that ethanol causes food prices to rise has already been retracted by the very same organization that started it: the World Bank. They retracted it in 2010, and again on two later occasions. In the absence of the World Bank's report there are no authoritative attempts to blame ethanol for food price increases.

As for Brazilian deforestation for sugar crops, that's incorrect. There has been deforestation in Brazil, but it's not to clear land for sugar cane production. Sugar is grown in other regions. It's similar to the U.S., where the best regions to grow crops for ethanol are not where the lumber industry gets most of their lumber.

I'm sure you'll respond and ask me to provide information that supports my comments. I look forward to hearing back from you, but if you'd like to save a step you can simply read my 60+ page report about it here:

https://www.theautochannel.com/news/2013/06/10/082044-gusher-lies-book-review-and-reply-to-robert-bryce.html

Follow Up From Raymond B:

Thanks for the comment. I will look up your report. The articles I've (previously) read state exactly what I said in my initial comment about ethanol. This is another very politically charged topic. Farmers have profited immensely from ethanol made from corn since corn prices have skyrocketed. They are firm supporters of the efficiency of ethanol fuel.

If all fossil fuels are banned by 2030 there will be no remaining large-scale agricultural industry in the USA or the world. Fossil fuels are needed for agriculture and 17% of all the energy we use is in agriculture. The USA feeds the world.

When we were dependent on the Middle East for oil, it made sense to make ethanol for gasohol. We now are a larger producer of oil than Saudi Arabia and don't need their oil. We don't even need Canada's shale oil which we have the refineries to refine.

I remain very skeptical of the claim that oil uses more oil energy than it produces. I'd want to see proof of that. I also want proof that making gasohol produces more energy than the oil energy used to grow corn, sugar beets or sugar cane and distill the alcohol from it.

I certainly am interested in knowing the truth and it is getting very hard to find truth in all the pseudoscientific claims being made on Internet news.

Reply from Marc

Ray, thanks for your reply. It sure is a politically charged topic; and given the amount of money involved in all aspects, how could it not be?

To say that farmers have "profited immensely from ethanol made from corn since corn prices have skyrocketed" is convolutedly incorrect.

Let's concede farmers have profited immensely. So what? Perhaps they're entitled to profit immensely, finally. If, as you write, America feeds the world, why shouldn't farming be very profitable? There has been no restriction on how much the oil industry can profit. I'd rather American farmers be rich instead of foreign dictators.

Corn prices have not skyrocketed, and during the times that corn prices have been pushed up, it was because of commodity speculators. The farmers aren't necessarily the ones who profited. As a matter of interest, corn prices this in June 2015 were about the same or slightly higher than they were in the 1970s, 1980s, and much of the 1990s. Recent prices are higher than in June, but they are about half what they were in 2011-2013.

I never wrote that ethanol "produces more energy than the oil energy used to grow corn, sugar beets or sugar cane and distill the alcohol from it." I wrote that gasoline was more energy-negative than ethanol. As you know, all man-made fuels require energy to produce them. This is due to the need to "convert" the raw material into its final usable form.

Links to the studies and reports related to gasoline being more energy negative than ethanol can be found in Part 5 of my report.

Follow Up From Raymond B:

I beg to differ. I own a farm and have rented 22 acres to a big farmer who raises corn. He told me corn prices tripled when government started the ethanol production programs and mandated gasohol. He even doubled the rent he was paying me because of the higher profits.

Maybe you are using the wrong time frame. Compare corn prices before the gasohol program and after the gasohol programs. In subsequent years I do not know what is happening to corn prices. I do know that corn production per acre has skyrocketed due to genetically modified corn and use of RoundUp and other herbicides that allow weed free crops and no-till farming. The increase productivity of all-American farms per acre has meant that least productive acres have been taken out of farming and used for other purposes. This may have brought corn prices back down. I just don't have the data.

If farmers are not making any additional money from growing corn for ethanol why are farmers and farmer-dominated states so intent on keeping the ethanol program alive? Politicians have wanted to kill the ethanol from corn program for years now and have been unsuccessful.

I don't understand "convolutedly incorrect." Are you a farmer or a farm lobbyist? To say farmers have not profited by higher corn prices is just not what the local farmers are telling me here in Maryland where I live. Are you saying middlemen are making all the profits or saying there are no large windfall profits from growing corn for ethanol? We both may be correct. Sometime articles only cover small time frames and portray a different picture than in larger time frames. I'd like to sort out what is true.

REPLY FROM MARC

Ray, other than being contrary, I'm not sure what it is that you believe that differs from what I've presented.

You question my use of "convolutedly incorrect." I used this because the intention of most farmers—like all businessmen—is to make a profit. Perhaps people became so used to hearing about farmers struggling financially they think it's alarming to learn they've made a profit.

Clearly, the advent of the ethanol blends in place of poisonous leaded gasoline and MTBE gasoline has benefitted farmers, but so what? If someone has been struggling for years to produce blue dye and the fashion industry decides that blue is the "hot" color, you wouldn't think ill of the guy for now being successful. Even if the dye manufacturer increased his prices because of the change in fashion trends, you wouldn't fault him. After all, that's what supply and demand is all about. So, you've "convoluted" the whole reason for being in business and added in incorrect data to boot.

Now, you might be trying to say that corn farmers took unfair advantage by tripling or quadrupling their prices, but that would be incorrect. The primary reason for corn prices going higher after the government decision to mandate ethanol-gasoline blends was due to commodity speculators pushing prices up. The farmers themselves didn't benefit to the degree the prices increased. And I would venture to say that many may have only marginally benefitted, with the only benefit being that they finally had a crop with a solid market.

The calendar dates that I refer to are stated in the paragraphs of my previous post. The figures I quote come from a website called Trading Economics, it's URL is

http://www.tradingeconomics.com/commodity/corn.

You can also find the history of crude oil prices at that website at:

http://www.tradingeconomics.com/commodity/crude-oil.

You'll be able to search a wide variety of commodities and prices over many decades and compare them against each other.

What you'll find is that there have been dynamic price changes since the ethanol mandate, but that the ups and downs correspond to the dynamic changes in the price of crude oil and oil-based fuels. In addition to the corn prices fluctuating because of commodity speculators, farmers and ethanol distributors were also affected as the price of diesel and gasoline skyrocketed.

I have come across people on message boards and at events who will challenge me by comparing a specific date when the price of corn was very low to a specific date when the price of corn was very high, and they will use that narrow example to prove how unreasonable the price of corn has become. That kind of comparison is like calling a baseball player "a lousy hitter" because he batted .147 during a particular weekend series, when he batted .347 for the entire season.

You do have to look at the larger picture, a picture large enough to take into account all the other benefits related to substituting foreign-dominated petroleum fuels with domestic non-fossil fuels.

Follow Up From Raymond B:

Marc, I have no problem with farmers making money. They are hardworking people and as long as government does not subsidize gasohol to make it competitive with gasoline it is fine with me. That has not been the case. I know you claim gasohol does not require more energy to manufacture than it releases, but that is what I have read in several articles over many years. I doubt gasohol would exist in a free market.

With that said, and realizing we may agree to disagree, I looked up corn prices and spoke with the farmer who rents 22 acres of my farm. Corn prices were pretty stable until 2007 when they rose from $100-$110/ton to $280-$335/ton.

The market for corn here in Maryland is different than the market in the mid-west. The farmer told me that three years ago corn prices in Maryland hit six-seven times the usual prices due to the drought in the Midwest. Then two years ago they were $5.50 a bushel. When the corn hit its high, fertilizer and herbicide and seed prices were still low, so farmers had exceptionally good years and made money. Then the seed companies, fertilizer companies and herbicide manufacturers raised their prices in response to the enormously high corn prices three years ago, and when corn prices dropped back the prices for these remained high and Maryland farmers barely broke even raising corn last year.

I discovered ethanol replaces about 1 billion gals of oil a year now. By 2009 ethanol market share in the US gasoline supply reached 8% by volume.

Gasohol is supposed to release slightly less CO2 than pure gasoline but E10 (10% ethanol gasoline mixture now marketed in the USA) really does not make a big difference with regard to CO2 emissions when you count the oil used to plant, pick and distill corn into ethanol. Decaying plant organic matter emits 60 gigatons of carbon into the atmosphere a year compared to eight gigatons of carbon emitted from humans burning fossil fuels, so you can see 10% ethanol really is irrelevant in reducing CO2 emissions.

The issue really boils down is why is the government subsidizing gasohol? It made sense when we wanted to stop dependence on Arabian oil, but now the USA is producing more oil than Saudi Arabia thanks to fracking. We have an oil excess and that has driven down gasoline prices and made gasohol even less competitive. I am sorry but I am just a retired middle-class fellow and I can't afford to pay more for gasohol just because some politicians from Iowa wants to help the farmers. It makes no sense. It takes money away from my pocket that I need to pay other bills, and if our government is going to pick winners and losers to subsidize, we are making a serious mistake in a free market.

Obama ordered that the US Navy use synthetic fuels rather than drilled oil. That doubled the cost of fuel for our navy. Again, U.S. taxpayers foot the bills and it makes no sense. If subsidizing different industries is the game of the future, well start with my synthetic sun machine that MAY extract sunlight from cucumbers as in Gulliver's Travels. I'll need about $10 billion in startup money, cash only, deposited into a numbered Cayman Island bank account. If I fail, I promise I did try really hard to make the project work, at least as hard as Solyndra's CEO tried to make Solyndra produce a single solar panel. I promise to also do as well as Fisker Car Co. that produced not a single electric or hybrid car at their US plant in Newark, DE despite the large government subsidies Obama gave them.

Reply from Marc

Ray, if you have no problem with farmers making money, why do you keep making a big deal out of corn prices? A $4 box of corn flakes contains about 6 cents worth of corn. I've already explained why there was a large increase in corn prices, if you have information that contradicts me, then present it.

If you aren't happy about subsidies, why don't you complain about the subsidies given to the very profitable oil companies, which greatly exceed anything given to all other energy resources combined?

Subsidies to the oil industry allow gasoline to be competitive with ethanol, not the other way around. If the oil companies had to pay for wars fought for oil, global defense of shipping and exploration and development, gasoline might be $15 - $20 per gallon in America.

For some reason some people who consider themselves patriotic Americans have confused relying on a foreign-dominated oil industry with supporting America. I think you've fallen into that trap. Moreover, ethanol has reduced the price of gasoline.

As for emissions, thanks to ethanol-gasoline blends we have less poison in the air and the skies over our large cities are the cleanest they been in decades.

There's plenty of reasons to dislike Obama and his anti-American policies, and I believe I am as harsh a critic of him as anyone—but giving away too much money is something that every president has done for at least 100 years, regardless of their political party.

December 11, 2015

Posted By Litesong:

Ethanol blends do NOT save oil supplies. In my last five, low 87 octane, low compression ratio (9:1 to 11:1) gasoline engines, using 10% ethanol blend (E10) lost 8%, 8%, 7%-8%, 7% & 5% mpg vs. 87 octane, ethanol-free gasoline(E0). Also, oil is burned to make ethanol, ship to blending sites & support the lying "ethanol in gasoline industry". The "ethanol in gasoline industry" produces an inefficient product & uses more oil than it can minimally save.

Reply from Marc

Hey Happy Holidays, Litesong –

I'm glad to hear from you. Frankly, I rely on people like you to help me keep my Ethanol Chronicles blog stocked with things to respond to, and the pickings have been meager these last few weeks. Either people are wary of crossing keyboards with me or they're worrying about where the next oil-funded terrorist act will take place.

I'm not completely sure what you are trying to say by referencing compression ratios, but you say your last five fill-ups did not demonstrate that ethanol reduces the amount of gasoline being used, and you are incorrect. If you replace one substance with another substance, you are automatically using less of the first substance.

The only way that you would not reduce the use of petroleum oil is to show that the manufacture of the ethanol requires so much petroleum oil fuels that it exceeded the original requirement. I recognize that you are trying to make that case by writing that "... oil is burned to make ethanol, ship to blending sites...," but taking into account whatever petroleum oil fuels are used in ethanol production it is not enough for your statement to be correct. In case you are not aware of it, all petroleum oil fuels that might be needed to make ethanol can be replaced by ethanol-based fuels, and this often occurs.

Moreover, virtually all fuels and usable energy sources require the use of other energy and fuels, so your comment is irrelevant if you are trying to imply that ethanol production suffers from some unique problem compared to gasoline and diesel fuel. Petroleum oil fuels are also used to find, ship, refine, and ship again crude oil and the finished petroleum fuels. In fact, the entire process of making gasoline causes gasoline to be more energy negative than the production of ethanol.

Getting less MPG by using an ethanol blend is not an issue of whether ethanol is helping to reduce America's dependency on petroleum oil. It is a personal issue of whether you are saving money by using E10 versus E0 (ethanol-free gasoline).

To determine if it is economically beneficial to use E10, you simply compare the price of E10 to E0. If you get 8% less MPG with E10 but save 10% on the cost of E0, then you have a net gain of 2%. In some regions of America, ethanol-free gasoline can cost $2 or $3 more per gallon than E10. So, if the cost of E10 is about $2.50 per gallon and you have to pay $3.50 for E0, then you are far ahead. In some areas E0 is only about 50 cents per gallon more than E10. This would result in a net savings of 12% by using E10. Since you find the loss of 8% MPG to be a bad, I presume you should be overjoyed in saving 12% or more on every gallon of fuel you purchase.

But let me take this issue further. Some people would argue that the overall reduction of petroleum oil that is displaced by E10 ethanol-gasoline is so insignificant that it's not worth the bother. To those people I say, "Fine—that's why America should be switching to E30, E40, E50 or E85 as the required blend level. This way every consumer would save even more money and we would be able to displace significant quantities of petroleum oil fuels."

The topic of ethanol versus gasoline is clouded in lies, but they are the lies invented and disseminated by the oil industry.

December 31, 2015

Posted By Greyswynd:

Does ethanol effect gas mileage negatively? My mechanic told me it does. My gas tank holds approximately 17.5 gallons of gas. I'm getting roughly 240 miles per tank, stop and go city/suburban driving. I drive a 2002 Toyota Sienna. I've heard a lot of hate stories about ethanol.

At first, I was siding with the other side about ethanol sucking water out of the air because I was biased, but when I heard your analogy of setting out ethanol in a glass and seeing whether or not the glass filled due to it sucking water out of the air sold me. The ethanol would of course, evaporate.

On top of that condensation isn't the same thing as hygro-whatever-you-call-it, like you explained. So why are you so pro-ethanol?

Reply from Marc:

Hey, Grey, I hope you had a great Christmas and that tonight will kick off a special New Year for you and your family.

Let me start with the 'ethanol in a glass' situation and its related issues:

Of course, evaporation would take place and that's one of the factors that makes the claim that ethanol sucks water out of the air so ridiculous. If your fuel was exposed to open air it would evaporate. This is true with gasoline as well, only slower. The point is, when would you have a situation in which you would just have your engine fuel left in the open? It doesn't happen in cars or motorcycles, except by accident; and it doesn't even happen in boats. Mercury Marine, the largest manufacturer of marine engines positively states that "ethanol does not grab water molecules out of the air." And they correctly attribute water collecting in boat fuel tanks as the result of condensation, not because the ethanol in the ethanol-gasoline blend sucked the water molecules out of the air.

Condensation is not the same as the hygroscopic process. People either ignorantly or intentionally confuse the two. An overly simplified way to look at what the hygroscopic process does is to say that a hygroscopic substance has a "wicking effect" on adjacent water. Cotton balls are a hygroscopic substance. If you place a cotton ball on your kitchen counter (instead of a glass of alcohol) the cotton ball doesn't suck moisture out of the air. But, if you place some water immediately adjacent to the cotton ball, the cotton ball will wick-up the water and "hold" it. This is essentially what ethanol does. Ethanol breaks down water molecules and it can hold the water elements. This is why Dry Gas (ethanol) is used in gasoline engines/tanks when there is water present because of condensation or other causes. The ethanol's hygroscopic characteristics allow the engine to fire and then expel the moisture in the exhaust.

You'd think that a certified auto mechanic would know the difference. Unfortunately, from experience in arguing the issue, I can tell you that there are far too many certified mechanics who are, I think, certifiable nuts, not certified mechanics.

Regarding your question "Does ethanol effect gas mileage negatively," let me break the question down. First, I'm sure that when you write "gas" you mean liquid gasoline, and not air-like fluid gas, such as natural gas or propane, right?

In that case, the question should really be "Does the presence of ethanol in a gasoline-optimized internal combustion engine negatively affect the output (mileage) of that engine?" The answer is yes, of course it does. The reason it does is that the gasoline-optimized engine is designed to run on the characteristics of gasoline. If the same engine was optimized to run on ethanol, then ethanol would produce comparable or better output (mileage) results than gasoline in a gasoline-optimized engine. The issue is optimization, not energy content or BTU rating.

This can also be proven by trying to use diesel fuel in a gasoline-optimized engine. Even though diesel has more "energy content" than gasoline (higher BTU rating), if you filled your gasoline vehicle tank with diesel you would not get more MPG, in fact you would get no MPG. This is because the gasoline-optimized engine is not optimized to run on diesel fuel. Similar to the ethanol-optimized engine scenario, a comparably sized diesel-optimized engine running on diesel fuel will get better MPG than the gasoline-optimized engine running on gasoline.

If ethanol or an ethanol-gasoline blend negatively affects the MPG of a gasoline-powered vehicle, why would anyone use it?

The answer has three parts:

1. The lower cost of ethanol usually makes up for the loss of MPG in a gasoline-powered vehicle. If you were to use E85 in a vehicle and get 15% fewer miles, but the E85 saves you 20% of the cost of ethanol-free gasoline, then you gain 5%. So, the use of ethanol doesn't negatively affect your vehicle's MPG. It would have a positive effect.

2. Using domestically produced ethanol helps make us

independent from the oil industry. Even domestic gasoline is controlled by OPEC in terms of availability and pricing. Using less gasoline, regardless of where it is produced, hurts foreign oil producers.

3. Ethanol is significantly cleaner burning. This helps with clean air and water. You'll notice I'm not referring to so called man-made climate change, I'm simply referring to having clean air and safer water supplies.

Regarding your 2002 Toyota Sienna: I have a non-flex fuel 2002 Ford Taurus. I use various blend levels of ethanol-gasoline, including E85. My car seems to run best at about a E30 to E40 level, which I achieve by 'splash-blending' my own mix (I pull up to an E85 pump and then to an E10 pump). The reason my car works better at E30-E40 than E85 is because the car's computer doesn't know how to handle the E85. But this is not an issue of lower "energy content," it's an issue of the car not knowing the proper air mixture (along with not knowing how to adjust timing and piston stroke length) necessary to run on E85.

If you have never tried anything higher than E10 in your Sienna, pump in one gallon of E85 the next time you fill up. See what happens. Most likely you'll not notice anything different. The next time try two gallons of E85, and so on until you experience a change in performance or MPG levels that you're unhappy with. You may never be unhappy, or you may find that a mid-range blend works fine, saves you a little money, and gives you satisfaction that you're doing something to help defeat terrorists and make America energy independent.

That's why I like ethanol. I can put it best by saying: I'd rather have my fuel money go to American farmers than to terrorists and foreign regimes that hate us.

If you have any further questions or comments, please let me hear from you.

Ethanol Chronicles
SPECIAL EDITION

David Are You There?

Originally published March 17, 2017

In 2015, I hosted a regular weblog on TheAutoChannel.com called "Ethanol Chronicles." I stopped at the end of 2015 because it became rather tiresome to respond to the same ill-informed and often inane questions and comments against ethanol, often from the same people. I guess they thought we wouldn't remember them, or that rephrasing an ignorant anti-ethanol comment would stump us. I'm not saying that there isn't a need to keep fighting back, but in 2016 I found other venues to present the truth about ethanol.

However, on March 16, 2017, Bob Gordon and I replied to a series of near-hysterical press releases issued by ethanol-haters. In response to one of Bob's replies, a reader named "David" wrote to my business partner on DISQUS to attack him and what he wrote. Bob's comments were included in a story titled "How Can Such Smart People Be So Stupid?"

https://www.theautochannel.com/news/2017/03/15/357517-how-can-such-smart-people-be-so-stupid-epa-to.html

Those who know us, know Bob and I are from Brooklyn. We discussed David's comments and it was decided that I would be the one to reply. Why me? Because it's never fair to pit a heavyweight against a lightweight, figuratively as well as literally. We didn't want the guy to be completely obliterated.

So, for the amazement, amusement, and edification of our regular visitors to www.TheAutoChannel.com, we resurrected the Ethanol Chronicles for a brief time to present "David's" comments, and my replies.

March 15, 2017

David To Bob Gordon:

I read the article you referenced. However, the writer of that article (you) is [stet] the stupid one.

You want to switch our fuel from gasoline with 10% alcohol to E85. The flex-fuel vehicles on the road today cannot run E85 without a computer retune and supporting modifications. Non-flex-fuel (most cars) cannot run E85 without extensive (and expensive) modifications. So, who pays for this?

What about the fleet of diesel trucks that transport nearly everything we use? They can't run E85. Ask any small engine mechanic and they will tell you alcohol-based fuel is bad news for your lawn mower.

You are wrong about the "facilities in place." Alcohol is NOT sent through the pipeline (causes corrosion), it's blended with gasoline at the time it is loaded on the tankers.

To make the switch as he suggests' would require massive amounts of land to be taken out of food production to provide his "dream fuel." Suppose the Midwest, our grain producing area, were to be hit with a 1930s type drought (the climate change lunies [stet] are predicting all sorts of calamities like this), transportation comes to a halt.

Since you referenced this article, did you also notice that the comments section on that article were CLOSED TO COMMENTS. Any wonder?

March 16, 2017

Marc Rauch To David:

Hi David - First, my business partner (Bob Gordon) and I would like to thank you for taking the time to read the article that you're complaining about. We like it even when unknowledgeable people visit TheAutoChannel.com because it shows up as a web-visit and there's always the possibility that you might click on one of the ads we get revenue from.

You have no idea what you are talking about. Vehicle owners around the world prove that late model non-flex-fuel vehicles can run on high ethanol-gasoline blend levels. I have done it personally for several years with a 2002 Ford Taurus. I've written about this on numerous occasions and testified about it twice in Congressional committee hearings.

Here's one link to a story about my Ford Taurus:

https://www.theautochannel.com/news/2015/08/10/138668-splash-blend-your-own-e15-e20-e30-or-e40-for.html

If you would like more, let me know.

It is true not all late model non-flex-fuel vehicles can go straight to E85, but we didn't say they should. Many of these vehicles can use E25 to E50 or more without any modifications.

However, those vehicle owners who would like to go to E85 may or may not require any modifications or might need a conversion kit (it is not particularly expensive). Expensive is the cost the EPA has forced on Americans who would like to convert their gasoline-powered vehicles to CNG; this costs about $10,000.

You ask, "Who will pay for it?", it being the kit to make non-flex-fuel vehicles capable of running on E85. Why is that a significant question? Those who wish to do it would pay for it—like when they decide which set of tires to use or what brand of oil to use; it's an optional choice. If all new vehicles were being built with the ability to run on E85, the cost is built into the sale price, and at that stage, it wouldn't really affect the price of a new vehicle at all.

You ask about diesel trucks and say they can't run on E85. You're correct, but no one is asking them to do that. They would run on biodiesel made from ethanol. And the best part is that no modifications are required for diesel engines at all. You should stop and think about what you're saying.

You ask about small engines and suggest that we speak with "any small engine mechanic." We have done that. We've talked to them and corresponded with them quite extensively. And we've published articles about ethanol and small engines. One example is this: "Why Do Small Engines Suffer from Ethanol Problems?" (It is in this book.)

Many auto mechanics (small and large engine mechanics) know as little about ethanol as you do; and like you, they believe all the bad propaganda spread by the petroleum oil industry. They have confused normal engines problems with problems they claim are caused by ethanol. In fact many of the problems are caused by gasoline. Read this:

"New Government Lab Research Confirms Ethanol Blend Fuels Are Better for Large and Small Engines." Here's the link:

http://www.theautochannel.com/news/2016/09/22/286340-new-government-lab-research-confirms-ethanol-blend-fuels-are-better.html.

Next you bring up the issue of transporting ethanol via pipelines. There are two things about this issue you don't understand. The first is that the ethanol business model should not be the gasoline business model. Ethanol should not be piped over long distances and would not be in a more robust ethanol-fuel environment. Instead, ethanol would be produced locally and 'short' hauled to retail outlets, following the dairy industry business model.

Then you assume the pipelines used to transport ethanol are not treated to be ethanol resistant. They are. In addition, the great mistake you and others have made is that you assume gasoline and oil are not corrosive, and that precautions weren't made over time to make pipelines and tanks resistant to them. However, petroleum oil and the finished fuels are corrosive, so an anti-corrosive ingredient is added to gasoline. It's really business as usual for those who understand how to do this. To people like you, it's a surprise. Then, you introduce the land issue to try to show you are a man of "environmental conscience." Very nice. Unfortunately, the land issue is really a waste of time because ethanol can be produced from so many different materials, including ocean algae and waste. As even you might imagine, using algae and seaweed requires no land whatsoever to grow the necessary crops—they flourish naturally. Using wood chips and other "cellulosic" materials require no extra land to grow anything.

Even if you would like to make this an academic argument for fun and consider only ethanol produced from corn, I suggest you read either David Blume's fabulous book *Alcohol Can Be A Gas*, or visit his website at *http://alcoholcanbeagas.com*, or both. I warn you that when you learn what David Blume will teach you about this issue, you will be very embarrassed for bringing it up.

Finally, you make the accusation that the comments section to the article was closed in order to avoid comments such as yours. If Bob or I were worried about comments like yours, I wouldn't spend this much time responding to you. The fact is we stopped using DISQUS on *TheAutoChannel.com* a couple of years ago.

I'm gonna' [stet] go one step further. I'm going to take your entire comment and my response to it and make a full story about it. It will appear on *www.TheAutoChannel.com*'s front page and then be available forever after via our internal search engine. It'll be posted in a few hours and I'll return to provide the link. If you have the guts to identify yourself with more than just a first name, I will be happy to include it along with your professional credentials, if you have any.

Best regards,

Marc J. Rauch

March 23, 2017

David To Marc Rauch:

Hello Mr. Rauch,

Guess you didn't expect to hear from me again. Your response to my comments concerning ethanol tells me that I must have pushed your hot button. I would have responded to you sooner had I not been out of town, away from my computer.

In reverse order, I am a retired business manager with education from (automotive) tech school; heavy equipment training in the Marine Corps; business major in college; and 70+ years of life experiences.

Early in my career, I, with two friends, opened an auto service business specializing in performance upgrades. We sponsored two of our customers' drag cars and, on occasion, I would accompany them to the track as part of the pit crew. We were very successful on the local tracks.

A later career change led me into the heavy industrial construction business, working for one of the USA's largest construction companies. I still maintained my love of the automobile and competed in local events when time permitted.

Now, to the details. I wrote that non-flex-fuel cars could not run on E85 without extensive (and expensive) modifications. First, here is, verbatim, from the owner's manual of my 2014 auto:

"Use only unleaded fuel or unleaded fuel blended with a maximum of 15% ethanol. Do not use fuel ethanol (E85), diesel fuel, fuel-methanol, leaded fuel, or any other fuel because it could damage or impair the emission control system.

"Use of any fuel other than those recommended may cause powertrain damage a loss of vehicle performance; repairs are not covered under warranty."

I also own a modified, high performance muscle car and discussed the use of E85 with my engine tuner. He told me that to correctly convert would not be possible without losing my emissions certification and that it would not pass state tests. "Racetrack only," he said. Even converting a race car to E85 requires different injectors, fuel lines and computer re-tune. So, when I wrote: "So who pays for this?", the feds sure won't. The cost must be borne by the owner. Even then, there is no benefit of extra mileage.

My comment concerning over the road trucks was because I understood your intentions was to replace America's petroleum-based fuels with E85. You clarified that with reference to biodiesel. However, biodiesel is still not widely available across the country.

You also stated that another alternative to diesel fuel was CNG and that the conversion costs about $10,000. Wrong. Ford offers a gasoline/CNG conversion option on the F150 for $350. Buyer only has to add a tank and related hook up items.

When you stated, "that many auto mechanics (small and large engine mechanics) know as little about ethanol as you; and like you, they have all believed all the bad propaganda spread by the petroleum industry." Here is what is in the owner's manual of my 2016 Kohler-powered lawn mower:

Fuel requirements: "Gasoline up to 10% ethyl alcohol/90% unleaded is acceptable. E15, E20 and E85 are NOT approved and should NOT be used."

So, are you saying that MANUFACTURERS know as little about ethanol as well???

The pipelines: You conceded that ethanol is NOT shipped through the pipelines. Several articles that I have read state that is because of the corrosive effect of ethanol.

I should have mentioned in my previous post that the recreational boating industry and the private aircraft owners have NO interest in alcohol-based fuels. That comes from owners of both.

You also listed several links to articles that I could reference to better my knowledge of ethanol. It was not necessary for me to read these as I have been following this story for many years and there are as many detractors as there are advocates.

Articles began appearing in the media as early as 2007 about the harmful effects of ethanol fuels. An article penned by Kenneth P. Green appeared on the American Enterprise Institute's website on July 29, 2008, gave us an interesting view of the ethanol vs petroleum debate: *https://www.aei.org/publication/ethanol-and-the-environment.*

A more recent article published in 2016 by the Yale School of Forestry and Environmental Studies titled "The Case Against More Ethanol: It's Simply Bad for the Environment" is also interesting reading and blows holes in your approach to cropland/seaweed use to make ethanol. The article can be found at:

https://e360.yale.edu/features/the_case_against_ethanol_bad_for_environment.

Argue with their "credentials!"

About your website: I had posted my comment via Disqus. The ONLY reason for linking to your website was the fact that a Bob Gordon had used the word "stupid" in referring to President Trump's impending trip to talk to automakers about the controversial EPA ruling. My first impression was of another liberal Trump-hating journalist using the internet to level a criticism of the President. And I don't think that was wrong. You say that your website does not use DISQUS, and that was the reason I was unable to post on your website.

The ethanol vs. petroleum debate is far from over. You mentioned in your bio that you had testified before Congress. I assume you meant about your preference for ethanol. Did you convince anyone of the benefits? One person who did like your idea was President George W. Bush. But isn't he the one you journalists all made fun of because you considered HIM a "lightweight"?

As businessman, the thing that almost always takes precedence over any decision to make changes is to consider whether the "new way" is more cost effective and does it improve the product or service. When this is imposed on the ethanol/petroleum debate, ethanol loses. It costs more to produce; is, and always has been, subsidized by the taxpayers; delivers LESS mileage than pure gasoline and is not a good business model; otherwise, there would be a big push to make your dream come true. This conclusion is in the two articles that I referenced.

In conclusion, I really don't care that you think of me as a "lightweight." Perhaps, because of my writing skills? But that is what YOU were trained to do.

Ben Franklin, writing in *Poor Richard's Almanac,* said: "The learned fool writes his nonsense in better language than the unlearned; but still 'tis nonsense."

Did I pass your "guts" test? I trust you to post this and INCLUDE THE TWO ARTICLES that I referenced.

Regards,

David

p.s. I was surprised that you, as a journalist, would have had typos in your response.

April 5, 2017

Marc Rauch To David:

On the contrary, David, I'm delighted you replied. It gives me more content to put online. I hope you had the chance to read our first exchange on TheAutoChannel.com and you will read this updated version of "Ethanol Chronicles: SPECIAL EDITION: David Are You There?"

You didn't push a hot button, we respond to people who are good enough to make comments, regardless of whether they are positive or negative. I wish those who I challenge would be as cooperative and considerate. They usually aren't.

Now to the details:

You tell me about your 2014 owner's manual, but you didn't tell me what kind of vehicle. That seems strange to me, but okay. You tell me that the owner's manual says anything above

E15 is unsafe. I already know that many owner's manuals say that. Some say don't use anything above E10, and there is even at least one non-flex-fuel car (a MINI) that says it's okay to use up to E25.

(See: *http://www.theautochannel.com/news/2016/02/01/190510-first-non-flex-fuel-vehicle-approved-for-e25-70-new.html).*

I don't care what the owner's manual says, I'm telling you what is true. The manufacturers have put limits in their U.S. owner's manuals in order to limit their warranty liability. But this doesn't mean the vehicle can't safely and economically use higher blend levels.

I'll give you an example. About a year and a half ago, I attended a media event to drive the new Volvo XC90 (beautiful car, by the way). During dinner, I asked the marketing director why the car is not being marketed as a flex-fuel car in America. He said, "Because it's not a flex-fuel car. You can't use anything higher than E10."

I insisted it was, he insisted it wasn't. This went back and forth until finally he said "Look, I'm the one who works for Volvo, I should know."

I had him where I wanted him.

I asked where the vehicle is manufactured, he said Sweden. I asked what's the difference between the XC90 made for use in Sweden and the XC90 made for use in the U.S., and he proudly said, "Nothing different." So I said, "But in Sweden your mandated regular fuel is much higher than E10." He didn't reply, I had him.

He did say that in Sweden they don't mandate E85, just E25 or E30. So I asked, why don't they market the car for the U.S. as being capable of running at E25 or E30. He said they don't do it because the engine and fuel system might have problems after five years.

I asked, "What's wrong with that?"

He said, "Because we make the engines to last for ten years."

I asked how long the engines and fuel system is warranted and he said three years. "In that case," I asked, "why isn't the engine and fuel system warranted for ten years?"

He said because it's too expensive.

So, it wasn't because the car can't use higher than E10, it's because they want a limit to their warranty exposure.

This is true with all cars sold unchanged in the U.S. and other countries. In Brazil, for example, the minimum ethanol-gasoline blend is E27. Many of the gasoline-powered cars and boat engines sold in Brazil are exactly the same as those sold in the U.S. If the engines can use E27 safely and economically in Brazil, they can do the same here. But they aren't sold that way, and their owner's manuals in the U.S. say not to use fuel over E10 or E15.

Now, if someone has a new vehicle that's still under warranty, and they're afraid to void the warranty, I always say that they shouldn't do anything that voids the warranty. However, there are millions of newish cars that are not under warranty any longer. Therefore, if a person wants to experiment, they should. And I say the same to you. Here's one of the typical stories I published over the years that addresses this issue:

http://www.theautochannel.com/news/2011/05/09/531654-using-e85-non-flex-fuel-vehicle-can-save-almost-1.html.

You mention your racing experience. Perhaps you should try racing in a different association, such as one of the associations sponsored by Ignite Racing Fuels:

http://igniteracingfuel.com/index.html.

If you visit this website, you'll find all kinds of stories about improved performance with their fuels (Ignite E85, Ignite E98, Ignite 108, and Ignite 114).

Regarding the question of engine optimization—yes, maybe changes must be made. That's part of what racing is about. If you normally fish for lake trout, but you go ocean fishing for sailfish, you may have to upgrade your equipment and it will probably cost you money.

Regarding bio-diesel: if it's not readily available it simply means that distribution has to be increased. It doesn't mean that bio-diesel won't work in every state. If people ask for it, or it's mandated by federal law, distribution will increase.

The CNG conversion option is for gasoline powered vehicles. I wasn't talking about diesel and I don't believe my comments were confusing on this issue. I wrote, "... the EPA has forced on Americans who would like to convert their gasoline-powered vehicles to CNG..."

Regarding the Ford CNG conversion option for about $300, this is for new vehicles with the conversion taking place in the factory at the time of assembly. This is exactly the problem. In the factory, at the time of assembly, the package is no big deal. But if you wish to convert an older car, let's say your 2014 vehicle, it will cost you thousands of dollars because of the fees levied by the EPA and some states. But it shouldn't cost that. Keep in mind that in some states, like California, you cannot convert any gasoline-powered car to CNG, regardless of cost. If you do the conversion outside of California and try to register it in the state, they will not allow it.

I owned a factory-converted CNG-powered Dodge Ram 1500 van for several years (I just recently sold it).

Regarding small engines and mechanics—I provided two links that deal with this issue. If you haven't already read this information you should do it now.

I stand by my statement that many auto mechanics know as little about ethanol as you do. As for the manufacturers, many don't know about ethanol, or they are doing what Volvo is doing, guarding against warranty claims.

As to the pipeline issue, I already answered it in detail, and I did state that corrosion could be a problem in untreated pipelines. However, when ethanol is shipped via pipeline, as they do in Florida, they use treated pipes. Ethanol is carried by tanker trucks and stored in tanks in the ground. These tanks use materials that are corrosion resistant, just like gasoline trunks and tanks. It's no big deal.

Regarding your comment on recreational boating and aviation, your impression is incorrect. Please revisit my comments about boats in Brazil. Mercury Marine has conducted significant tests and has no problem with E10, in fact they say it may be better to use E10 than E0. But, again, Mercury sells its marine engines in Brazil, and Brazil's minimum mandated fuel is E27, so we know that Mercury Marine engines sold in America can use higher blend levels than E10. This link will take you to the extensive ethanol webinar conducted by Mercury Marine:

https://youtu.be/GbkmY7TMKMg

As far as aviation is concerned, it's a different issue. In order to save small aircraft manufacturers from the added cost of re-tooling their engines, the FAA granted a waiver for the U.S. They considered that the amount of tetra-ethyl lead and other bad aromatics emissions from light aircraft was not important enough to worry about.

Embraer Aircraft produces ethanol-gasoline powered airplanes, and they have no problems.

(http://www.embraer.com/en-US/ImprensaEventos/Press-releases/noticias/Pages/Embraer-celebra-dez-anos-do-Ipanema-movido-a-etanol.aspx.)

Here's a video I shot of Greg Poe and his ethanol powered stunt plane:

https://www.youtube.com/watch?v=ierFrs5FkAU.

Regarding your experts and their published articles, you provided links and said to me "Argue with their credentials!"

Kenneth P. Green doesn't know what he's talking about. The link you provided was from 2008. It contains information that was out of date and irrelevant at that time. Today, the true information is, in many instances, miles ahead of him. I have to wonder if he was paid by the oil industry to give the presentation. I wrote to Mr. Green last week and received a reply. I responded to his reply with additional information, but he hasn't yet gotten back to me. You can read all about my correspondence with him here:

http://www.theautochannel.com/news/2017/04/03/364606-uncovering-gas-roots-contemporary-ethanol-opposition.html

The second person you refer to is C. Ford Runge at the University of Minnesota. His paper "The Case Against More Ethanol: It's Simply Bad for the Environment" was published last May, eleven months ago. I responded to Mr. Runge in August 2016. Here's my published story:

http://www.theautochannel.com/news/2016/08/28/275763-case-against-ethanol-opponents-they-are-simply-incorrect-video.html

And so, David, I did argue with them, and they are no match for the facts.

You write that you think that my partner and I are left-wing Trump bashers. Here's something you might find interesting: I am probably the most conservative right-wing guy you'll ever meet. My partner and I both supported Trump, and I wrote my first editorial supporting him five years ago. Because of my editorial I was invited to be a guest on Michael Savage's radio show. I used to be a frequent guest on a local Libertarian Party TV show, but I stopped doing it because they weren't as strong as I thought they should be. But this isn't a Republican vs. Democrat issue, or a liberal vs. conservative issue, this is an issue of the oil industry against the public.

The issue with DISQUS is exactly as I told you. Comments weren't closed because we wanted to avoid your comments, they were closed because we stopped using DISQUS.

The ethanol vs. petroleum debate is over. The strength and money of Big Oil allows gasoline to continue to be our primary engine fuel, but real arguing about facts is long gone. Should you come across someone who would like to debate me in a live, open forum, please let me know immediately.

Lastly, it's surprising that if you found some typographical or spelling errors, and saw fit to mention it, that you didn't reveal what they were. In any event, I admit to being mortal and to occasionally allow my fingers to type too fast, as well as to overlook typos. Because I am a professional journalist, and TV/Film producer, I know that spelling errors and typos exist in nearly all published works. I like the Ben Franklin quote you provided. Fortunately, it's not relevant to me because he was referring to fools.

By the way, you didn't pass the guts test—you didn't tell me your name and where you're from.

Gasoline vs. Ethanol
Ten Pictures That Tell the Story

Undecided as to whether we should change over to higher ethanol-gasoline blends for use in our vehicles? Take a look at these photographs.

Originally published November 10, 2017

Dolphins covered in oil.

Pelican covered in oil.

Dead Americans from Iraq.

Fracking.

Tar sands.

Or we can have this:

U.S. corn fields.

Which do you think is better for your lungs; for the environment; for all living creatures?

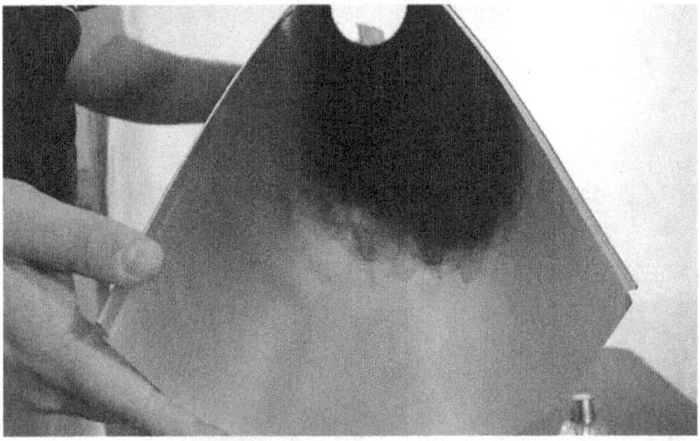

Burning gasoline residue.

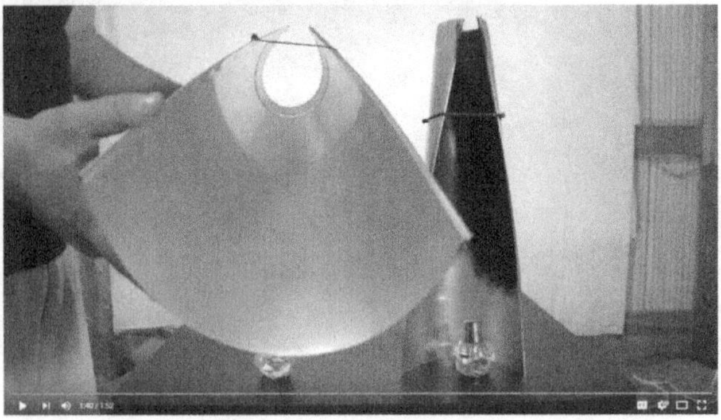

Burning ethanol residue - none

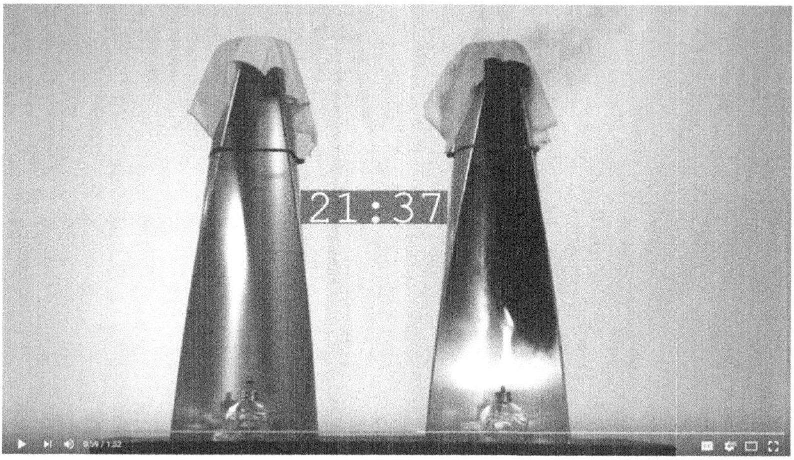

Comparing burning ethanol and gasoline.

Isn't it time we told Big Oil to go frack themselves?

Bonus Picture

With an ethanol spill the worst you get is a 'happy' seal.

Mugs, Pugs. Thugs, Nitwits, Halfwits, Dimwits, Snipers, Vipers, Con Men and Oil Men

No further explanation needed!

The Auto Channel Fights for the Truth about Ethanol Versus Gasoline

Lies and Misconceptions about Ethanol Continue to Keep it from Public Acceptance

Originally published May 25, 2010

A couple of months ago Bob Gordon and I posted some stories and a RoadTrip video we produced about ethanol. The video featured interviews with actor and environmental activist Ed Begley, Jr., and David Blume, author of *Alcohol Can Be A Gas*, the best and most complete book ever written about producing and using ethanol.
(https://www.youtube.com/watch?v=TshLDrREHk8)

Bob and I studied the issue of alternative fuels and energy and we've become very enthusiastic supporters of all the alternative technologies. We like CNG; we like propane; we like algae biofuels; we like electric; we like wind energy and solar energy, and we even like the concept of powering vehicles with compressed air. But most of all, we like ethanol. Why? Because ethanol can be used right now, anywhere in the U.S., and by most vehicles—without any engine conversions. Virtually every fuel-injected automobile can use E10, E15, E20, E30, and E50 without conversions and without being official flex-fuel vehicles. Some can even use E85. And with an inexpensive, easy-to-install device, every fuel-injected vehicle could use up to 99.9% alcohol (E100). Best of all, you can make alcohol fuel right in your own kitchen, garage, basement or backyard. You can't do that with any other alt-fuel.

In discussions about alternative fuels, it is often said that there is no "magic bullet" and that a variety of solutions will have to replace gasoline. Bob and I disagree with this philosophy. We think that one alt-fuel could become a primary replacement for gasoline, and that that alt-fuel is alcohol (ethanol). Gasoline was never the only engine fuel in use during the past 80 years. Diesel was required for powering large vehicles and equipment; and for those machines, something other than a straight alcohol or alcohol-gasoline blend is required. But all standard gasoline-powered vehicles could use just one fuel—alcohol. Domestically produced, environmentally friendly, terrorist free, and considerably more economical than gasoline.

To be sure, there are many questions surrounding ethanol, but the point is to tell you about our fight to overcome the lies and misconceptions about ethanol promulgated by the oil and gasoline industries.

In response to the stories and videos we produced, we received emails challenging our information. The challenges were based upon information contained in two anti-ethanol videos on YouTube. One video contains comments made by David Fridley, a scientist at the Lawrence-Livermore Laboratory in Northern California, and the other was a John Stossel story featuring Jerry Taylor, a Senior Fellow at the CATO Institute in Washington, DC. Stossel is a TV consumer reporter and advocate. The two videos can be found at:

(http://www.youtube.com/watch?v=DeVT7jMYZlo&feature=related

https://www.youtube.com/watch?v=j9QQcP_Y1Il)

Bob and I consider information contained in these two videos gross misconceptions and possible lies circulated by people working for the oil and gasoline industries. These dubious views are supported with millions upon millions of dollars oil industry apologists pour into politicians' campaign coffers and hand to others in the hip pockets of the oil/gasoline industry.

After watching the Stossel/Taylor and Fridley videos, I sent both emails expressing how appalled I was at their lack of knowledge and the fact that they were, perhaps unknowingly reciting a litany of lies promulgated by the industry. I offered various facts contradicting their statements. Mr. Fridley sent a very nice reply in which he restated what he had said in the video. He had no answers to the points I raised. Mr. Taylor replied a bit differently. He wrote:

"You will have to be more specific with regards to my appalling ignorance. But before you pipe up, make sure you're not demonstrating any of your own." He included a link to a report he wrote for the Federal Reserve Bank of St. Louis, titled "An Economic Critique of Corn-Ethanol Subsidies."

(https://object.cato.org/sites/cato.org/files/articles/jerrytaylor_aneconomiccritiqueofcornethanolsubsidies_2009.pdf)

I have to admit taking on a scientist at the prestigious Lawrence Livermore Lab and a guy described as "among the most widely cited and influential critics of federal energy and environmental policy in the nation" is pretty heady stuff for a simple little man like me who hails from the backwoods of Brooklyn, NY. The good thing is that when you catch a rascal with his hands full of misinformation, it's really not so daunting a task.

After reading Jerry Taylor's report to the Federal Reserve Bank of St. Louis, I sent him the following email:

> Hi Jerry -
>
> Thank you for having taken the time and having the interest to respond to my email. I also appreciate you sending me the link to the report you wrote for the Federal Reserve Bank of St. Louis.
>
> I read it and will comment on it below, along with addressing the specific problems I have with your responses to John Stossel in the video interview. To make sure you know the video interview that I am referring to, you can find it at:
>
> *(http://www.youtube.com/watch?v=j9QQcP_Y1II)*

It may be available at other places on the web as well.

In my initial email to you, I wrote that I was "absolutely appalled at how little you know (or knew) about the subject (of ethanol). I hope you learned something since you did that video." I then asked, "How can the Cato Institute be for individual liberty, free markets and peace if you spout all the lies that OPEC and the gasoline companies have fabricated about alcohol/ethanol?"

The reason for my being appalled is that you are touted as being "among the most widely cited and influential critics of federal energy and environmental policy in the nation." Now I understand that being widely cited and an influential critic doesn't require that you actually know anything about the given subject, it may mean that you just have a good press agent. But since those that cite you, such as Mr. Stossel, do so in a manner that would portray you as having some expertise, then I personally feel that you should have some knowledge of the subject.

I'll start with two issues in generalities: Alcohol as an engine fuel, and then subsidies.

Basic Facts About Alcohol/Ethanol

Alcohol (ethanol) as an engine fuel can be made from a variety of source materials; corn is only one, and from a productivity from yield standpoint, it is not necessarily one of the most efficient and economical sources. To take up the issue of ethanol by limiting one's argument to corn ethanol alone—regardless of the perspective—is ignorant and/or dishonest. It's dishonest if the position being argued is against the use of all ethanol. John Stossel approached the subject of ethanol in his video story as if it could only be made from corn. Your interview segment of the video only concerned corn. Given your overall predilection to be against alternative fuels, your responses were dishonest. If you were honest you would have brought to Stossel's attention that there are other source materials that might be far more efficient and economical.

John Stossel is a practitioner of TV sensationalism masquerading as journalism. Television, being the content ravenous beast that it is, requires a steady ingestion of stories. They don't have to be good stories, or important stories, they just have to be stories. The talent of the presenter to sensationalize or embellish the story is all that is needed. Stossel has become pretty good at this. Sometimes his stories are interesting and true, and sometimes they're interesting and highly sensationalized or embellished to give them an air of credibility. My guess is that he may have purposely left out the inclusion of other raw source materials for ethanol in order to create a sensationalized story. But this is not an excuse for you, if you were honest or objective.

You smugly pose circumstances that you think prove ethanol's impracticability by couching its production and distribution along lines of a business model used by the oil/gasoline industry. Since your bio on the Cato Institute website doesn't list any entrepreneurial business experience, I assume you have had none. This would certainly account for why you are ignorant of how business works in the real world. It might be polite to think that your lack of business understanding is the reason for your poor business interpretations, and it would have been hopeful if your report to the Federal Reserve Bank demonstrated that you had learned something about either real business or ethanol. In the video interview you did with Stossel, alas, that was not the case.

In in a real-world situation, the production and distribution of ethanol would rely on a business model significantly different from the oil/gasoline model for production and distribution. There would be no long-distance shipping of ethanol over the road or in container ships; it would never be piped from the Midwest to the coast. For you to present this possibility in the Stossel interview was beyond belief. Those really involved in the study and practical production of ethanol rely on a proven business model that has been successful in every country in the world. It can best be described as the "Dairy Business Model." The ethanol would be produced locally, then picked up and distributed to retail outlets in the same manner as milk.

On this subject of local production, corn might be the source material used in a given region, provided that it was the most economically viable source for that local market. Cattails, switchgrass, cane sugar, beet sugar, seaweed, garbage, discarded paper, wood chips and wood remnants might be more economical and plentiful in other regions. Moreover, cattails, switchgrass and seaweed require no expensive or environmentally harmful fertilizer (the requirement for fertilizer to help grow corn is another canard often used by those seeking to discredit ethanol). In fact, when it comes to cattails, switchgrass and seaweed (garbage too, for that matter), municipalities go out of their way to limit their abundant availability. Even if you think what you say was true, the argument that the U.S. can't produce sufficient ethanol from corn is moot.

Opponents of ethanol will argue that ethanol has negative EROEI (energy returned on energy invested) because of the energy it takes to produce and run the farm equipment that is required to grow corn. This is preposterous since it relies on the assumption that there is no ethanol-ready equipment currently in use by farmers to plant, harvest and store corn, and that an entirely new collection of mechanized machinery and storage facilities would have to be manufactured from scratch. It also fails to account for the ethanol fuels used to run the equipment at lower cost—with less pollution than gasoline or diesel.

Any claims that ethanol (alcohol) can cause damage to vehicle engines is totally false. Virtually any modern vehicle (1990 to present) can use a gasoline ethanol blend of proportionately high levels of alcohol with no problem, even if they are not officially flex-fuel vehicles. For modern vehicles that might have a problem with E85 or higher alcohol, an inexpensive easy-to-install device renders the vehicle capable of running 100% alcohol. Older vehicles, after being fitted with fuel injectors and replacing any rubber that is not tolerant of ethanol, can also function at full capability without any damage to the engine. Ethanol cleans engines, it does not harm them.

Government Subsidies

The argument against using government subsidies to increase the production capacity of ethanol and to encourage its use is ridiculously puerile. It's amazing that anyone would use it, regardless of where on the political spectrum they may fall.

I guess the only reason for doing so is that most politicians and cited analysts know that the general public will never research an issue and challenge such egregious statements as "If ethanol made a lot of sense, it wouldn't need to be subsidized." By the way, if this statement sounds familiar, it's because it's what you said to John Stossel. You may have thought you were safe from rebuke, but "Tag, you're it!"

There is very little that would have gotten accomplished in America (to the good and bad) if not for government subsidies, incentives, allotments, protectionist laws and taxes. No energy sector was developed and in continuous play without substantial subsidies, including the construction of hydro-electric dams, nuclear power plants, and oil exploration and refineries. This is as true today as it was when the first pools of petroleum oil were tapped in Pennsylvania in the 1850s.

Products made from petroleum oil, such as kerosene, would have never come into popular use for in-home heating and lighting if not for onerous taxes placed upon alcohol. Despite the smell and harmful fumes of kerosene, consumers were driven to it because its cost was so much lower than alcohol: kerosene production was taxed at just 10 cents per gallon, while ethanol was taxed at more than $2 per gallon. As the new petroleum oil companies prospered and grew, they were able to buy the political support they needed to protect their development of new oil-based products and keep alcohol production too costly. They finally made it completely illegal from 1920 to 1933, handing the oil companies the monopoly on producing vehicle fuel.

From 1973 to 2003, the Federal government paid $74 billion for energy subsidies to support nuclear power and fossil fuels, compared to just $26 billion in subsidies during the same time frame for renewable energy technologies and energy efficiency. In August/September 2008, Congress authorized $25 billion in subsidies to go to the U.S. automakers to help advance future fuels and technologies. Not one of the three Detroit automakers stepped up to claim some of that money. One month later, the CEOs of the three companies went to Washington to beg for subsidies to allow them to continue making gasoline-powered vehicles. The government gave GM and Chrysler more than twice what they offered for future fuel and technologies subsidies. The only people who benefited from this were the oil companies and OPEC. The oil companies and OPEC should have bailed out Detroit, not the American people. Gifting GM and Chrysler with billions of taxpayer dollars—and allowing them to bleed shareholders and debtors of untold billions more—is as much an oil/gasoline subsidy as handing all that money directly to the Saudis, Iran and Hugo Chavez.

It's amazing that opponents to alt-fuel subsidies think it's okay to subsidize oil and gasoline R&D, but that it's unpatriotic or anti-free market when it comes to subsidizing something that might wrest power away from OPEC and the oil industry.

Your Report to The Federal Reserve Bank

The report you wrote for the Federal Reserve Bank presents many of the same incorrect and misleading points that you made in the interview. However, the report goes much further in making outrageous claims (by you directly or via the use of quoted citations) with greater expansion on those statements made in the John Stossel interview. I've listed several of them below, with my comments. I stopped at nine examples for the sake of brevity and because the redundancy of your claims in opposition to ethanol—and your false statements—don't require reiteration.

1. You wrote, *"If ethanol is truly economically competitive with gasoline absent government preference—as many of its supporters seem to believe—then private investors will produce ethanol for the market regardless of whether government lends a hand (Tyner and Taheripour, 2008)."*

After a hundred-plus years of government support and subsidized funding to the oil/gasoline industry, it's absurd to think that an under-funded fuel/energy alternative of any kind could compete with gasoline without significant government assistance. The benefits of developing a domestic alt-fuel industry that causes less pollution and frees us from external forces makes the effort worthwhile, even if it was no more economical than petroleum oil fuels. For private investors to produce ethanol in the quantity that it needs to retire gasoline, the investors and the alt-fuel industry need the same government intervention that was used to make oil/gasoline the primary fuel. It's time to level the playing field.

2. You wrote, *"If ethanol is not economically competitive with gasoline, then subsidies distort the market by steering investment away from economically attractive gasoline and toward economically unattractive ethanol."*

This clever ruse hides the real issue. It's gasoline's price that is the distortion. So much in subsidies and incentives are given to the oil/gasoline companies that the price of gasoline is artificially held down. If they had to pay to defend their oil wells, pay to protect their lines of distribution, pay for all their own exploration, and bailout makers of gasoline-powered vehicles whenever they run into trouble (thereby protecting their primary user of their gasoline and oil), gasoline prices would be much higher.

Furthermore, if the government removed their unjustifiable limits on adding alcohol to gasoline, and mandated E15, or E20 or E50 for all fuel-injected vehicles, producers of ethanol would be able to attract more private investment, build more efficient plants, amortize expenses more economically, and take advantage of volume pricing as it might relate to the production and acquisition of raw materials.

3. You wrote, *"Economists broadly agree that, as a general rule, leaving production and consumption decisions to market actors proves more economically efficient than leaving the same to governmental planners. Only if some unique and fundamental failure occurs that prevents gains to trade in a given market is there room for the argument that government intervention improves the functioning of those markets (Cowen, 1988, and Cowen and Crampton, 2003)."*

By "economists," I assume you mean that great body of people, most of them never having been involved in the entrepreneurial building of a business. Those whose knowledge of market actions and demands come from textbooks or a laboratory-protected environment never had to face real supply and demand or competitive issues because they were able to simply buy their way into a prominent market position. The price of gasoline has nothing to do with the common definition of supply and demand. It is a highly regulated and rigged market. It is no more subject to legitimate supply and demand forces than is the diamond industry under the dictatorial control of DeBeers.

My biggest objection to what you wrote is that without the government stepping in to correct the unfair advantages given to the oil/gasoline industry since the mid-1800s, "market actors" will never get the chance to prove that ethanol can compete with gasoline and overcome it.

4. You wrote, "*The oil industry's reluctance to use high blends of ethanol in gasoline absent a government mandate, build ethanol delivery infrastructure to supply service stations, or provide E85 pumps are often marshaled as evidence that oil companies are unfairly strangling an economic competitor in its bed. The existence of this self-serving oil cartel is said to explain why this otherwise commercially attractive transport fuel – ethanol – requires government subsidies and consumption mandates. Yet, as of 2007, 38 percent of the retail fuels market was composed of independent service stations, not vertically integrated franchises, and another 13 percent of grocers and other hypermarkets. Only 49 percent of retail fuel was sold by stations associated with major oil companies. Likewise, 56 percent of the refining market was composed of independent, vertically de-integrated refining companies (Lowe, 2008). Big Oil is simply incapable of keeping ethanol out of service stations if profits are to be made by selling ethanol to motorists.*"

At best, this is just another example of your naïve understanding of commerce and any market. But since you are "among the most widely cited and influential critics of federal energy and environmental policy in the nation," I must conclude that your view is another baseless attack upon ethanol.

Why, and how could fueling stations sell anything but the product that is provided to them, and for which they have customers? You make it sound as if station owners can control the marketplace. Perhaps in 1915 an abundance of fueling stations with a preference for one fuel over another could have influenced which vehicles their local neighbors might buy, but those days are long gone.

Gasoline companies no longer need to own filling stations to control the market. They've done it by buying politicians and car company CEOs. Real competition between gasoline brands has been eliminated through mergers and mandated gasoline formulas: there is virtually no difference between them. Service stations no longer offer free drinking glasses, dishes or trading stamps because they no longer have to compete for business.

5. You wrote, *"The EIA calculates that federal oil and gas subsidies outside the electricity sector total $30,000 per million British thermal units (BTUs). Biofuel subsidies outside the electricity sector, however, ($3 billion of the $3.2 billion of which are directed at ethanol via the blenders' tax credit), work out to $5.72 million per million BTU (EIA, 2008, Table 36). Using EIA figures for oil and gas subsidies and estimates of the cost of the blender's tax credit from Koplow (2007), economist Douglas Tiffany (2008) calculates that oil subsidies in 2007 were slightly less than $0.15 per gallon of gasoline while ethanol subsidies totaled $0.588 per gallon. Whether we embrace a narrow or broad definition of subsidy, the conclusion is the same; oil subsidies are relatively trivial while ethanol subsidies are relatively substantial."*

The 2007 gasoline subsidy that you cite most likely did not include the cost of fighting a war in Iraq and a war in Afghanistan. It probably did not include the cost of stationing air force units in Saudi Arabia, Kuwait and other oil producing nations. The subsidy probably also did not include the cost of our navy protecting and escorting oil tankers (even those not registered to the U.S.). I'd be surprised to learn that the subsidy included the cost of beefing up the protection of American embassies around the world. And the 2007 subsidy definitely did not account for the cost of the lives of the U.S. servicemen who died while fighting these wars to protect the oil/gasoline assets. Protection subsidies like these have been paid by our government every year since at least the First World War.

There is nothing trivial about the subsidies that have been and are given to gasoline. Your estimation of them as "relatively trivial" is an indication of your lack of a moral compass.

6. You wrote, *"Proponents of ethanol subsidies argue that if the price of oil included the cost of our "oil mission" in the Middle East, the wars that the U.S. military engages there to protect oil supplies, the costs associated with our need to "kiss the ring" of Middle Eastern oil producers, the economic damage by terrorists from the flow of petrodollars into their coffers, and the harm done to U.S. interests by oil rich states like Iran, Venezuela, and Russia, then oil*

consumption would be far less than it is now. Alas, it is believed that those national security externalities are not embedded in gasoline prices and, as a result, gasoline consumption is heavily subsidized. Ethanol consumption is thus suboptimal and ethanol subsidies are an appropriate remedy. Economists, however, are far less worried about the national security costs of America's reliance on oil (foreign or otherwise) (Bohi and Toman, 1996) and with good reason: Economists understand oil markets far better than do foreign policy elites. The alleged national security externalities associated with gasoline consumption are for the most part a figment of an imagination unmoored from a good understanding of market reality."

If the cost of the wars and military actions were covered in the price of gasoline and/or the Federal taxes placed on gasoline, our debt would not have risen so high since Operations Desert Shield and Storm. This proves the cost of gasoline is heavily subsidized by the government.

Also, economists are worried about national security costs. Whether they understand anything better than anyone else is debatable, but for you to say that they are not concerned about those costs is outright nonsense. News reports and video clips that we have seen in real-time since Iraq first attacked Kuwait—right up to the latest reports on CNN or FOX News, are not a figment of anyone's imagination. American forces were and are fighting to protect the Saudi royal family's oil assets.

7a. You wrote, *"Many believe that reliance on foreign oil requires the United States to militarily defend friendly exporting states and to ensure the safety of oil supply facilities and shipping lanes. Those marching under banners declaring "No Blood for Oil" seem to believe that is the case, as do most mainstream foreign policy analysts. Delucchi and Murphy (2008) offer a rigorous attempt to quantify the public dollars associated with the "oil mission." They suggest that if motor vehicles in the United States did not consume Persian Gulf oil, the U.S. Congress would have likely reduced military expenditures by $13.4 to $47 billion in 2004 (one of the two years examined in the analysis)."*

The U.S. sent no troops to fight in the long Iran/Iraq war, even though both countries produce and export oil. Our non-involvement had virtually no effect on our way of life and there doesn't appear to have been any ill-effects to freedom for first-world countries. Not participating didn't cause us to spend extra billions to defend and prop up one or the other country. From 1980—the start of the Iran/Iraq war to 1988 (the end of the war)—the price of gasoline didn't change. It actually fell slightly. In the effort to defend the Saudi family and Kuwait (in other words, to subsidize them) we experienced skyrocketing gasoline prices and an exploding national debt. The only reason we went to the aid of Saudi Arabia and Kuwait, and then back into Iraq was to protect their oil assets. It's very possible that if we did not engage in the first and second Gulf Wars that prices of gasoline might not have gone to the levels that it has, and we wouldn't have spent trillions on the fight. We also would not have lost thousands of servicemen.

As far as I know, from 1980 to the present day, the United States has lost no servicemen in the effort to protect and defend ethanol production.

7b. You wrote: *"If U.S. motor vehicles did not consume any oil at all, military expenditures would have, oddly enough, gone down by far less: by $5.8 to $25.4 billion in 2004. The "best guess" of this analysis is that, if U.S. gasoline consumers were forced to pay for the U.S. oil mission, gasoline prices would increase by $0.03 to $0.15 per gallon. Simple economics, however, suggests that the oil mission—however large it may be—is unnecessary, regardless of what Congress may think. Oil producers will provide for their own security needs as long as the cost of doing so results in greater profits than equivalent investments could yield."*

If the oil producers are so capable of providing their own security, why was it necessary to stage Operation Desert Shield and then Operation Desert Storm? Why have the United States and other countries not been reimbursed 100% by the Saudi and Kuwaiti governments? Since we acted as a mercenary force, why didn't we make a profit by doing so?

For you to say that oil producers are willing and able to provide for their own security, after what has gone on, you would have to be deaf, blind and dumb.

7c. You wrote: *"Because Middle Eastern governments typically have little of value to trade except oil—oil revenues, for instance, are 40 to 50 percent of Iranian government revenues and 70 to 80 percent of Saudi government revenues – they must secure and sell oil to remain viable (EIA, 2006). Given that their economies are so heavily dependent on oil revenues, Middle Eastern governments have even more incentive than do consuming states to worry about the security of oil production facilities, ports, and shipping lanes (West, 2005). In short, whatever security our military presence provides (and many analysts think that our presence actually reduces security; see Jervis, 2005) would be provided by incumbent producers were the United States to withdraw. That Saudi Arabia and Kuwait paid for 55 percent of the cost of Operation Desert Storm suggests that keeping the Strait of Hormuz free of trouble is certainly within their means. The same argument applies to al Qaeda threats to oil production facilities. Producer states have such strong incentives to protect their oil infrastructure that additional Western assistance to do the same is probably unnecessary. Although terrorists do indeed plot to disrupt oil production in Saudi Arabia and elsewhere, there is no evidence to suggest that producer-state security investments are insufficient to protect their interests. The U.S. oil mission is thus best considered a taxpayer-financed gift to oil regimes (and, perhaps, the Israeli government) that has little, if any, effect on the security of oil production facilities or, correspondingly, the price of oil. One may support or oppose such a gift, but our military expenditures in the Middle East are not necessary to remedy a market failure."*

My response in 7b answers all this. But in addition, if all of our efforts and expenditures were unnecessary, as you set forth, that doesn't mean that we didn't spend billions/trillions on this particular oil and gasoline subsidy, it just means that it was another ill-advised subsidy going to the oil/gasoline industry.

8. You wrote: *"Does Western reliance on oil put money in the pocket of Islamic terrorists? To some degree, yes. Does that harm Western security? Probably not - at least, probably not very much. Before we go on, it is worth noting that only 15.5 percent of the oil in the world market is produced from nation-states accused of funding terrorism (Lundberg Survey, 2006). Hence, the vast majority of the dollars we spend on gasoline do not end up on this purported economic conveyor belt to terrorist bank accounts. Regardless, terrorism is*

a relatively low-cost endeavor and oil revenues are unnecessary for terrorist activity. That a few hundred thousand dollars paid for the 9/11 attacks suggests that the limiting factors for terrorism are expertise and manpower, not money. This observation is strengthened by the fact that there is no correlation between oil profits and Islamic terrorism. In Taylor and Van Doren (2007), we estimated two regressions using annual data from 1983 to 2005: the first between fatalities resulting from Islamic terrorist attacks and Saudi oil prices and the second between the number of Islamic terrorist incidents and Saudi oil prices. In neither regression was the estimated coefficient on oil prices at all close to being significantly different from zero. During the 1990s, inflation-adjusted oil prices and profits were low. But the 1990s also witnessed the worldwide spread of Wahhabi fundamentalism, the buildup of Hezbollah, and the coming of age of al Qaeda. Note, too, that al Qaeda terrorists in the 1990s relied on help from state sponsors such as Sudan and Afghanistan – nations that are not particularly known for their oil wealth or robust economies. Producer states do use oil revenues to fund ideological extremism. Saudi financing of madrassas and Iranian financing of Hezbollah are good examples. But given the importance of those undertakings to the Saudi and Iranian governments, it is unlikely that they would cease and desist these activities simply because oil profits were down. They certainly were not deterred by meager oil profits in the 1990s."

I must question your sanity. You start by writing that even if Western reliance on Middle East oil does put money in the hands of terrorists, it's "not very much," that's like saying "It doesn't mean anything." You point out that the attacks on September 11, 2001 only cost a few hundred thousand dollars – hardly enough to mention.

To a sane person, any money given to terrorists is too much. A sane person says, "Let's not do business at all with an oil producing regime that sponsors terrorists." A person with a working brain says, "We must do everything to reduce our dependence on the product that keeps these murderers in business. We must become energy independent."

Not you. In the Stossel interview, you declare that we will never be energy independent, and that domestic alt-fuel sources will never solve the problem. Ultimately you may be correct, if you and others continue to twist and misrepresent the facts regarding ethanol and other alternative fuels.

9. You wrote, "*Regardless, ethanol production cannot displace significant amounts of gasoline consumption (Akinci et al., 2008). Even if the entire U.S. corn harvest were dedicated to ethanol production, only 3.5 percent of current gasoline consumption would be displaced (Eaves and Eaves, 2007). All available cropland in the United States would have to be dedicated to corn production if all U.S. vehicles were powered by fuel composed of E85 ethanol. By 2036, all rangeland and pastureland would have to be added to that total to maintain adequate production. By 2048, all land outside of urban centers would be required for corn production (Dias de Oliveira, Vaughan, and Rykiel, 2005). Thus, no matter one's opinions about the dangers of oil dependence (foreign or otherwise), corn ethanol cannot displace enough oil to matter.*"

Ethanol can be produced in sufficient quantity to replace not just foreign oil, but all oil. After the period of time it will take to transition away from the use of gasoline—which should be mandated to be not more than five years, petroleum oil in America should be used exclusively to make non-fuel products that require it as a chemical ingredient.

About two years ago I wrote a paper on this topic, titled "No New Gasoline-Powered Vehicles By 2014…Can It Be Done?"

It can be found here:

https://www.theautochannel.com/news/2008/06/23/090705.html

I am appalled by your ignorance, your lack of morale fiber and shocked that any person or entities would seriously take advice from you. You should be ashamed at yourself, and the CATO Institute should be embarrassed to have to as an associate.

Cordially,
Marc J. Rauch

(I never received any additional reply from Jerry Taylor)

Thank Limbaugh, Beck, Hannity, Prager and Savage for Rising Gasoline Prices

High Profile Talkers Go Pro-OPEC To Attract the Most Buck for Their Bang

Originally Published December 28, 2010

The only reason oil and gasoline prices are rising again is because there is no serious competition to oil-gasoline. It's not that there aren't legitimate alternative fuels that could compete—of course there are. But the oil-gasoline industry has done such a masterful job by convincing politicians to see it their way with generous campaign donations and doing the same to big media by providing them with billions in ad revenue and have thereby kept alternatives off the starting line. Supporting oil-gasoline you have Republicans, Democrats, Conservatives, Liberals, Communists, and Libertarians; they're all there. If there's one thing that can be said about most politicians, it's that they can be bought. It's irrelevant whether they've spent their lives fighting for the concept of open free markets, or in pursuit of a glorious socialist wet-dream—for many; if there's a dollar bill to be had and they can get it, slap a sold sticker on their foreheads.

(Every U.S. President and U.S. Congress since WWI has dropped the ball on energy independence and is guilty of allowing the oil industry to con the American people.)

Mainstream media is the same. It doesn't matter if the publication or broadcast outlet is supposed to be on the right or the left, they are "pro-green." Not "pro-green" as in the green environmental movement, they are pro-green as in pro-greenbacks. The bean-counters who are responsible for making the decisions in the media, and the mouthpieces they retain to present their positions are only concerned with where they can get the most buck for their bang.

There are politicians on the right who want to do away with subsidies, claiming that subsidies are wasteful. So how do you explain Diane Feinstein, the powerful liberal senior senator from California? Feinstein can typically vote in favor of anything that gives taxpayer money away. Recently, however, she opposed the continuation of subsidies for ethanol production. Who'd ever have thought that she would come down on the same side of a spending issue as Rush Limbaugh? She received no money from the ethanol lobby. Is that why she voted against it?

Then we have the stalwart icon of the liberal environmental and socialist agenda: PBS.

If there's an insect to be saved from extinction, a human condition to be championed, PBS is your advocate. So how do you explain what happened to the television series San Francisco's PBS station produced with David Blume about ethanol?

Blume is definitely one of the most knowledgeable persons in the world on ethanol and the author of the exhaustive book *Alcohol Can Be A Gas*. After just one episode, the series was cancelled. The entire series was already in the can, ready to go. It was cancelled because the oil-gasoline contributor to the PBS station threatened to withdraw its funding if the show continued. This oil-gasoline company is one of the advertisers on TV and in print, promoting their support of finding alternative fuels and energy solutions. In other words, PBS can only support social issues if the gasoline companies allow them to do so.

We must not forget the self-anointed guardians of America's sovereignty and moral values: Rush Limbaugh, Glenn Beck, Sean Hannity, Dennis Prager and Michael Savage (there are others, but the list would be too long). They want to do away with out-sourcing jobs; they want to strengthen our borders; they want to reduce the level of imported products, and they want hip-hoppers to pull their pants up. But when it comes to oil and gasoline, anything goes: Suddenly, there's nothing we can't do to aid our brothers in Saudi Arabia, etc.—you know, the place where terrorist funding and all those terrorists come from.

Rush Limbaugh has actually referred to the oil-gasoline companies as "our gasoline companies." But we, if he means America, don't have any gasoline companies. Sure, there are gasoline companies that may have offices here in America, with

(l-r, top) Rush Limbaugh, Glenn Beck, and Sean Hannity.
(l-r, bottom) Dennis Prager and Michael Savage.

pay taxes. But these aren't "American" companies; they are global companies that participate in a global oil economy that is controlled by regimes that are enemies of America. When was the last time you heard of an "American" gasoline company refusing to follow the dictates of OPEC's pricing—even for gasoline refined from "American" petroleum-oil? NEVER!

Increased drilling for oil on American property will have no mitigating effect on the energy situation until and unless guarantees are put in place to ensure the oil/gasoline won't be sold to other countries, and the price at the pump will reward Americans for permitting our natural resource to be used by the oil companies.

If subsidies are no good, why do the radio-templars support all the subsidies that are given to the oil-gasoline industry, a highly profitable industry that requires no financial help?

If Rush, Glenn, Sean, Dennis, Michael and the others followed the principles they espouse, they would speak out against gasoline and support domestically produced alternative fuels. They would stop repeating and spreading the lies about ethanol, CNG and propane fabricated by the gasoline companies. They would stop pretending there was a division between "American" oil companies and foreign oil companies. If they hate the idea that "American" jobs and money have left the country, they should hate any money that goes out of the U.S. to prop up oil-producing foreign companies. The patriotic position to take is to support domestically produced, domestically priced alternative engine fuels. Since their voices are the loudest speaking for America, they ought to be championing America. If you stand for America, then stand up for America.

NOTES

NOTE #1: I like listening to these guys and have done so for years. I agree with many of their positions; but compared to them, I'm a true patriot.

NOTE #2: Obama is not off the hook for being responsible for this mess. He's proven to be as incompetent as his severest critics feared; he dissembled about everything he said in his energy policy speech in August 2008; and we'll be done with him in six years. Some or all of the radio guys could be with us for decades.

An Open Letter to Tim Searchinger and All Other Gasoline Apologists

Originally published February 11, 2011

Hi Tim -

In reading your editorial in the *Washington Post (http://www.washingtonpost.com/wp-dyn/content/article/2011/02/10/AR2011021006323.html),* I'm reminded of the famous movie line, "Just when I thought I caught up with the rat race, the rats got faster."

For a few years, whenever those opponents of ethanol launched a new attack on its widespread acceptance or expansion, we would hear about how the use of corn was affecting the overall price of food and causing worldwide starvation.

It took some time, but finally ethanol advocates started to prove that corn use for ethanol had no relationship to the rise in corn prices or general food prices, and that its use to make biofuels did not contribute to either a shortage or world hunger—primarily because the world outside of America doesn't eat corn, they eat wheat. That is why the World Bank's statement in the summer of 2010 reversed their earlier claim that ethanol was the reason for corn price rises and shortages.

Obviously, from your story, you know that the cost of the corn in a box of corn flakes is a minuscule amount, roughly 1.5% of a $4 retail price.

So, to use the rat race analogy, you came up with a new way to cast the same old worn-out story. Instead of corn flakes, which you agree that the world doesn't eat like we Americans do, you substitute corn meal for corn flakes. But you didn't do your homework very well.

Corn meal is used for many other things besides feeding starving people around the world. Corn meal is used as a plant fungicide, for weed prevention, insect repellant and to protect wooden floors. Why not blame the rise in corn meal prices on people who wish to use a non-chemical solution to their garden and pest problems? What? Chemical fertilizers and insecticides are not good enough?

Why not be critical of people who like to dance on wooden floors? You mean it's okay for a family in Somalia to starve just because some ballroom in New York doesn't want to refinish their wood floors so often?

Corn meal is used by seafood chefs as a simple, low-cost way to fatten clams and mussels before cooking them in any number of favorite recipes. Why not blame seafood lovers and restaurants for causing a run up of corn prices and shortages? I'll bet that there are far more people who regularly eat clams and mussels than use ethanol.

Expanding on this chain of thought, let's look at another significant use of corn: as a sweetener in soft drinks, candy and other snacks. Here we are, in the midst of a fast-food-obesity epidemic, and you're blaming ethanol for too much use of corn. Wouldn't it be better for everyone concerned if you, a fellow with a degree at a prestigious think tank, made the argument that we shouldn't be eating so much junk food? It would be better for the health of non-starving humans, while leaving more "corn" for the starving humans.

Next, since even the worst scenarios merely depict a situation in which corn supplies will only be low until the next harvest, not exhausted, then there isn't really any reason for corn prices to rise...except for one reason, and one reason only: commodity speculators. Even you must know that the previous rise in corn prices had been errantly blamed on ethanol production and was due to commodity speculators.

Most importantly, blaming ethanol production for the real or imagined starvation of peoples around the world has absolutely nothing to do with corn ethanol production and use. Blame it on the massive wheat failures in Russia and China. If either or both Russia and China would improve their farming methods, instead of screwing around with deviant forms of government for the past eight decades or so, they probably wouldn't have had to face crop failures, again. So why do we have to make up for their failures? Why are we responsible? Why should we have to sacrifice, once again, national aspirations for energy independence because of Russia and China's farming failures? When and where have you written a paper that lays the blame for food price increases or shortages on them?

It seems your education was wasted. If your goal in life was to figure out how to suck up to the oil industry, why didn't you open a gasoline service station and be satisfied with providing clean restrooms for your customers? Why take up space in a classroom? Why tarnish the name of an institution like Princeton? Why write and publish a report that isn't fit to wrap fish in?

If you got this far in the letter, thanks for your time. I hope that it has some impact on you.

Marc J. Rauch

(I never received a reply from Tim Searchinger.)

Big Oil's Anti-E15 Campaign Filled with Gross Exaggerations and Misinformation

The People Who Rammed Lead Down Our Throats for Nearly A Century and Denied It Was Harmful Have A New Story to Peddle

Originally published May 16, 2012

In 2010, the EPA issued a study and held a teleconference to announce, that after testing, E15 engine fuel (15% ethanol mixed with 85% gasoline) had the same effect on spark-induced internal combustion engine vehicles manufactured after 2007 as E10 engine fuel (10% ethanol mixed with 90% gasoline). There were no deleterious effects. NONE.

Listen to the complete October 2010 EPA press conference here:
https://www.theautochannel.com/news/2010/10/13/501225.html

Oil and gasoline industry executives and apologists, politicians and media personalities became apoplectic. They issued press release after press release denouncing the EPA repeated their tired rhetoric about how harmful ethanol can be to automobile engines. They challenged the EPA's authority to take on the study and announce that E15 was ready for prime time. Big oil claimed the EPA did not do sufficient testing and their results were premature. Among other things, they threatened to sue the EPA.

Four months later, in January 2011, the EPA extended their E15 announcement to include all passenger cars and light truck gasoline-engine vehicles manufactured for model years 2001-2006, giving them the green light to use it.

Read more about EPA's January 2011 waiver here:
https://www.theautochannel.com/news/2011/01/21/515393.html

The oil-gasoline lobby had fits and conniptions. They recirculated the press releases they used four months earlier, added some new vitriolic condemnations, and made good on the promise to sue the EPA.

Again, Big Oil claimed millions of vehicles would be damaged, that the EPA waiver was made without sufficient testing, and that it was premature.

In May 2012, the American Petroleum Institute and two auto manufacturer associations held a teleconference to announce the finding of a study they released several days earlier. The study, conducted by a private organization called the Coordinating Research Council, studied the use of E15 in eight vehicles and found that two of the vehicles suffered from pitted valves, leaks, and loss of compression. A former member of CRC joined the Big Oil panel to present these results and explain the technical aspects. Using the study's results, the group unanimously stated that their earlier criticism about the EPA results was correct, that the EPA was premature in issuing the waiver.

https://www.youtube.com/watch?v=CSqBZi8kexo

I listened to the live press conference and participated in it. In my opinion, this study and its findings are irrelevant, immaterial, off-base, wrong, and/or entirely fraudulent. The study and its findings are yet another episode in the long history of oil industry misdeeds that have always been, and still are, designed to keep America, and most of the world, enslaved to gasoline.

The EPA Waiver

The EPA took on the E15 study because they were asked to do so by members of the ethanol industry. They didn't do it because some New World Order socialist conspiracy to controls our lives. Perhaps because the request came from a non-governmental source, or maybe because the EPA lacks the authority to mandate the use of E15, they did not issue an E15 mandate. They simply observed that E15 could be used without undue harm. They recommended additional warning labels be placed at filling stations so consumers would understand that the fuel from a marked E15 pump was E15 fuel, not E85, not regular gasoline (E10), not diesel, plain water or compressed air. There are similar labels on other station pumps, so this label wouldn't be out of the ordinary.

The EPA's study was not done in haste. The original request for testing was submitted in March 2009. The preliminary findings were released in September 2010, 18 months later. That amount of time is at least as long as the time of the study conducted by CRC to dispute the EPA findings.

In the anti-E15 press conference, Big Oil stressed that CRC's study was conducted by experts. They used the word experts again and again, as if to say the EPA relied on the opinion of monkeys. The EPA's study was based on work done by the government's testing laboratories: The National Renewable Energy Laboratory and the Argonne National Laboratory. I tried to ask the panel if they felt the two U.S. labs were less expert or if they didn't have experts, but I was cut off before I could ask the question.

Rest assured that both government laboratories have experts. It makes sense that when the EPA did the test, they also tested E20 to see if E15 represented an invisible line that, if crossed, would have disastrous effects. E20 proved as harmless to the tested engines as E15. I know this because during the October 2010 press conference, I asked the EPA experts if they tested any other blend levels at the same time. You can hear it for yourself if you listen to the press conference via the link above. (Argonne and NREL have been studying ethanol for years.)

Why didn't the EPA recommend E20, too? I asked that question. They (the EPA) said they were only asked to test E15, so that's all they responded to. The E20 was used as a control of sorts. The significance here is that if the EPA wanted to force the evil ethanol monster on us, as API contends, they would have gone straight to E20. That information is now available because the question was asked.

To support the findings of the government labs, Ricardo Laboratories, perhaps the world's most prestigious private testing lab, conducted their own independent study of E15 and arrived at the same conclusions as the EPA and broadened their acceptance of E15 to include vehicles manufactured as far back as the 1994.[1] Read about the Ricardo Study here:

https://www.theautochannel.com/news/2010/09/15/496452.html

Ethanol is not a new engine fuel. It was used in the very first internal combustion vehicle in the 1820s. Until General Motors scientists invented leaded gasoline, ethanol was considered the better fuel. What the GM scientists sought to achieve with the invention of leaded gasoline was a fuel that mimicked the performance qualities of ethanol. Gasoline didn't become the dominant fuel because it was better than ethanol, it became the dominant fuel because GM, the world's largest auto maker, found a way to make billions of dollars of profit by promoting their new patented product. GM was given mighty assistance in achieving fuel dominance by partnering with John D. Rockefeller's oil-gasoline companies. Rockefeller then delivered the *fait accompli* when he gave millions of dollars to politicians to vote for Prohibition—virtually wiping ethanol from the field of competition.

Ethanol and various blends of ethanol-gasoline have been studied, tested and used in America and around the world for more than 150 years. To say that studies using ethanol are premature is a joke as unfunny as the one they told when they said leaded gasoline was not harmful to living things.

Pitting Valves, Leaks and Loss of Compression

The anti-ethanol crowd likes to say ethanol damages engines. They say it as if gasoline has no negative effect on engines. The anti-E15 panel members in the teleconference repeated, "It is gasoline that damages engines" ad nauseum. Gasoline causes pitted valves; gasoline can leak; engines running on gasoline can experience compression loss. Drive by any engine shop and see if they are working on engines that haven't been using ethanol. Those engines are damaged because engines deteriorate over time, and gasoline damages engines.

If ethanol would not be mixed with gasoline in higher compression engines, (all modern engines), the engines would blow apart because of the knock caused by straight gasoline. During the years that tetraethyl lead was added to gasoline, the knock was quieted. When leaded gasoline was finally legislated out of existence, the options were to go back to mixing in ethanol or methyl tert-butyl ether, aka MTBE. When they finally realized that MTBE was another poison, it, too, was eliminated. The only solution was the original solution—ethanol.

I've run across a few knuckleheads online who promote freedom of choice by saying Americans should be allowed to choose gasoline without ethanol. They don't get that if their gasoline didn't have ethanol in it, their engines would knock themselves apart. This freedom of choice has nothing to do with the Open Fuel Act, which would give us a choice in deciding which fuel to use.

Gasoline Engines Not Made for E15

The only comment I agreed with was made by one of the speakers who said gasoline engines were not made to run on E15. That statement is correct. However, it's irrelevant and immaterial.

A gasoline-optimized engine that can run on any fuel other than gasoline is remarkable. It's a testament to how fantastic ethanol is. Just think, if we were ever faced with an apocalyptic future, as envisioned, for example, in the *Mad Max Road Warrior* movie, we wouldn't have to hijack remnants of gasoline to run our gasoline-powered motor vehicles, we would just make ethanol out of almost any plant life, regardless of where we are. We wouldn't need to drill thousands of feet below the surface with expensive equipment—we could simply ferment whatever grows and use stone-age technology to distill it into alcohol.

A point most people overlook is that ethanol isn't the perfect fuel for a gasoline-optimized engine because the engine is optimized to run on gasoline, not because ethanol is not as good or better. In an engine that is optimized to run on ethanol (with proper fuel injectors, appropriate spark timing, and greater piston compression), ethanol will perform equal to or better than a gasoline-optimized engine. If you tried to use pure gasoline or E10 in the ethanol-optimized engine, your results would probably not equal the results of pure ethanol or a high ethanol-gasoline blend in a gasoline-optimized engine.

What we need is a fuel whose source and availability are not dependent upon foreign interests; a fuel whose success doesn't contribute to the war chests of terrorists and terrorist regimes; a fuel that doesn't poison us; a fuel that can be produced inexpensively; a fuel that can be produced in any geographic region.

We need a fuel like this right now, not 50 years in the future. Ethanol is that fuel. Ethanol's cousin, methanol, also meets these qualifications, particularly if the methanol is produced from easily sourced waste materials (not from CNG). Being able to power our vehicles from a fuel produced from an increasingly endless supply of human excrement is better than any *Star Trek* fantasy we can imagine, because it's doable right now.

As The Auto Channel opined in June 2008, we need legislation to mandate the end of new gasoline-powered vehicles. If you live in fear of a New World Order controlled by the people who occupy the United Nations, you should be equally afraid of the people who occupy the OPEC headquarters in Geneva and their minions housed in the API office building in Washington, DC.

Liar, Liar, Pants on Fire Update (July 16, 2018)

If you read "The Hypocrisy of Big Oil and API" in this book, you know that ethanol-gasoline blends of up to E30 were used for many decades in countries around the world. There were virtually no stories blaming ethanol for any problems to vehicle engines, fuel tanks or fueling stations. Everything claimed by the moderator in the audio API press conference was a lie. Each statement of alarm and caution against using ethanol was a lie. That's why apologists exist—to make lots of money feeding us a crock of malarkey.

Notes

[1]Why did Ricardo Labs only okay E15 for vehicles manufactured since 1994? In the early 1990s nearly all auto manufacturers changed certain metal and rubber components that were susceptible to corrosion from ethanol, in order to allow ethanol to be used without any harm. Typically, any ethanol-gasoline blends with higher ethanol levels are not recommended for use in cars manufactured prior to 1991-92. The great preponderance of vehicles on the road today were manufactured after the mid-1990s.

Biofuels Are Worth Every Effort and Cost
An Open Letter to Economist Richard W. Rahn

Originally published July 24, 2012

Newsmax.com reprinted an anti-ethanol article originally published by *The Washington Times*, written by Richard W. Rahn.

This is the kind of story that puzzles us (we know the answer, but it's still very puzzling) because it was written by someone who says he's pro-American and pro-capitalism. The article was published by two media outlets also claiming to be staunchly American and capitalist, but they take a very un-American position. The *Newsmax.com* story can be found at:

https://www.newsmax.com/Rahn/biofuel-gas-corn/2012/07/24/id/446374/

My response to Richard Rahn:

Hi Richard -

I just finished reading your article, "Biofuels Not Worth the Trouble or Cost" on *Newsmax.com.*

As with the positions taken by your colleague Jerry Taylor* at Cato Institute, I shook my head in bewilderment wondering how you arrived at your conclusions. The article seemed to be written by someone with absolutely no idea what he is talking about—or by someone in the employ of the petroleum oil industry. (After reading my letter, I hope you will be able to enlighten me as to which you are.)

I researched your name. I read that you're an esteemed economist; that you served as member of a national monetary authority, and that you are an economics professor. While I would agree that those credentials might entitle you to comment on how someone else accomplished a real goal, or to guide students through a curriculum that studies processes, those are not the credentials that would allow you to understand what the whole biofuel picture is about.

In my opinion, in order to understand the dynamics of what you attempted to explain, you would need true business experience. Perhaps, in the past, you may have tried to sell lemonade as a child to your neighbors or perhaps you flipped burgers during your high school summer vacations and started to get an understanding of what business and supply and demand are really about. But if you did, you got lost when you entered academia and became just another zombie-economist prattling on about chalkboard business formulas and equations that mean nothing in the real world.

Almost all manufacturers make products they believe will make them the most amount of money. This is the key to the entire capitalist system. I shouldn't have to point this out to you, since Cato Institute is supposed to be a bastion of capitalism and the free market economy. Unfortunately, your article indicates you don't understand this principle. Before you continue reading this letter, you might want to consult with your colleagues and confirm that manufacturers do, indeed, try to make products that make them the most amount of money.

Farmers are like manufacturers. The only difference is that they "manufacture" food instead of refrigerators or cars or television sets. But they have the exact same underlying goal: grow or produce items that will make them the most amount of money. If a farmer can make more money growing apples than potatoes, the farmer should, and probably would grow apples. This is not rocket-science economics; it is basic business sense.

Consequently, if a farmer can make more money growing corn than apples, why wouldn't he do that?

According to your article, farmers are villains for choosing to grow more corn than other products. You don't consider them savvy businessmen; you make it seem as if they are greedy and out to hurt people. What the heck is up with that?

If the U.S. government should happen to provide some financial inducement for a farmer to grow corn instead of some other crop, why is it wrong for a farmer to take advantage of the financial inducement? Would you vilify a building construction company for seeking government housing contracts during the national home-building depression? I don't think you would, so use some good old common sense in evaluating the decision/reason to grow corn.

I haven't yet addressed the central point of your article that biofuels are "Not Worth the Trouble or Cost." Before I do, I wanted to point out that since you don't understand the basic economic issues, you are not competent to evaluate the value of biofuels.

You wrote, "The argument for mandating ethanol in motor fuel was to help make the United States energy-independent and reduce carbon-dioxide emissions." Now the inconvenient facts:

- Recent studies show that the total carbon-dioxide emissions from growing, harvesting, processing and burning corn as ethanol are much greater than those from oil and gas production and use.
- Ethanol reduces gas mileage in cars because it is less energy-dense than gasoline, and it causes more wear and tear on engines.
- Without subsidies, ethanol is more costly than oil and gas."

Here are the facts you missed or misrepresented:

- Recent studies do not show that the production and use of ethanol creates more carbon-dioxide emissions than the production and use of gasoline. Stories circulated by the oil industry have attempted to dissect the entire process of both fuels and then cherry-pick certain elements of gasoline production/use and then compared them to cherry-picked elements in ethanol production/use which may appear favorable. These are not "recent" stories. They are re-worked lies and misinformation the oil industry has circulated for decades.

- To say that ethanol reduces mileage in cars because it is less-energy dense than gasoline is ludicrous. Energy density has nothing to do with the difference in mileage. The typical spark-induced internal combustion engine will get better mileage using gasoline because the engine is optimized to run on gasoline. If the engine was optimized to run on ethanol, it would get at least the same mileage, with more power.

  In order to optimize an engine for either ethanol or gasoline you need three key components: the fuel injectors, the timing of the spark, and the length of the piston stroke. None of these three components rely on "energy-density." The argument of energy-density, predicated upon BTU-ratings, is not relevant to internal combustion engines. BTUs are a good measure for steam-powered engines.
- Ethanol does not cause more wear and tear on engines; it helps them run more smoothly. It is why ethanol must be added to gasoline. Without it, poisonous lead or other poisons, the engines would shake apart from engine knock. Ethanol cleans engines. It removes the gunky deposits that form in gasoline-powered engines.
- Ethanol costs less than gasoline. Even without pump subsidies, E85 costs less than gasoline. Even when taking into account any reduction in mileage from using E85 in a gasoline-optimized engine, the lower cost for E85 makes up for the reduction in mileage. Therefore, there is a net advantage for the consumer.

Furthermore, since ethanol can be produced by anyone, anywhere, from a very wide assortment of raw materials that can cost as little as zero, ethanol will always be easier and cheaper to produce than gasoline—until and unless the oil industry can get the government to make alcohol production illegal again. I hope you are not an advocate for Prohibition.

The real issue is not even price related. The real issue is being independent from foreign interests who control our energy and our economy. If you do not support the idea of a New World Order in which the members of the United Nations tell us what to do, then how can you support an Oil World Order that controls us?

In simple terms, I'd rather have my fuel money go to American farmers than foreign terrorists!

I challenge you to respond to this letter. I dare you to step up to the plate and take a swing. You don't need two good eyes to see the truth. Heck, you don't even need one.

You're up!

(I never received a reply from Richard Rahn.)

AAA Blunder on Ethanol Sets Off Firestorm of Criticism

FOX Business News Worsens the AAA Blunder with a Barrage of Stupidity

Originally published January 8, 2013

On January 3, 2013, we received a copy of a video interview produced by *FOX Business News* with Melissa Francis, host of FOX's *Money* show, and Lauren Fix, who bills herself as "The Car Coach."

The topic was a recent editorial published by AAA that calls for the national suspension of E15 fuel sales because AAA claims that gasoline-ethanol blends can damage vehicle engines. The AAA editorial was published on December 12, 2012. This editorial came after a similar AAA editorial two weeks earlier that made the same claims. The two editorials can be found at the following links, respectively:

http://newsroom.aaa.com/2012/11/new-e15-gasoline-may-damage-vehicles-and-cause-consumer-confusion/

http://newsroom.aaa.com/2012/12/suspend-sale-of-e15-gasoline/

It is unknown why AAA felt the need to reiterate the same information, under a slightly different wrapping, just two weeks apart. Perhaps they did it because they felt there wasn't sufficient public attention paid to the first editorial, and it needed to be spiced up with a more dramatic headline. Perhaps the first release was a test to see if the editorial would create backlash; and not receiving any great blow-back, they decided to take a harder stand. Or perhaps, if I were really cynical, which I often am, I would say that maybe the petroleum oil industry paid AAA more money to repeat the accusations of engine damage.

When the first of the two AAA editorials was made public via a press release on November 30, 2012, TheAutoChannel.com published the story without countering information, because it seemed to be another attempt by Big Oil to fight the

EPA's recommendation that E15 fuel (a blend of 15% ethanol with 85% gasoline) be allowed to be sold for use in vehicles manufactured since 2001.

Background

On May 16, 2012, the American Petroleum Institute (API) staged a telephone press conference to present a litany of negative assertions related to the Environmental Protection Agency's decision on E15. API claimed everything from insufficient testing to the decades-old oil industry lies about ethanol being inferior to gasoline and that it damages vehicle engines.

On that same day, *TheAutoChannel.com* published a comprehensive story covering the API press conference, generally describing the API assertions as "Gross Exaggerations and Misinformation." This short description was a polite overall critique compared to the specific comments set forth within our report. (The story, along with the complete audio recording of the API press conference and the earlier EPA press conference can be found at:

http://www.theautochannel.com/news/2012/05/16/036507-big-oil-s-anti-e15-campaign-filed-with-gross-exaggerations.html

In addition to the May 16th story, we published several other opinion pieces responding to claims made against E15 and the EPA decision by third parties, apparently on behalf the oil industry. Links to three of these items can be found immediately below. The first story link covers the U.S. Court of Appeals decision against the oil industry lawsuit to stop E15. In its decision, the U.S. Court of Appeals stated there is "No proof that any engine has been or would be damaged by E15."

http://www.theautochannel.com/news/2012/08/17/046592-u-s-court-appeals-smacks-down-anti-ethanol-lobby.html)

http://www.theautochannel.com/news/2012/02/17/025717-afpm-warns-adverse-effects-e15-use.html

http://www.theautochannel.com/news/2012/07/12/042492-kansas-drivers-should-beware-new-e15-fuel-gasoline-pumps.html

We took the November 30th AAA editorial as a regurgitation of old anti-ethanol lies. Because we have a significant body of work done over the last several years about the benefits of ethanol, we didn't bother confronting the AAA allegations.

Back to The Present

With the publication of the second AAA editorial and the follow-up interview produced by *FOX News,* we decided to delve deeper into the issue. From comments and related links that we've received, it's clear that statements made by Melissa Francis and Lauren Fix met with wide scorn and disagreement.

After watching the interview, Ron Lamberty, Senior Vice President of the American Coalition for Ethanol, wrote to me:

> I was so stunned by the inaccurate remarks that I tried to count them (46—but I may have missed one or two), and then I started going through the video to document each misstatement and provide the correct information. I'm up to 15 of them at a minute and a half.

The video interview can be watched in its entirety at *http://video.foxbusiness.com/v/2000862202001/warnings-not-to-use-e15-gas-in-your-car/*

The first thing that came to mind when I saw the video was, "Why isn't someone from AAA being interviewed on the story?" But that issue was quickly overshadowed by the barrage of stupidity that followed. In the email that I sent to FOX News, Ms. Francis and Ms. Fix, I wrote:

> The information provided by Lauren Fix about E15 is almost completely untrue. Lauren's explanation of phase-separation and the food-price argument about corn are preposterously puerile. In fact, if you live in a cold climate and your fuel tank and lines have a tendency to collect condensation (water), which could

> freeze and cause real damage, the solution is to put Dry Gas in the fuel tank. Dry Gas is alcohol. Alcohol is ethanol. Ethanol "absorbs" the water moisture.
>
> Fuel left in an unused engine for an extended period of time can break down and cause a starting or running problem. This is true of all fuels, especially gasoline, which leaves gummy varnish-like deposits. Sta-Bil, another product used to stabilize the gasoline left in dormant engines, also contains alcohol to help prevent the gummy build up. And again, alcohol is ethanol. (See Sta-Bil footnote at bottom of this chapter.)
>
> E15 will not damage the engines of vehicles older than 2012. It has been extensively tested. It can be safely used in all modern gasoline-powered vehicles manufactured since the early 1990s, whether they are "flex-fuel" vehicles or not. When the EPA conducted their tests on E15 and gave it their 'clean bill of health," they also tested E20 and reached the same positive conclusions.
>
> I have been test-driving vehicles for 25 years and have regularly used various blend levels of gasoline and ethanol since the time they were made available in California, with no negative reactions. I own and drive a non-flex fuel 2002 Ford Taurus I run on high blend levels of gasoline and ethanol. My vehicle suffers no problems normally associated with all-gasoline-powered vehicles.

I waited a day for their individual or collective rely. Then I went back to the November 30th AAA editorial and also studied the December 13th AAA story that triggered the FOX interview.

On January 4th, I sent this email to Michael Green, Manager of AAA Public Relations. Michael Green is listed on the AAA webpage as the contact person for more information about the stories, although the author of the two editorials is identified as Robert Darbelnet, CEO and president of the AAA.

Yesterday, I was made aware of a video story produced by Melissa Francis and FOX News based either upon the above titled AAA editorial written by your organization's CEO, or a very similar earlier AAA editorial released on or about November 30, 2012. (Please feel free to share this email with Robert Darbelnet and other AAA staff members.)

In the video, Ms. Francis introduced Lauren Fix to comment on and explain the 'warnings' in your editorial. Although neither Ms. Francis nor Ms. Fix identified Ms. Fix as an official AAA spokesperson, she seems to have acted in that capacity.

I found almost everything Ms. Fix said about ethanol to be either wrong information, a possible lie, or a recitation of the typical bad propaganda spread by the oil industry and its lackeys over the past 80+ years. In a separate email, I made my opinion known to both Ms. Fix and Ms. Francis.

Michael, unless you can confirm Ms. Fix as an official spokesperson for AAA, I am sure you will not have any remarks to make regarding her comments, and that's fine—since the point of this email is not to get the AAA reaction to her comments. I'm simply including this episode as background for the questions I have regarding the AAA editorials.

My questions are:

"1. What oil-industry-independent 'research-to-date' was Mr. Darbelnet citing that '...raises serious concerns that E15...could cause accelerated engine wear and failure, (and) fuel system damage?'

"2. What information do you have, other than unsupported oil-industry claims, that the EPA did not conduct tests sufficient to determine the safety of using E15 in gasoline-powered passenger vehicles manufactured in the past two decades?

"3. Why does the AAA consider independent E15 testing conducted by Ricardo (findings released

September 2010) to be insignificant confirmation of E15 as safe for all modern gasoline-powered vehicles?

4. In paragraph 8 of the editorial, Mr. Darbelnet states, 'Some of those supporting E15 admit the fuel may cause damage,' and gives this example: '...[S] ome underground storage tank systems, both new and used, exhibited reduced levels of safety and performance when exposed to E15.' Given that all fuel underground storage tank systems routinely experience problems, what information do you have—other than oil-industry anti-ethanol-biased research—showing that E15 underground storage tanks experience problems that are greater and/or more frequent than underground storage tanks used for diesel, E10 gasoline, E85, or gasoline that contain no ethanol?

5. Regarding paragraph 8... how does this potential problem relate to vehicle engine damage? And, wouldn't it be fair to say that combining the two points is to create an irrelevant red-herring warning?

"6. Does AAA agree with the overall warning *FOX News* issued, one they based on the AAA editorial about E15—or did they overstate your concerns?

I received this reply shortly thereafter:

Lauren Fix is neither an employee or spokesperson of AAA and her views are independent of ours. We did not work with either Fox or her prior to the interview. AAA's views are expressly stated on our website.

In regard to your specific questions, I have passed them along to our engineering team to review. In regard to question four, I can say that the warning regarding storage tanks was made by the Renewable Fuels Association to fuel retailers and was not from research conducted by AAA. They would therefore be in the best position to say why they made that recommendation.

In response to this reply, I then wrote:

Thanks for your quick reply and clarification confirming that Lauren Fix has no official relationship with AAA.

While I look forward to receiving the AAA engineering team's comments, I'm troubled by your response to my question #4 regarding the storage tanks. You wrote, '...the warning regarding storage tanks was made by the Renewable Fuels Association to fuel retailers and was not from research conducted by AAA. They would, therefore, be in the best position to say why they made that recommendation.' But the AAA story called for the suspension of the sale of E15, claiming it damages the engines of most gasoline-powered vehicles, and offered as part of your proof that the ethanol industry concedes that there are problems.

Let's face it, engine damage is the crux of the story; it is, in fact, the salient part of the entire denigration effort by the oil industry and anyone associated with them to stop ethanol—their (false) claim that ethanol damages engines.

RFA didn't issue a warning that consumers shouldn't use ethanol as an engine fuel and then cite a storage tank issue as the reason. The storage issue has absolutely no bearing on ethanol's capability as an engine fuel. AAA took the RFA warning out of context and made a leap that should never have been made. It would be like someone using the warning printed on plastic bags (the suffocation warning) to claim that carrying groceries in the bag makes the items being carried dangerous to eat. One thing has absolutely nothing to do with the other. It is entirely possible that a fuel can be the safest, most efficient and economical fuel to use in an engine but requires some degree of care when storing. Would AAA recommend that people stop drinking milk because if it's not refrigerated it could render a person seriously ill?

As I noted previously—gasoline also requires care when storing and is far more dangerous than alcohol. Why not issue a recommendation that all gasoline sales be suspended until the explosion/fire/storage/pollution problems related to gasoline are solved? If AAA or Mr. Darbelnet are unclear as to the issues regarding ethanol underground storage so that you can't respond to my question, it should never have been included in the story, whether ethanol damages engines or not. If AAA is objective on the overall issue of ethanol versus gasoline, as is alluded to in the editorial, then a big mistake was made.

Michael Green's reply was:

Our engineers will take a look at your questions and I appreciate your feedback. I must say that your very strong advocacy of E15 seems uncharacteristic of an independent news source.

Our main concern from the beginning is regarding consumer warranties. Nearly every automaker is on record saying that your car's warranty will be voided if you use E15. This is of serious concern to AAA and should be of concern to anyone who owns a car. AAA believes that ethanol-blended fuels have the potential to provide motorists a clear choice at the pump that supports jobs, promotes energy independence and reduces fuel costs. Both E10 and E85 provide options for consumers at this point, and AAA would support a motorists' right to choose E15 once basic thresholds have been met regarding consumer protections.

You rightly note that the U.S. Department of Energy rigorously tested E15. This research focused primarily on exhaust emissions and associated components such as catalytic converters. While this research was consistent with the agency's mission, it never fully examined whether E15 might damage engines and fuel systems. While current research is not sufficient to oppose E15 as a product, the significant lack of consumer awareness and protection highlights the need for additional study.

Both the EPA and the ethanol industry recommend that motorists check with the manufacturer on whether their vehicle's warranty would cover E15 usage. AAA did this and found that automakers have approved less than five percent of cars on the road to use E15. Five manufacturers (BMW, Chrysler, Nissan, Toyota and Volkswagen) state their warranties will not cover E15 claims. Eight additional automakers (GM, Ford, Honda, Hyundai, Kia, Mazda, Mercedes-Benz and Volvo) state that E15 does not comply with fuel requirements specified in most owners' manuals and may void warranties. The only vehicles on the road approved by manufacturers to use E15 include flex-fuel models, 2001 and newer Porsches, 2012 and newer GM vehicles and 2013 Fords.

The automakers' position leaves a huge gap between the limited number of vehicles that automakers will cover and the vehicles the EPA ruled can use the fuel. This sort of conflicting information puts motorists at risk, and AAA believes it is inappropriate to sell E15 while these issues remain unresolved. It is for this reason we made public our call for additional education and more effective labeling at the fuel pump before we would support the sale of E15. AAA recommends the EPA, fuel producers and automakers collectively develop a long-term plan that promotes public education, while implementing improved labeling and warnings at the pump. Additional research also is necessary to better understand the full consequences of using E15 in older and newer vehicles.

After considering these comments, I wrote back:

I appreciate your taking so much time to address my comments and questions. And I especially appreciate your recognition of our strong advocacy of ethanol. I say ethanol, in general, as opposed to just E15 because our support is for ethanol in general, not just a particular blend level.

However, should you have the occasion to do so, you'll find that The Auto Channel is also a very strong advocate of all domestic alternative energy solutions, and we have spent considerable time covering all possibilities. We have evolved to favor ethanol because of its ubiquitous production possibilities (it can be made almost anywhere, by almost anyone, from a wide variety of resources), and because it can be used immediately by virtually any modern gasoline-powered vehicle (to one extent or another) without expensive engine conversion and without requiring changes in local or national laws. The same is not true for CNG or propane, which we otherwise consider to be excellent alternatives.

Our primary objective in the adoption of an alternative fuel solution is not predicated upon environmental concerns, but upon economic opportunity. That's not to say that we don't appreciate an alternative fuel's environmental friendliness—we love clean skies and water as much as anyone—we just don't think that energy independence and a healthy environment need to be mutually exclusive.

Regarding the issue of The Auto Channel's media independence vs. advocacy, there's no inconsistency.

We look for the truth and try to report it as truthfully as possible. When and if we learn that our position is incorrect or founded on incorrect information, we change our opinion. We're not wedded to any financial remuneration that would spell our demise by changing our position.

We have repeatedly asked for information that would prove our position wrong—if, indeed, it is wrong. But that information is not forthcoming What we get, if we get anything, is repetition of the oil industry party line, same-old-same-old lies, distortions, and nonsense.

Setting aside our feelings for domestic alternative fuel and energy solutions, we do not alter our reporting on gasoline or diesel-powered vehicles, unless there is a particular 'energy' reason for doing so. We can still fawn over a Maserati or Ferrari, no matter the fuel it uses and what its MPG is. We're as fascinated as the next guy with electric solutions—we're just realistic in how we appraise electric's potential in the next few decades.

TheAutoChannel.com contains over one million pages of content and our overwhelming position can only be defined as objective. Unlike many alt-fuel advocates, we do not favor small vehicles versus large vehicles. We think that people should be able to buy whatever size vehicle they want and burn as much fuel as they want. The only concern we have is that the fuel should be domestically produced, be inexpensive, and not enslave us to foreign producers and questionable regimes. Moreover, we don't jingoistically think that only America should benefit from an enlightened domestic fuels program, we think all countries should do the same with their respective resources.

We are aware of the warranty issue and the reluctance of auto manufacturers to accept responsibility for vehicles using a fuel that they don't explicitly endorse. But the manufacturers are always reluctant to change and fight anything that is not part of their status quo—unless they perceive a significant marketing or financial advantage in doing so. If the manufacturers had their way, they never would have abandoned leaded gasoline; they would never have adopted the use of catalytic converters; they might not have voluntarily accepted universal seat belts and airbags, or many other safety features, etc., etc. Their reluctance is less indicative of qualitative or safety reasons than it is because of negative financial implications arising from competitive concerns or insurance claim vulnerability.

Moreover, OEM (original equipment manufacturer) personnel are as likely to be ignorant of alternative fuel features and specifications as non-industry personnel. Even those specifically tasked with marketing non-traditional fuel-powered vehicles are, typically, only semi-knowledgeable about their specific energy format and will often repeat the egregiously incorrect information about alternative fuels. Consequently, their misunderstanding of ethanol's benefits is easy to understand.

The answer, for us, is a national paradigm shift. Ethanol is the best engine fuel for most passenger and lighter commercial vehicles. History and the test of time have proven this. Even if the use of ethanol in any blend level did create engine damage, the answer is not to abandon ethanol; the answer is to immediately mandate that all new passenger vehicles and light trucks sold in America be built with engines optimized to run on ethanol (not gasoline). The government should provide a trillion dollars (since the administration seems so happy to dish out economic stimulus money) for whatever refitting is

> necessary for all vehicles produced since the mid-90s. The really good news is that because ethanol does not damage engines and can be used in very high blend-levels with little or no re-fitting, the government may not have to lay out very much money.
>
> The result would be a fantastic multi-year economic stimulus. Concerning the warranty issue, let the government mandate that five or ten cents of every gallon of ethanol sold be set aside in a fund to reimburse the manufacturers for costs related to "ethanol-caused" repairs, if any are reported and confirmed. Auto dealers would love for their service departments to have the extra work; they just don't want to have to foot the bill."

I didn't expect a reply and I didn't get one. During my correspondence with Mr. Green, I copied Lauren Fix, Melissa Francis and FOX News on everything. I wanted to see if they would comment on the AAA back and forth, or at the least my original comments to them. As of January 8th, I had not received any response from Francis, Fix or FOX, and nothing further from AAA.

I did, however, receive an email from Christina Martin, Executive Vice President of the Renewable Fuels Association, on the point of contention that I raised with AAA about E15 underground storage. Ms. Martin forwarded this from one of her colleagues:

> You are correct, our statements were precautionary and stated in a way to prompt station owners to investigate their tanks and fuel compatibility—not just take a blanket statement by a handbook as sufficient. Page 10 of the retailer handbook states not all tanks may be compatible with E15...there was no testing that supported that statement, in fact DOE (Department of Energy) just finished testing on UST

(Underground Storage Tanks) system materials late last summer.

Here are some facts to use:

• According to EPA Underground Storage Tank Program, there are 590,000 active underground storage tanks currently in America; there is a 70.9% compliance rate to prevent and detect leaks at these active USTs. That indicates there are 29.1% of the current tanks that are leaking or have serious potential to leak.

• Further, EPA UST program states there has been a steady decline in annual UST releases since 1990; the very same time period when ethanol blending, storage and dispensing has reached an all-time high.

• The Department of Energy's Oak Ridge National Laboratory has studied ethanol's potential to affect underground storage tanks and published their results in July 2012. Oak Ridge concludes that "in general, the materials used in existing UST infrastructure would not be expected to exhibit compatibility concerns when moving from E10 to E15." Oak Ridge admits housekeeping issues like water contamination and microbial growth are more likely to have a negative impact on USTs than adding ethanol to gasoline.

As I made clear in my first email to Michael Green, AAA took a small piece of information out of context, blew it out of proportion, and used it to help support the fallacious oil industry position that ethanol damages engines.

Also commenting on the AAA editorials was Bobby Likis, host of the nationally syndicated Car Clinic Radio Network. He wrote:

I'm both surprised and disappointed in the entire premise of the AAA E15 article…In the 41 years I have

> been—and still am—hands-on in the diagnosis, service and repair of cars, I find the repeating of an already dispelled myth that E15 ruins engines distasteful and contrary to the results found in extensive EPA and university studies, as well as in my first-hand experience with over 175,000 cars that have rolled through my service shop. In 41 years, we have not diagnosed a single 'ruined' engine due to ethanol.

David Blume, perhaps the world's leading authority on ethanol production and author of *Alcohol Can Be A Gas*, had this to say about the AAA editorials:

"Of course, this isn't the first time AAA sold themselves to the American Petroleum Institute. The first time, in the 1930s, had them conducting a phony test on phase separation in an attempt to prevent wider-scale use of alcohol."[1]

The oil industry used all of these anti-ethanol arguments years ago to discredit the government's decision to deploy E10 gasoline blends—the gasolines that are our standard regular and premium gasoline fuels. The claims weren't true then, and they are not true now. We also do not hear of consistent, universal problems from other parts of the world where higher ethanol-gasoline blend levels have been used for more than 50 years.

The oil industry has one goal: to keep us enslaved to gasoline and reliant upon foreign entities. AAA's participation in this scheme, whether willingly or unwittingly, damages America's economic health. As The Auto Channel has pointed out in the past, AAA is not the only "American" organization acting as an apologist for the oil industry. The contamination runs deep.

Even if the use of ethanol in any blend level did create engine damage in a gasoline-optimized internal combustion engine, the answer is not to abandon ethanol. The answer is to immediately mandate that all new passenger vehicles and light trucks to be sold in America be built with engines that are optimized to run on ethanol (not gasoline)—and the government should provide a trillion dollars (since the administration seems so happy to dish out economic stimulus money) for whatever refitting is necessary for all vehicles produced since the mid-90's.

(1) Additional information about the ethanol tests conducted by AAA in the 1930s, and the ensuing controversy from the questionable results, can be found in *The Forbidden Fuel – A History of Power Alcohol* written by Hal Bernton, William Kovarik and Scott Sklar. Pages 19-20 and 91-93.

(2) Sta-Bil Footnote: After publication of this story we received a call from the PR firm for Sta-Bil's manufacturer Gold Eagle Co. They would like us to advise our readers that contrary to popular belief, and although there are fuel additives on the market that contain alcohol, that Sta-Bil does not. From my recollection as a Sta-Bil user, I recall that it did contain some type of alcohol. Various references on the Internet seem to confirm that a previous Sta-Bil formula did contain alcohol. But we could all be wrong. In any event, the official response from Gold Eagle Company concerning the Sta-Bil formula is: "STA-BIL® Brand Fuel Additives are a complex blend of chemistries. There is no one additive that accomplishes all of the water handling required to prevent water damage in an engine's fuel system. We are unable to divulge the actual chemistry as it is a closely held trade secret."

The following is a list of various engine fuel additives we believe contain some type of alcohol. The point is alcohol cleans gasoline engines and removes water from the fuel system. If alcohol damages engines in the manner suggested by the oil industry (and AAA), then why would anyone sell or buy these products?

- Gumout Winter Fuel Injector Cleaner - contains isopropanol alcohol 30-50%, plus detergent mixture, petroleum distillate(s) and trade secret formula.
- Mercury Marine Fuel System Cleaner - contains

15% isopropanol alcohol and naphtha solvent.

- CD2 Engine Tune-up - contains methanol (60-70%), Isopropanol (20-30%), 2-Butoxyethanol (10-20%) and Diacetone alcohol (10-20%).
- Turbo Octane Boost 108 - ingredient list includes only Methanol and t-Butyl alcohol.
- Neutra Fuel Stabilizer - primary ingredient = N-Butyl Alcohol.
- Evinrude BRP/OMC 2+4 Fuel Conditioner - contains 95% alcohol.
- K100 G Marine Gas Treatment, K100 MD, K100 MG, and other similar K-100 products: Primary ingredients = 2 Butoxyethanol, Ethylene Glycol, Monobutyl ether.

If anyone finds this information is incorrect, please let us know.

Oil Industry Stages Another Sham Anti-Ethanol Press Conference

Originally published January 29, 2013

On January 29, 2013, the American Petroleum Institute held a telephone press conference in their continuous attempts to undermine alternative fuel solutions and keep us addicted to gasoline.

The two presenters took on the surprisingly conciliatory pretense of "having nothing against ethanol," and went so far as to say that there are fine "blending qualities" about ethanol—but they did what API has done for decades—they misrepresented facts and presented disinformation and mischaracterized ethanol. To listen to them, anyone would think that vehicles running on gasoline never experience engine breakdowns or system failures, and that fuel pumps, fuel lines, pistons and cylinder walls only show signs of wear if alcohol somehow gets into the fuel system. They warned consumers would experience roadside breakdowns and costly repairs if they used E15. Yes, they suggested that millions of vehicles that have broken down over the past 100 years broke down because they used fuel other than gasoline.

The API is full of malarkey.

The API stated, yet again, that the EPA prematurely approved the use of E15 in vehicles manufactured since 2001. API says that there hasn't been sufficient study. They conveniently ignore the fact that alcohol/ethanol has been used in vehicle engines since the mid-1850s; that virtually every independent study of gasoline vs. ethanol shows ethanol is the superior engine fuel; and that even the automobile industry's top scientists (including the General Motors guys who invented leaded gasoline) believed that ethanol is the better fuel. They also ignore the fact that there are other countries on this planet that have relied on ethanol and various ethanol-gasoline blends for years, with no problems other than those customarily experienced by internal combustion engines.

They glossed over the fact that the government's testing laboratories have studied ethanol and different blend levels for many, many years. Ricardo Laboratories—the world's leading and most respected private lab dealing with fuel issues—did its own study showing that E15 can be used in all vehicles going back to the early 1990s.

API again mischaracterized the EPA's E15 waiver; making it sound like it wasn't just a recommendation, but a mandated imperial order. They referred to a recent AAA warning about E15 causing engine damage as if AAA had conducted their own conclusive testing of the fuel. But AAA didn't; they relied on the lies fed to them by the oil lobby and went on to invent their own gross exaggerations to make the warning seem more urgent. It's just more malarkey; hot, stinky, grotesque batches of malarkey.

But perhaps the single most senseless misstatements the two presenters made was when they characterized ethanol as being dangerously corrosive. Let me remind everyone that ethanol is alcohol. Alcohol can be consumed; it can be rubbed on your body; it is used to clean and disinfect sensitive medical instruments that are used on internal organs of the human body; and you can leave an open container of alcohol in a closed room without killing any people who may be in the room. Try doing these things with gasoline. And, if you only use gasoline and you experience water related problems such as freezing, you add alcohol to the gasoline to solve the problem. In my estimation, if there is something that's bad in ethanol, it's the gasoline added to the alcohol to denaturize it.

The petroleum oil industry is controlled by foreign regimes and terrorist groups who want to keep us addicted to gasoline. Regardless of where and how the oil is found, it is controlled by OPEC. There are no "American" petroleum oil companies; they are international companies who may have offices here. They hire uninformed PR apologists to spread their lies and make huge donations to politicians to keep us using their poison.

Even if ethanol caused some or all the engine problems the oil industry pretend that it does, the solution is not to ban ethanol. The solution is to ban the use of gasoline as an engine fuel and to mandate that all engine components be manufactured to be compatible with alcohol use. The ethanol should be produced in America by Americans.

Simply put, I'd much rather have my fuel money go to American farmers than foreign terrorists.

Why Do AAA and Big Oil Feel They Must Lie to America?

Doesn't Lying About One Thing Taint the Veracity of Everything Else They Say?

Originally published May 23, 2013

The week before Memorial Day 2013, I received an AAA press release concerning travel over the weekend. The release carries the usual information issued about projected numbers of Americans traveling via automobile during the holiday. This bit of fluff becomes a highly valuable fact used by hundreds of radio DJs and news readers around the country after they've told us which celebrities are celebrating their birthdays. But the release contained information that is becoming more visible in AAA press releases: Misinformation and lies about ethanol.

The press release stated:

> As travelers plan to journey an average of 690 miles, those filling up on high ethanol-gasoline blends like E85 (up to 85% ethanol), or mid-level ethanol blends such as E15 (15% ethanol), will notice that they'll need to stop at the pump more often. Ethanol contains 33% less energy than traditional gasoline (and is more expensive when adjusted for this), meaning consumers will get fewer miles per gallon. Unfortunately, 95% of consumers surveyed by AAA were not familiar with E15, meaning that travelers could unintentionally fill up on the fuel, and end up spending more on gas and potentially wrecking their engines.
>
> More and more ethanol is being forced into U.S. gasoline every year as a result of the Renewable Fuel Standard (RFS)—a policy that is diverting 40% of all U.S. corn to use in fuel instead of food. The RFS

> is also increasing prices for meat, poultry and dairy products, and will make Memorial Day cookouts more expensive. Since the RFS was expanded in 2007, prices for cereal and bakery products have risen 77%; prices for meat, poultry, fish and eggs have increased 78%; and prices for vegetable oil and fats (e.g., butter) are up by 444%.

What was the purpose of including this information in a pre-holiday AAA travel press release? It's like giving some upcoming motion picture preview information and warning America about how much toilet paper will be used and wasted in the theater rest rooms. It was a puzzling combination of issues, but I later learned why there was this mash-up of divergent messages.

This is the heart of the matter: anti-ethanol information provided in the press release are either lies, gross exaggerations, misstatements, or all three.

1. The press release implies that E15 will be used by a significant number of Americans. If this was true, it would be great news to the ethanol industry. So far, E15's availability is so limited in America that using it over the Memorial Day Weekend is very close to none if viewed on an overall fuels-to-be-purchased pie chart. And even if E15 had the negative effects that they claim it has—which it doesn't, the overwhelming majority of holiday travelers are more likely to encounter a visitor from Mars.

2. While it is true that ethanol has less energy per gallon than gasoline when calculated according to BTU-style measurements, the energy content has nothing to do with why a high ethanol level blend might deliver less miles per gallon of fuel. The vehicles will get less mileage because the engines are optimized for gasoline. If the same engines were optimized for ethanol they would get as much or more miles by using ethanol. Even flex-fuel vehicles are optimized to run on gasoline. Although FFV engines might have some component parts that are supposedly more ethanol-friendly, they are not ethanol optimized so a flex-fuel engine will deliver less miles with E85 than regular gasoline (E10).

Another way to explain and prove this point is that diesel fuel, also made from petroleum oil, has more energy per gallon than gasoline. But if you try using diesel in your gasoline-optimized vehicle you will get fewer MPG—the engine might not even start. The use of the energy content issue is hocus-pocus designed to keep fear alive in the car owner.

3. But wait, it's even worse than ornery hocus-pocus misdirection. The press release uses the 33% less energy figure as a blanket measurement to tell you how much mileage you will lose and then uses the lower mileage figure against the lower cost per gallon of a higher ethanol blend fuel to tell you that the cost savings isn't worth the effort. But the 33% loss is a theoretical number. The true on-the-road figure is probably closer to just a few percentage points, perhaps as much as 10% depending on the blend level. If you buy a gallon of E85 that costs 25% less than regular gasoline, but you lose 5% or 10% in miles, you still have a nice net gain. Consequently, using ethanol would be less expensive.

4. Then AAA says that in their study 95% of all those surveyed are not familiar with E15, which means that they could unintentionally fill up with E15 "...and end up spending more on gas and potentially wrecking their engines." This is an irresponsible warning and an insult, compounded with a falsehood. As I said in #1 above, first the drivers have to find E15.

The press release also suggested that car owners don't know how to read. E15 equipped pumps and hoses are marked as clearly as diesel pumps and hoses. Yes, some "very happy" holiday revelers will undoubtedly mistake a water hose for a tire air-hose, and some will leave their Big Gulp drinks on the roof of their cars as they drive away, but the damage of losing a $2 Mountain Dew to carelessness is far greater than any damage that will happen to a gasoline-powered engine using E15 fuel—whether E15 use is unintentional or not.

Then there is AAA's threat of damage. The press release tells the public E15 will result in "...potentially wrecking their engines." They don't state how the "wreckage" might occur; they don't indicate the nature of the damage, and they offer no statistical information concerning the likelihood of any such damage. It is simply an over-stated, irresponsible, unproven warning.

U.S. government testing has proven E15 (and E20) will not damage gasoline-powered vehicles manufactured after the year 2001. Testing conducted by Ricardo Laboratories, the world's most respected private fuel testing company, has proven that E15 will not damage passenger vehicles manufactured since the early 1990s. Historic testing conducted on pure ethanol and various blends of ethanol-gasoline for more than 100 years have proven ethanol does no more damage to an internal combustion engine than gasoline, and in fact, ethanol is safer to use than gasoline.[1]

6. The press release also trots out the old, discredited, food-vs.-fuel argument to blame higher food prices and world hunger on corn-based ethanol. American farmers grow as much corn as they do because there is a market for ethanol. If there was no market for ethanol, the farmers would not grow as much corn; it's that simple.

Anti-ethanol forces create the image that corn targeted for consumption at Memorial Day or 4th of July picnics or for starving third-world people is suddenly redirected to greedy ethanol producers. This is just not the case. The overwhelming majority of the corn is grown because of known demand from ethanol producers.

Food prices have increased because of "transportation costs" but not because corn is being used to produce ethanol. Food prices have increased because of higher oil-based transportation fuel costs: gasoline and diesel. Food costs have also increased because of marketing costs related to packaging cost increases. One of the reasons for packaging cost increases is higher printing costs due to ink costing more. The ink is manufactured from petroleum oil. Increases in the price of oil, not ethanol production, are pushing up food prices. In any event, ethanol can be produced from many more raw materials than corn, so if AAA had any concern for the American public, but still didn't like corn, they would advocate the use of other items—many of them are non-edible, readily available materials.

The Same Old Story

This isn't the first time AAA has carried water for the oil industry. AAA's efforts to mislead the public about ethanol dates back to at least the 1930s, when they staged comparative testing and doctored the results.

In winter 2013, AAA issued two E15 warnings. These warnings became the subject of a FOX News broadcast that exacerbated the lies and misinformation contained in the AAA warnings. The Auto Channel challenged FOX and AAA on these warnings in an editorial published on *TheAutoChannel.com* on January 8, 2013 and subsequently widely republished on other websites. ("AAA Blunder on Ethanol Sets Off Firestorm of Criticism.") Read it at *https://www.theautochannel.com/news/2013/01/08/062115-aaa-blunder-ethanol-sets-off-firestorm-criticism.html*

Despite assurance from AAA that they would respond to the points we made in that editorial, we have received no further information from them. On April 26th, just a couple of days before I went to Washington DC to give a presentation about E15 to a Congressional Staffer meeting, I reminded AAA that I was still waiting for some explanation for wrong information contained in the winter warnings. The same head of AAA public affairs said he would get back to me in two weeks. Four weeks went by since that promise, and more than four months have passed since the earlier promise was broken.

FOX News never replied.

And now, the update:

After we published this editorial on a Thursday morning, I was informed the AAA Memorial Day press release didn't come from AAA. It was issued by the American Fuel and Petroleum Manufacturers, a big oil pro lobby headed by Charlie Drevna. Apparently AFPM took some information from AAA and used it in a press release to make it look like it was from AAA, thereby giving the overall information an air of objectivity. Sounds like the kind of misrepresentation and sleazy PR stunts I've known Drevna and his group like to pull just as Americans head into a holiday weekend.

Hey, Charlie! Thanks for confirming your lack of ethics!

For more on Drevna and AFPM see "AFPM Issues Panic-Stricken Warning Against Ethanol to Keep America Addicted to Foreign Oil."

https://www.theautochannel.com/news/2012/02/17/025717-afpm-warns-adverse-effects-e15-use.html

So, I say to AAA, AFPM, the American Petroleum Institute and all the other entities the oil industry pays to create and disseminate negative ethanol information:

If this is a case where you have been caught telling lies in an effort to keep America chained to gasoline and to keep ripping off the public, you have no need to reply.

If you really think you are not lying, and you have serious and reliable information to substantiate any of the negative things you claim about ethanol, please let us know what it is. We are not on anyone's payroll. If gasoline is the best fuel—the best for our vehicles and for our country—then tell us what it is, and we will shout it from the roof tops. But please don't continue to urinate on our legs and tell us it's raining!

Additional Update - May 25, 2013:

Shortly after publishing the above editorial, I notified both AAA and Edelman Public Relations (the PR firm sent out the press release) about my published response. On the following Friday I received replies from both. AAA denied any knowledge of the press release or relationship with the oil lobby and stated that their only interest is the American motorist.

Edelman PR said the email they sent wasn't a press release but a "media pitch including news and data points." The agency representative went on to say that she had not been "speaking on behalf of AAA, AFPM or anyone else and that it was by no means an official statement." The original release from this person at Edelman PR offered to arrange for me to speak with "a representative from the American Fuel and Petrochemical Manufacturers or representatives...or consumer advocacy groups who can provide further insight into how government ethanol laws are creating problems for consumers this weekend."

Edelman PR is the public relations firm for both American Petroleum Institute (the primary mouthpiece for the oil industry) and Saudi Arabia (the world's #1 producer of oil).

These were my responses:

TO AAA: "Thank you for your reply and your comments this morning. Knowing for certain that AAA was not involved clears up certain aspects of the text, while making the effort by Edelman Public Relations all the more insidious (in my opinion).

If AAA is on the side of the motorist/consumer and you have no allegiance to the oil industry, you owe it to the motorist/consumer to get facts correct. The two press release warnings that AAA issued in November and December went far beyond advising consumers about potential warranty issues. The releases made definitive claims about the negative effects of using ethanol that were not true. The effects of those false claims were then amplified by the ridiculous story done by FOX News. You may not have endorsed the FOX story or the persons talking about the AAA warnings before they produced the show, but unless you can say AAA subsequently contacted FOX and Lauren Fix and took them to task for taking the AAA warnings out of context, then your silence is tantamount to endorsement.

I agree with your comment that "A large AAA/oil-industry conspiracy to conspire against ethanol makes good...headlines." But if it's not your intention to get that kind of publicity, I suggest you don't issue false and misleading information. I didn't write the AAA warnings press releases, AAA did. And if I am incorrect in the statements that I made about the AAA statements, then you should have had your engineers get back to me in a timely manner. Now that you are aware of how AAA's "Memorial Day Weekend" projections were taken out of context by the oil industry to further denigrate ethanol, you should make them aware you don't appreciate their deceitful tactics in lying to motorists and consumers."

To Edelman PR:

Sorry if I misconstrued the meaning of the email you sent me.

Since you included several pieces of AAA information included in AAA-issued press releases and offered to put me in contact with people and groups related to the claims made in the email text, including AFPM, it sure looked like a press release to me. After all, Edelman Public Relations is not working for The Auto Channel, and I didn't request information from you prior to receiving your email. To the best of my knowledge, Edelman does represent at least one oil industry organization. So, based upon my 40+ years of personal hands-on experience in marketing and advertising, I would call your email a "press release."

Knowing that the email was just intended to be a "media pitch," I'm curious as to why you, or others at Edelman, chose to disguise the comments about E15 as if they were written by AAA. I'm also curious to know where you got the misinformation about E15. If your email wasn't an official statement from one of Edelman's clients, who on Edelman's staff concocted this stuff? And after reading my editorial responses to the egregious points you made, why haven't you circled back to whomever it was that wrote the false and misleading information and told them of my comments?

Notes

[1]Some rubber and metal components used in vehicle engines and drive trains manufactured prior to the early 1990s may be susceptible to alcohol erosion and corrosion. It can also be true for other non-automobile engines manufactured before and after the 1990s. It is for this reason higher ethanol-gasoline blends are not recommended for vehicles manufactured before the early 1990s and for other types of engines. I know of no one involved in the ethanol industry who doesn't adhere to this caveat and instruct users accordingly.

Bear in mind, gasoline is also corrosive and is far more damaging to many materials used in automobiles than ethanol. Ordinary water is also highly corrosive and can damage automobile parts.

Ethanol Honesty Is the Best Energy Policy

Originally published January 30, 2015

It's always troubling when a person touted as having a high proficiency in one area tries tackling something in which they have zero proficiency. I think this may be the situation with Debbie Carlson, a professional writer who typically pens financial articles. On January 28, 2015, *TheGuardian.com* published her story titled "Energy hypocrisy: Ethanol isn't a good fuel, but it's not going away anytime soon."

(https://www.theguardian.com/business/2015/jan/28/ethanol-biofuel-corn-expensive)

Ms. Carlson apparently failed to investigate the subject sufficiently.

To be fair, if "fair" is the correct word to use, the genesis of Ms. Carlson's article may not have come from an original inspiration; it may have been initiated via a solicitation from an interested subjective party. These kinds of solicitations usually come to freelance writers and speakers with suggested phrases and key points that must be addressed in the presentation. Because these solicitations often emanate from what seem like respectable middlemen and public relations companies, the freelancer often does little or no independent fact checking, choosing to rely on the veracity of the information provided.

Given the large number of inaccurate statements in Ms. Carlson's article, her investigative skills probably weren't at fault.

My problem with Ms. Carlson's article starts with the title. There are two reasons for this. First, it's untrue. Ethanol is an excellent fuel. Second, she included nothing in the story to support calling ethanol a bad fuel.

Top automotive people have always considered alcohol to be the best fuel. This was true of Henry Ford, Charles F. Kettering, Thomas A. Midgley and T.A. Boyd (GM's top scientists in the early 20th century). It's still true today, with the continued use of ethanol or methanol in the most advanced race cars and high-performance street cars. Kettering, Midgley and Boyd were responsible for inventing leaded gasoline, which became the dominant engine fuel because GM was searching for a formula that would imitate ethanol's anti-knock characteristics. It became the favorite fuel of GM when bean counters at GM realized that they could patent the invention and make billions from it. GM was the world's largest auto maker even then, and when combined with patents owned by Standard Oil and DuPont, they had a virtual global monopoly for a liquid fuel that could be used in high-compression engines. (Competition from ethanol in the U.S. was virtually non-existent because of Prohibition). Gasoline became the primary engine fuel because of greed, not because of its inherent quality as a fuel. The world's biggest car maker, oil producer and chemical company virtually bought the position by buying political power.

Beyond the headline, the information presented by Ms. Carlson doesn't get any more honest as she thoughtlessly rattles off hackneyed, long disproved criticisms of ethanol like a high school cheerleader who doesn't understand the rudiments of the game she's cheering for.

Ms. Carlson wrote that while ethanol was supposed to help reduce our dependence on foreign oil, combat climate change, be a gateway for more renewable fuels technology and reduce gasoline prices because it was cheaper, it hasn't done any of these things.

She is wrong, it has done all those things.

If just one gallon of gasoline was displaced by one gallon of ethanol, it has lessened America's dependence on foreign oil. This is simple mathematics. If America used 13 billion gallons of ethanol in 2014 to help power our passenger vehicles, that means we reduced our dependence on foreign-controlled oil by 13 billion gallons.

If one is to argue that man-made climate change is real and that vehicle emissions play a large part in creating the problem, then the use of ethanol in place of gasoline does combat man-made climate change.

Ethanol is a cleaner burning fuel; it produces fewer harmful emissions than gasoline.

Paid critics of ethanol often use oblique elements of the total ethanol production process to arrive at the net conclusion that ethanol is a "dirtier" fuel than gasoline. But if the same methodology of including secondary considerations was related to gasoline production, it would reveal the true extent of gasoline's negative impact on the environment and show horrifying differences between gasoline and ethanol, with gasoline being the definite loser.

Among the secondary considerations of gasoline production is the amount of fuel required to militarily defend and protect the oil industry, including the production of the vast amount of military materiel needed and expended, and the thousands of lives lost every year. Yet to date, not one American serviceman or woman has been killed defending domestic ethanol production.

Ethanol's value as a gateway to more renewable fuels is incalculable. Its mere presence demonstrates that a substance other than gasoline or diesel can be used to power an engine. Regardless of how negatively a critic tries to paint ethanol, we see there is a solution to myriad problems and concerns related to petroleum oil—including issues of long-term, uninterrupted availability. On a tangible level, acceptance of crop-based ethanol led to breakthroughs and advancements in the development of cellulosic ethanol, methanol and methane fuels.

Studies released in 2009 and 2012 by Professors Xiaodong Du and Dermot J. Hayes[1] showed that ethanol use did reduce gasoline prices. Despite rebuttals by Professors Christopher Knittel and Aaron Smith, Professor Hayes successfully defended the findings of his and Du's work. Knittel subsequently acknowledged, "That ethanol most likely reduced gasoline prices," while stating the reduction was less than that claimed by Hayes and Du.

However, a reduction is a reduction—and in a world where a three-cent reduction in the price of gasoline is considered national news, I can't see how Debbie Carlson was able to categorically deny that ethanol has not reduced the overall price of gasoline. All the studies, claims, and counterclaims by Professors Du, Hayes, Knittel, and Smith were concluded long before Carlson wrote her story. So, if Carlson was out to publish an objective story about ethanol, this information was available to her. Instead, it's clear her story was not meant to be objective; it was meant to be a bash-ethanol-at-all-costs story so that she could receive some type of financial reward from the oil industry.

Let's say ethanol provides no reduction in the cost of filling your fuel tank. Let's say the entire reason for replacing gasoline with ethanol is just a wash on the effective price per gallon. Let's say the only benefit is that America gets to retain the one billion dollars that leaves America every day to pay for foreign oil. These dollars may never make their way back to America. If we can avoid sending the $1 billion a day ($360 billion per year) out of the country and use those dollars to pay salaries here and buy domestically produced goods, those dollars become stimulus for the economy without having to print additional dollars and devalue those already in circulation. If those dollars are spent on American workers and American products, we actually wind up recycling the dollars and multiplying its stimulus effect several-fold. It could be possible to recycle the $360 billion three, four, five or six times, giving us a trillion dollar-plus stimulus every year for every year we do without gasoline. If the $800 billion stimulus package in 2008-9 was significant in helping our economy, imagine what this stimulus could do for America.

Ms. Carlson goes on to decry the amount of corn being used for ethanol. She attempts to re-ignite the preposterous flames of the "food vs. fuel" argument. She writes, "We're putting nearly 40% of the US corn crop in our gas tanks."

Central in trying to make this an alarming statistic is the imagery that just as the corn is about to be distributed to millions of corn-on-the-cob-deprived starving people around the world, greedy ethanol producers swoop in and buy up all the food to be turned into fuel. That is not how the system works.

There is no question that more corn being grown in America today is being used for ethanol production than as compared to, say, 10 years ago. But the reason for this is that the corn is specifically grown to be used for ethanol. There is demand for the crop, so farmers grow more. This means that farmers (American farmers) can grow something that is profitable. Moreover, it means that they can grow something without having to turn to public assistance.

In 2000, U.S. corn production was 251,854 metric tons. In 2013, U.S. corn production was about 353,715 metric tons. Despite the increase in the amount of corn grown between the two years, the actual amount of corn available for human consumption remained the same. Although most of the world outside of the western hemisphere does not eat corn the way we do, world corn production reached record highs in 2014. Therefore, it's safe to say that there were fewer starving Africans being deprived of non-nutritious high-fructose corn syrup products. Considering the obesity problem that we have in America, if we were depriving someone of corn chips, we might be doing them a favor.

Ms. Carlson tries to take another swipe at ethanol's impact on gasoline prices by writing that "since mid-December, ethanol prices have risen above reformulated gasoline prices because of the sharp drop in crude-oil and gas prices, along with a rise in corn prices. As of 26 January, Chicago Board of Trade ethanol futures were holding around $1.448 a gallon, whereas New York Mercantile Exchange reformulated gasoline futures prices were at $1.3167, giving the renewable fuel a 13-cent premium."

There are two problems with this statement and its implication. She's assuming the current surprisingly low price of crude oil will remain surprisingly low, and she fails to calculate in all the other marketing and distribution costs associated with the retail price of gasoline. There are no public filling stations in America regularly offering gasoline at $1.31 to their retail customers. Moreover, in a check around my neighborhood, at those stations where E85 is available, it is available at about 40 cents less per gallon than regular E10 and 50 or 60 cents less than premium E10.

With the above real at-the-pump prices in mind, her entire explanation of how the Renewable Fuel Standard may he causing higher ethanol prices is rendered immaterial and is incorrect.

Other comments made by Ms. Carlson—such as the "2012 drought decimating corn supplies" are untrue or misinterpretations. Even with the drought conditions in 2012, American had a corn surplus that year.

If ethanol producers are struggling to make a profit, it's because they are limited to how much ethanol they can sell, not because (or solely because) of oil's current low price. There is no reason whatsoever for passenger vehicles on the road today not to be using E15, E20, E30, E40, or any other blend individual owners would like to put in their vehicles. These increased blends would be a boon for domestic ethanol producers and farmers, without having to raise the price of ethanol. They would keep billions and billions of dollars from leaving our country to support terrorist regimes and anti-American incidents.

As always, I'm happy to hear any criticisms, objections or clarifications. I did write to Ms. Carlson directly and submitted this article to her for comment. I never received any reply from Ms. Carlson.

Notes

[1]Studies released in 2009 and 2012 by Professors Xiaodong Du and Dermot J. Hayes:

"The Impact Of Ethanol Production On Us And Regional Gasoline Markets" *(https://www.sciencedirect.com/science/article/pii/S0301421509002584)*

"The Impact of Ethanol Production on US and Regional Gasoline Markets: An Update to May 2009" *(https://www.card.iastate.edu/products/publications/pdf/11wp523.pdf)*

"The Impact of Ethanol Production on U.S. and Regional Gasoline Markets: An Update to 2012" *(https://www.card.iastate.edu/products/publications/pdf/12wp528.pdf)*

"Response to 'Ethanol Production and Gasoline Prices: A Spurious Correlation' by Knittel and Smith" *(https://www.card.iastate.edu/products/publications/pdf/12wp529.pdf)*

"Second Response to Knittel and Smith" *(https://www.card.iastate.edu/products/publications/pdf/12wp532.pdf)*

"Ethanol Production and Gasoline Prices: A Spurious Correlation – Christopher R. Knittel and Aaron Smith" *(https://arefiles.ucdavis.edu/uploads/filer_public/2014/06/19/knittelsmith_130510.pdf)*

Why Is Jay Leno Misrepresenting Ethanol?

Originally published March 9, 2015

There are two things everyone knows about Jay Leno: He's a great comedian, and he's a serious automobile enthusiast. Generally, when you become great at something, you learn a lot about that subject. Even if you don't want to learn about the subject, and you want to be good at engaging in the activity, it's impossible to not become a great student of the history and mechanics of that subject.

I have no doubt that Jay is a master of comedy history, along with the mechanics of what is funny and why it's funny.

I've watched enough video of Jay and his vehicles to believe that he is equally a master when it comes to knowing about his vehicles and the history of how they were designed. I also know that Jay has been a proponent of alternative fuels and advanced technologies. In fact, we even have a few dozen stories and videos of Jay on The Auto Channel website that feature him discussing these things.

See:*https://www.theautochannel.com/search/results.html?q=Jay+Leno*

There are at least two other media pieces done by, or with, Jay enthusiastically discussing his E85-powered 2006 Corvette. One piece was a *Popular Mechanics* text story published in 2008.

(https://www.popularmechanics.com/cars/a3213/4261712/)

The other was a video produced in Las Vegas at SEMA 2007.

(https://www.youtube.com/watch?v=JZTnDWZJzEM)

In both stories, Jay has a favorable opinion of high-level ethanol gasoline blends versus gasoline without ethanol, or even just E10 gasoline (10% ethanol/90% gasoline). Among other benefits, Jay cites ethanol's higher-octane rating, cooler operating engine temperatures, lower harmful emissions, and ethanol's engine cleaning characteristics that leave behind no nasty gasoline residue and gunk that clog key engine components, such as pistons and valves.

Then it was brought to my attention that Jay had authored a new story that appeared in the March 2, 2015 issue of *AutoWeek* magazine. The article, titled "Can't We Just Get Rid of Ethanol?" *(https://autoweek.com/article/car-life/jay-leno-hates-ethanol)* basically proposes that the United States end the "Renewable Fuel Standard" (RFS) because of issues related to the use of ethanol fuels in older vehicles. At the close of the story, Jay exhorts readers and automobile enthusiasts to write to their legislators to demand action against ethanol.

Clearly, there is a difference between old cars and new cars—between classics and antiques and late model vehicles that make up the majority of vehicles on the road today. It is understandable for Jay to express two different opinions about ethanol as it pertains to old cars versus new cars.

(For those of you keeping score at home, the average age of all cars and trucks on the road in America is about 10 years. Since the early 1990s, all gasoline-powered passenger cars and trucks manufactured for America have used engine and fuel-system components resistant to alcohol's solvent properties.)

The problem is that Jay did not say, "Write to your legislators to demand more freedom of fuel choice to give us old car owners easier access to ethanol-free gasoline." Instead, he called for less freedom of fuel choice. As much as I hate to say it, Jay used untrue and misleading information to sway the argument. And so, since I think that Jay does know better, he is spreading misinformation in the *AutoWeek* story.

For example, Jay states that "ethanol will absorb water from ambient air...causing corrosion and inhibiting combustion."

Ethanol doesn't absorb water from the ambient air. This lie is one of the oldest and most malicious created by the oil industry to denigrate ethanol. The only thing new is how Jay used the word "ambient." I've not seen that before. I've seen quotes that use the word "thin" to denote ordinary air that we normally breathe, but not "ambient." It doesn't matter. That's not what happens.

Many years ago, some clever oil industry person learned that ethanol (alcohol) is a hygroscopic substance, and that the general dictionary definition for a hygroscopic substance is that it can attract moisture from its environment. What the oil industry wag then did was to substitute the word "attract" with "absorb," and "air" for environment. Thus, attracting moisture from its environment magically became absorbing water out of thin air.

Here's an Abbott & Costello routine that presents a startlingly clear analogy of how silly Jay's hygroscopic statement is:

Costello tells Abbott that a loaf of bread is the mother of the airplane. Abbott tells Costello that he's crazy. Costello asks Abbott if he agrees that necessity is the mother of invention. Abbott replies yes. Costello then asks if bread is a necessity; Abbott says yes. Costello asks if the airplane is an invention; Abbott says yes. Therefore, exclaims Costello, if bread is a necessity and the airplane is an invention, then a loaf of bread is the mother of the airplane.

What I'm getting at is just because you can play semantic games with the definition of "hygroscopic," it doesn't mean the result of the game is relevant and correct.

To prove that alcohol will not absorb water right out of the thin or ambient air, I always offer this simple at-home experiment: Fill any open container halfway with alcohol and place it on your kitchen counter. Allow it to sit for one or more days. If alcohol absorbs water right out of the air, when you check the level of liquid in the ensuing days you would find it has risen. If the level of the liquid in the container has risen (without any manipulation, change to the environment of your indoor kitchen, or interference in the natural process) and you can document it, I will pay you $1,000.

Cotton is also a hygroscopic substance and offers additional proof that being a hygroscopic substance doesn't mean it absorbs water right out of the air. Place a ball of cotton on the other side of your kitchen counter and see if it gets saturated with water from sitting out in the open. It won't.

One of Jay's other points was that if you poured a gallon of water in your gasoline tank your vehicle will probably have great difficulty starting, but that's not how water gets in your gasoline tank. You can get water in your fuel system because of condensation. If you have water in your gas tank, do you stick a straw in and suck it out? No! You add a product like Dry Gas. Dry Gas is alcohol, ethanol, —you use ethanol to solve the problem. That's right, ethanol solves the problem!

Ethanol doesn't actually absorb the water; water molecules fill in around the ethanol molecules, which gives the appearance of it being absorbed so ignition and combustion of the gasoline can take place. The water molecules are vaporized and expelled in the exhaust. In other words, ethanol aids combustion and does not inhibit combustion.

Jay goes on to say, "It gets worse. Ethanol is a solvent that can loosen the sludge, varnish and dirt that accumulate in a fuel tank. That mixture can clog fuel lines and block carburetor jets." The sludge, varnish and dirt that Jay is referring to is caused by gasoline. So ironically, the cleaning characteristic that Jay is now criticizing is the same beneficial cleaning characteristic that he previously championed when discussing the benefits of ethanol.

Then Jay writes, "Blame the Renewable Fuel Standard (for these problems)." But that's not where the blame lies. The blame lies with gasoline, the liars in the gasoline industry, and the politicians who forced us to use gasoline. Ethanol cleans the gunk; gasoline causes the gunk.

Even if ethanol is never introduced into a fuel system, the time will come when the engine must be cleaned. The engine repair industry wasn't invented with the advent of E10 or E85 gasoline. Engine repair, maintenance and replacement is a natural result of the internal combustion process. If we can have a fuel that (as Jay previously wrote) burns 100 percent, leaving behind no nasty residue and leftover gunk that clogs key engine components, why shouldn't that fuel be readily available? Shouldn't that fuel be our primary default engine fuel?

Jay talks about damage that ethanol has caused to the fiber diaphragms in the fuel system of one of his Duesenbergs. He's probably correct about this. But is this why America should abandon the RFS and return to gasoline that contains poison?

If you watch the video that Jay did in Las Vegas in 2007, he says something very interesting. In response to the question of how he selects which vehicle he is going to drive to work on a given day, with great humility, he acknowledges that there are greater problems in the world to worry about. That was a correct, very modest response. So, in keeping with that recognition, I suggest that the fiber diaphragms in his or anyone else's Duesenberg have no significance to determine what engine fuel the nation should use.

As a former owner of a classic car, I can appreciate his concern for his vehicles, but I suggest he suck it up as noblesse oblige, a sacrifice that he must make.

For some inexplicable reason Jay also throws in negative comments about ethanol producers and the "food vs. fuel" argument. Perhaps Jay didn't write this article himself. Perhaps it was written by some API stooge and Jay signed off on it. Jay refers to some ethanol producers as "giant agri-businesses," and its mention is couched within a paragraph meant to demean producers and the effort to make America energy independent. Yes, some ethanol producers are large corporations, but compared to giant oil companies, they are like mom and pop businesses.

For example, in the same year ExxonMobil reported their fiscal fourth quarter profit as $40 billion, Archer Daniels Midland (ADM) reported their fiscal fourth quarter profit of $372 million. Although $372 million is nothing to sneeze at, it's less than 1% of ExxonMobil's profit. If there's a picture being painted about huge greedy companies looking to take advantage of the American consumer, the illustration is ExxonMobil, not ADM. Remember, too, that the ethanol used in America is produced in

America by Americans. No American military man or woman has ever died defending domestic ethanol production and distribution.

As for "food vs. fuel," Jay might as well claim that the Earth is flat. About 10 years ago, the World Bank issued a statement claiming increased corn-based ethanol production was causing food prices to rise. Since that time, the World Bank has twice rescinded that earlier claim based on new and better studied information. The fuel-related culprits they acknowledge as causing food prices to rise are gasoline, diesel, and other petroleum oil products used in packaging.

Jay was irresponsible for writing, or signing off on, this article. The bulk of the responsibility for letting this misinformation come to light belongs to *AutoWeek* magazine. No matter what Jay Leno had to say, they should not have allowed it to be published, or they should have published it with disclaimers. I guess *AutoWeek*'s decision was predicated on the hope of increased advertising support from the oil industry; and the reason they chose to embellish Jay's story in the online version of the story with the clickbait headline "Jay Leno hates ethanol," was to make sure they were kissing oil industry ass. I assume Jay made the same decision to potentially get oil company sponsorship for his new automotive content ventures. If that's the case, there is little reason for this article to have been written and published in the first place.

See also: "Can't We Just Get Rid of Ethanol Ignorance?" by Bobby Likis, a true automotive expert, at:

http://www.ethanolrfa.org/exchange/entry/cant-we-just-get-rid-of-ethanol-ignorance/

Lauren Fix Takes Ethanol Opposition to A New Level

Originally published August 28, 2015

In January 2013, I wrote an editorial lambasting several critics of ethanol. The primary antagonist of the piece is AAA, but a close second is the automotive media personality Lauren Fix, who uses "The Car Coach" as her tagline. (See article earlier in this book.)

The editorial has its roots in a couple of AAA press releases that denigrate ethanol. Then Melissa Francis, host of the *FOX News' Money* show, added gasoline to the fire by inviting Ms. Fix to explain and expand upon the factually erroneous AAA press releases.

When The Auto Channel published the original story on this matter, I sent Ms. Fix several emails questioning her statements and alerted her to my editorial. I gave her abundant opportunity to respond and provide supportive information for her comments or to recant them. She never responded. And although we often received press releases or stories from Ms. Fix, she stopped sending us anything.

I've been at the same automotive press events as Ms. Fix, and we were often in very close proximity (sometimes only one seat away). She has never taken the opportunity to respond or explain herself. Since she's "The Car Coach," if I was wrong, shouldn't she coach me on the truth?

More than two and-a-half years later, Ms. Fix was recently a guest on a streaming radio show, *The Jacki Daily Show,* hosted by Jacki Pick.

Ms. Pick has an impressive resume. An attorney, she's served on Congressional committees in Washington. She wields two decades of experience related to energy and the environment.

The Jacki Daily Show episode is "Pipeline Benefits and Ethanol's Drawbacks," and you can listen to the entire show at: *(https://soundcloud.com/jacki-daily/fracking-lifts-hs-drop-out-rate-real-oil-safety-record-pipelines-rail-ethanol15-kills-warranties)*

I didn't hear my name nor was there any veiled reference to corresponding with anyone, including me, who was critical of her past comments about ethanol.

The show began with the recorded disclaimer that Jacki Pick uses all her experience to present an independent view that brings "America great energy security," and follows by stipulating that her opinions are her own and free from any outside influence.

Ms. Pick led off by reading an email from Brian Williams, Legislative Director of The National Center for Policy Analysis, where Jacki Pick is the Chief Operating Officer.

She reads, "I ruined a weed trimmer and a chainsaw on this stupid E15 ethanol gasoline."

The Legislative Director said he ruined a weed trimmer and chainsaw by using E15! Where did he find E15, and what made him want to use E15 in his weed trimmer and chainsaw? If he's based in Dallas where the "National Center" is, E15 is scarce. It's not especially easy to find in the DC area, if that's where he's based. Any pumps that do offer E15 have warning labels clearly stating E15 should NOT be used in gasoline-powered equipment, such as weed whackers and chainsaws.

With Jacki Pick being schooled in energy, she should have laughed or said something teasingly flippant about her colleague's ridiculous statement and misuse of E15. She didn't.

She went on a tear against "bad" ethanol and went on to explain she doesn't really know anything about ethanol, because she's "a fossil fuel person." But she tells the audience, she did do some "research." Now, armed with that research, she tells us that she learned that E15 is not even approved for 95% of the gasoline-powered vehicles on the road... that's how bad she says that E15 is. Unfortunately, she must have done her research at the same school where Lauren Fix earned her Master's.

On an official basis, E15 is approved by the EPA to be used in all gasoline-powered passenger vehicles manufactured in or after 2001. The vast majority of gasoline-powered vehicles on the road in America were manufactured in or after 2001. This accounts for 75% of all cars and trucks on the road today. This means that the 95% statement is woefully incorrect.

It's two minutes into the show and three incredibly vacuous statements had already been made.

Ms. Pick then informs her audience that the use of E15 will cause mechanical trouble. Then she creates a fear-mongering tale about how, as a single woman, she doesn't want her car to breakdown in some desolate rural area or grungy urban neighborhood. She follows the statement with a warning that your boat could get stuck in the middle of the Gulf of Mexico if it's using E15. And that would cause the Coast Guard to waste its valuable time rescuing you. Does she really believe gasoline would protect her vehicle from engine and/or fuel system in "inconvenient" neighborhoods?

The real question is, "Where did Jacki Pick conduct her research? What were her resources? She's a lawyer; lawyers are supposed to know how to do research.

Lauren Fix Hasn't Been Introduced Yet!

Of the five U.S. states that share coastline on the Gulf of Mexico, only Florida and Alabama offer E15. This means that if you live in Texas and want to take your boat on the Gulf using E15, you'd have to first travel through Louisiana and Mississippi to get to Alabama. Let's say you live in Florida or Alabama, what do you think the chances are of finding an E15 pump at a local marina? I don't know the answer, but I'm guessing zero percent. How likely are you to get stuck in the middle of the Gulf of Mexico because you put E15 in the fuel tank? My educated guess? ZERO percent.

Ms. Pick closes her monologue by casting doubt on whether anyone in the government is an expert on the subject. This was her preface to Lauren Fix's introduction, wherein she calls Fix "America's Top Woman in Car Care, Education and Auto Industry News; as well as an "Automotive Expert."

Ms. Fix immediately states, "This is a very serious subject" and she's been "fighting it [ethanol]for years." Considering her on-air comments in the past, on *FOX News*, we must conclude she's been purveying misinformation about ethanol for years.

I have no clue about what Ms. Fix knows about car care; but if you use her knowledge of ethanol as the benchmark of what she knows about the automotive industry, then Lauren Fix may be the most laughable automotive expert in the entire English-speaking world. How can she be an "expert" when she doesn't know what she's talking about?

Lauren Fix says, "Ethanol is so damaging to your vehicle... we don't use it in race cars."

How come Lauren Fix, "America's Top Woman in Care, Education and Auto Industry News and Automotive Expert," doesn't know ethanol has been used in race cars for almost 150 years. NASCAR, America's #1 motor sports association, uses only E15 in their Sprint Cup Series and SuperTruck racing series. Ethanol fuels are also used in IndyCar racing, drag racing and boat racing.

Ms. Fix says we have ethanol-gasoline blended fuels "because some politician has a deal going with another one."

No, Ms. Fix. It has taken this long to rid us of the poisons added to gasoline by a patented GM formula to stop engine knock—because politicians had a deal with the oil industry and General Motors. This was the status quo until the evidence against tetraethyl lead was so overwhelming it was banned. So Big Oil persuaded politicians to allow another poisonous ingredient, MTBE, made from petroleum oil to stop engine knock. Learning MTBE was also poison, it, too, was banned, and non-poisonous ethanol was brought in to do the job all the oil and energy people knew was the solution in the first place. Since then, ethanol-gasoline blends are playing a significant part in clearing the air in cities like Los Angeles and New York.

All of the above happened only five minutes into the show. If I wasn't already bald, I would have pulled my hair out.

Ms. Pick asks Ms. Fix how much damage ethanol will do to modern passenger cars and trucks. Ms. Fix says it will cost thousands of dollars because ethanol is very, very corrosive.

Of course, no studies are cited to support her erroneous conclusions. Neither one of these supposed geniuses mentions the private studies by the most prestigious laboratories that repeatedly prove ethanol-gasoline blends will not damage modern gasoline-powered passenger cars and trucks.

Yes, ethanol is a corrosive liquid, but so is gasoline. Solar rays are corrosive, air is also corrosive, and of course water is one of the most corrosive liquids on Earth. The not-so-secret secret is to use materials that are resistant to the substance you are dealing with.

Take water for instance: We can't live without it and we consume it as if it has no corrosive characteristics whatsoever. We swim and bathe in water, including the most corrosive of all water, saltwater. It's not that the water we consume and bathe in is not corrosive; it's that our bodies are not very susceptible to water corrosion.

You can't just put gasoline in any old container, it must be a container that is not susceptible to gasoline corrosion. And when tetraethyl lead was added to gasoline to stop engine knock, the General Motors and Standard Oil people had to come up with another ingredient to mitigate the highly corrosive nature of tetraethyl lead (they used ethylene bromide, another highly poisonous substance). Conversely, alcohol (ethanol) such as scotch, rum, vodka, whiskey and brandy can be left in a in a glass or plastic bottle, or metal flask in your home for years and there will be no degradation of the alcohol nor the containers during that time. There will also be no so-called "phase separation," which is caused by gasoline's inability to safely "hold" water the way ethanol does.

Rubbing alcohol presents another good example of how compatible alcohol is with difference materials, regardless of what the rubbing alcohol is made from. Some rubbing alcohols are just regular grain alcohols with a denaturizing ingredient to render it non-drinkable. And if you have rubbing alcohol in your home, chances are that it's in a plastic bottle. The question to ask is why doesn't the alcohol eat away at the plastic bottle? It's because plastic is resistant to ethanol's solvent characteristics.

During Prohibition (when alcohol was supposedly not available in America) and in the decades subsequent to the end of prohibition, automobile parts makers had to contend with gasoline corrosion, not ethanol corrosion.

In those regions of the world (like England) where ethanol-gasoline blends were available for decades, they didn't experience any greater problems using ethanol-gasoline blends because the ethanol wasn't the problem, it was the gasoline. It wasn't until the late 1950s, when DuPont invented a new rubber material called Viton that gasoline's corrosiveness was contained.

Beginning in the mid-1990s, when vehicles were being built to meet E10 standards, very little was done differently.

Another question: How come these two "automotive" geniuses didn't have access to reliable information? Why doesn't the expert know these things after reading our material and materials produced by other credible sources? Ms. Fix was provided with links to reliable, greatly respected sources. She never refuted any of our claims, never offered evidence contrary to our findings, and never proved her points.

Instead, Ms. Fix blames ethanol phase separation that she says takes place when you drive around town doing errands. Fact: It doesn't happen.

She foists that old canard about ethanol grabbing water out of thin air so that it collects in the bottom of your gas tank and screws up the system. As we have repeated too many times already, water in the tank is formed by condensation, and ethanol gets rid of it by absorbing it and spitting it out through the exhaust system

Before E10 gasoline became the norm, when water was formed in your fuel tank from condensation, you poured ethanol into the tank to get rid of it. Yes, that's right, condensation occurred long before E10, E15, or E85 fuel blends. Condensation has been occurring for billions of years.

Mercury Marine, one of the world's largest manufacturers of smaller marine engines, said this about Ms. Fix's, and others' claims about ethanol phase separation and absorption of water from the air: "There is no active transfer mechanism for ethanol molecules to reach out and 'grab' water molecules out of the air.

...E10 may actually be a superior marine fuel as it tends to keep low levels of water moving through the fuel system, keeping the system 'dry'."

Mercury Marine recommends that if you're going to leave your engine idle for extended periods, other than removing all the fuel, fill the tank with E10 to reduce the amount of exchange with the air that may bring create condensation. The Mercury Marine information is good because too many people say ethanol-gasoline blends in boats cause greater problems than they supposedly do in road vehicles.

Lauren Fix says it will cost thousands of dollars to fix the problems caused by ethanol. I nod my head in semi-agreement, but not for the reason she thinks.

It might cost hundreds or thousands of dollars to fix problems attributed to ethanol because some car mechanics are as uninformed as she is. They invent the problems; and if they are unscrupulous as well, they can really sock it to you.

Near the six-minute mark of the show, Ms. Fix announces oil companies use lower octane gasoline to dumb-down ethanol-gasoline blends because of the excessive corrosion caused by ethanol's naturally higher levels of octane.

No.

Ethanol has an octane rating over 110. If you blend ethanol with 87-octane regular gasoline, the octane rating of the blend would naturally be higher than the regular gasoline by itself. E10 gasoline should have an octane rating of about 89. But if the gasoline companies allow E10 to have 89-octane, they lose money on every gallon of 87-octane fuel containing 10% ethanol and would lose the opportunity to sell a mid-level 89-octane fuel at a higher price. They intentionally dumb down the octane level of the gasoline in the blends so that they're not giving away the extra two octane points for free. It's a case of the oil companies trying to earn a few pennies more per gallon. It is not a corrosion issue.

You can use 89-octane gasoline in a car designated as an 87-octane car and the higher octane will not cause your engine or fuel system to corrode simply because of the higher octane. The same scenario is true when you step up to premium 93-octane gasoline. The oil company doesn't give you 90% of 93-octane gasoline blended with 10% of 110-octane ethanol. They use gasoline with a lower octane rating to save a few pennies per gallon.

An automotive expert should know this information. If you're a "Top Person in Car Care, Education and Auto Industry News" because you're really good at knowing how to clean white wall tires, maybe you wouldn't know about octane levels and such. In that case, don't pass yourself off as an expert and don't let other people falsely introduce you as an expert.

Moving on, and we're now only at about the seven-minute mark, Lauren Fix talks about receiving lots of letters from people who had their warranties voided because they used the wrong fuel. She's implying that the vehicle owners used something other than E10 in their new vehicles and developed engine or fuel system trouble. When they went to the dealer to get it fixed, the dealer refused to do the service under the warranty.

If the vehicle is still under manufacturer warranty, it means the vehicle had to have been made sometime in the last few years. If that's the case, it means the vehicle is warranted for E10. If the vehicle is flex-fuel, it is warranted for anything up to and including E85. For Ms. Fix to claim she received such letters from vehicle owners is ridiculous. The only possibility is that the complaining vehicle owners used E15 or E85 in a non-flex fuel car. If they claim they used E15, I want to know where and how these people found E15 because it is scarce. I believe Ms. Fix's claim about receiving many letters about this problem is a gross exaggeration.

I challenge Ms. Fix's claim that new vehicle owners with valid warranties experienced any problems by using E15, assuming they could find it. Studies conducted by the most prestigious laboratories have clearly stated that E15 will not have any effect on any modern gasoline-powered passenger cars and trucks. Moreover, I know from years of personal experience in using splash blended E20, E30, E40, and E50 in non-flex fuel cars, that it causes no ill effects. (Splash blending is when a consumer manually does his/her own blending by pumping a quantity from an E85 pump and then pumping in a quantity from an E10 pump).

If Lauren Fix is talking about people who filled their non-flex fuel vehicles with E85, then the only problem they might have experienced would have been that the Check Engine light lit up and/or they experienced some rough running because the on-board computer didn't know how to handle the E85. Both these problems are temporary, and the vehicle would automatically resume normal operation after refilling with the correct fuel.

Is it possible that a vehicle owner accidentally used E85 instead of E10 in their non-flex fuel vehicle? Sure, but it's also possible for a person to accidentally use diesel or plain water instead of E10. In instances like these the problem is not the fuel; it's the individual.

Ms. Fix repeats her nonsense about ethanol causing corrosion problems in new and old cars. She cites cars going back to the early 2000s; and for older cars, she talked about cars from the 1960s and earlier. But by the mid-1990s, cars made for America had to use parts that were not susceptible to ethanol corrosion.

For older cars, it's possible that ethanol caused a problem. But then everything can cause a problem, they're old cars. She makes it sound as if in the pre-RFS ethanol days in America vehicles didn't regularly require repairs, maintenance, and replacement parts. She makes it sound as if ethanol-free gasoline is a soothing, life-giving elixir that gently encourages an engine to run. Gasoline is a highly combustible liquid. Using gasoline in a vehicle causes thousands of small explosions to occur every minute. Parts wear down; they break under stress; explosions cause stress. Why does a so-called automotive expert not know this? Why does a so-called independent automotive expert, and a so-called independent energy consultant have to invent derogatory stories about ethanol?

At 11 minutes into the show, Ms. Fix brings up the old canard about ethanol having 30% less energy than gasoline so that you get far fewer miles per gallon on ethanol or an ethanol-gasoline blend than you would from ethanol-free gasoline. In my opinion, using the BTU energy-content issue to explain the difference in MPG when you use ethanol is the litmus test that proves a person's ignorance.

Using ethanol or an ethanol-gasoline blend in a gasoline optimized vehicle will deliver fewer miles per gallon. However, the lower MPG is not because of the difference in BTUs; it's because the engine is OPTIMIZED to run on gasoline. A comparable engine OPTIMIZED to run on ethanol will deliver the same or better MPG. If you don't know this, you don't understand the issue. Ignoring the facts and continuously stating the wrong information is dishonest and deceitful.

Then, Ms. Fix promotes products that "stabilize" ethanol and makes your engine run better. Here's what Mercury Marine had to say about ethanol additives:

"No fuel additive can prevent ethanol from acting as a solvent. No chemical agent can be added to E0-E10 gasoline, in a reasonable quantity, that will recombine a phase separated layer. Fuel cannot be rejuvenated."

As I mentioned earlier, I'm using the information from Mercury Marine because it has often been claimed that boats suffer from ill effects of ethanol far worse than cars and trucks. But if E10 is actually better for a boat engine and fuel system than E0, then it must be better than E0 in a late model car or truck.

Lauren Fix and Jacki Pick begin a discourse on political reasons for using corn and growing corn. If Ms. Fix and Ms. Pick are incapable of understanding how internal combustion engines and fuels work, they have no business explaining the economics of farming. When Ms. Fix mentioned that she is reaching out to an unnamed national politician who supports E30 so that she can educate him, you just have to wonder how she gets through the day without bumping into walls.

The entire interview with Lauren Fix, showed how little she knows about ethanol issues. That interview took place more than two and a half years after the FOX News interview, and Ms. Fix learned nothing about the facts since in the interim. If you're interested in fuels, energy, vehicles, business or farming, and believe in ethanol, you should listen to this interview to see how people can twist the truth.

See also: Auto Repair Expert Bobby Likis: "What Really Needs to be Fixed Re: Ethanol" at:

https://www.theautochannel.com/news/2015/09/03/140638-auto-repair-expert-bobby-likis-what-needs-to-be-fixed.html

Open Letter to Mark Levin About Ethanol, Donald Trump and Ted Cruz

Originally published January 19, 2016

I'm the co-founder and co-owner of The Auto Channel, the Internet's largest automotive information resource and a small broadcast television network. I am a free-market, balls-to-the-wall capitalist, entrepreneur, and American patriot. I have owned my own businesses since my late teens, and I'm now in my mid-60s.

I was listening to your radio show while driving. I heard your comments about ethanol, Donald Trump and Ted Cruz.

Your position on ethanol is incorrect; not a little incorrect, completely incorrect. If you are against subsidies, you should be howling mad at the oil industry. They receive many, many times more subsidies than the ethanol industry, and this has been going on for over 100 years.

You say ethanol subsidies should be removed; and if ethanol is so good, let it stand on its own. But the truth is that all subsidies to the oil industry should be removed and then let's see how ethanol can stand on its own. If all oil subsidies were removed, gasoline would probably cost $15 or more per gallon.

Our country's enslavement to petroleum oil has been responsible for the death of hundreds of thousands of American servicemen. To date, not one American soldier, sailor or airman has been killed defending the production and distribution of domestic ethanol fuel.

You say ethanol negatively affects the food market and prices. This is false. The originator of this theory was the World Bank in a report they published in 2006. They officially retracted that report—twice, at least. The cost of petroleum oil and its finished fuels that cause food prices to rise.

You also said ethanol damages engines. This is a gross over-statement. Ethanol cleans internal combustion engines; it prolongs the life of internal combustion engines. Ethanol allows gasoline to be safely used in internal combustion engines. If ethanol was not blended with gasoline for use in automobiles, some other antiknock agent would have to be used. The choices are tetraethyl lead, MTBE, or a cocktail of so-called aromatics. We already know that lead is poisonous, and after six decades, it was finally banned in most instances. It was replaced by MTBEs, another poison that was banned—and replaced by ethanol. The concoction of aromatics used in ethanol-free gasoline are as poisonous as tetraethyl lead, and Ethanol-free gasoline is more expensive than gasoline blended with ethanol. If no anti-knock agent is used in an automobile engine it would literally knock itself apart.

You complain that ethanol-blended gasolines are not free-market. This isn't true, because ethanol-free gasoline is available. Being forced to use ethanol-free, poisoned gasoline for 60 years was not the act of a free-market. Ethanol is truly free-market, because anyone anywhere can produce it out of a wide variety of materials.

It is impossible to be a patriotic American and dismiss the ethanol industry in favor of the oil industry. The money and taxes we pay for fuel should go to American farmers, not to totalitarian dictators who spawn terrorism.

Please take the time to learn the truth. When someone with your influence and power takes such a dogmatically incorrect position, it casts a great pall over everything else you stand for. If you are wrong about ethanol, how many other things can you be wrong about?

I was surprised to hear of Trump's support for ethanol, but I am delighted about it.

I'm available to answer questions or debate any part of this issue at any time.

Regards.

Open Letter to Jillian Kay Melchior and the National Review About the Iowa Primary

Originally published January 24, 2016

I read your recent article in *National Review* titled "Trump's Support for Ethanol Is Bad for Taxpayers and Their Cars" *(https://www.nationalreview.com/2016/01/donald-trump-ethanol-subsidy-support-bad-taxpayers/)*, and I'm perplexed as to why you would take on the subject. You wrote the article as if you knew something about the subject, but the information you present shows you know nothing about it. Was this simply your contribution to the *National Review*'s effort to scuttle Donald Trump's presidential campaign? Did the editors at *National Review* think no one would read your article?

What ethanol subsidy are you complaining about? The blender subsidy, which went to the blenders (not the ethanol producers), ended in 2011. Federal tax credits ended in 2014. If you are against subsidies, you should be howling mad at the oil industry. They receive many, many times more subsidies than the ethanol industry ever did; and oil subsidies and tax credits have been in effect for over 100 years.

Ethanol opponents always say ethanol subsidies should be removed; and that if ethanol is so good, it should be able to stand on its own. It does. But the truth is that if all oil industry subsidies are be removed, gasoline would probably cost $15 or more per gallon.

You write that (ethanol) tax credits allowed between "1978 and 2012 cost the Treasury as much as $40 billion." In that same time frame, oil tax credits cost the Treasury several times that amount. Over the past century, our country's dependency on petroleum oil has been responsible for the death of hundreds of thousands of American servicemen and women, along with millions of injuries and permanent disabilities. Meanwhile, not one American serviceman or woman has been killed defending the production and distribution of domestic ethanol fuel. Does the wellbeing of American servicemen fit into the *National Review*'s definition of patriotism?

The mandate that you refer to doesn't mandate only corn be used to make ethanol. It mandates that a renewable substance be used. Ethanol is simply the best, safest and least expensive renewables end-product to use.

You are wrong when you say that ethanol negatively affects the food market and prices. The originator of this theory was the World Bank in a 2006 report that they have since retracted—twice. The cost of petroleum oil and its finished fuels, as well as the collaterals involved in bring food to market, caused food prices to rise.

You then wrote about the BTU canard that has been disproven dozens of times by scientists. You wrote, "Vehicles can drive fewer miles per gallon using ethanol blends than they would with pure gasoline, so Americans end up spending an extra $10 billion per year for fuel." This is false. The lower cost per gallon of ethanol-gasoline blends versus ethanol-free gasoline makes the ethanol-gasoline blends more economical. Ethanol-free gasoline can cost anywhere from about 50 cents to $3-$4 more per gallon than an ethanol-gasoline blend. If a vehicle gets 5% fewer miles per gallon by using E10, but the ethanol-free gasoline costs 20% to 100% more than E10, then the use of E10 saves the consumer money. Higher level ethanol-gasoline blends like E85 result in bigger savings.

For the record, there is no such thing as "pure gasoline." Gasoline is not like extra virgin olive oil; it's a combination of ingredients. For gasoline to work safely in high compression internal combustion engines, it must have an anti-knock ingredient.

You wrote, "Ethanol also guzzles 40 percent of the U.S. corn crop, and the resulting scarcity drives up the price of food." This statement is pure drivel. Clearly, you have no concept of how commerce works. American farmers grow as much corn as they do because they have sales orders to do so. If there was no market for corn ethanol, they wouldn't grow so much corn.

In 2000, U.S. corn production was 251,854 metric tons (MT); with 226,668 MT going to feed people and livestock. In 2013, because of the market for corn ethanol, U.S. corn production was estimated to be 353,715 metric tons. If 40% went to produce ethanol, it left approximately the same amount for human and animal consumption, and there was still a surplus. Moreover, a significant byproduct of ethanol production is distillers-grain, widely used for animal consumption. Not a single American went without corn chips, corn flakes, or corn-on-the-cob because of a corn shortage.

You then launch into the oil-industry-invented lie that ethanol causes engine problems and top it off by writing, "In fact, many vehicle manufacturers will no longer offer warranties when ethanol comprises 10 percent or more of fuel." In fact, that is not the case. Ethanol does not cause engine problems; it solves engine problems. Ethanol cleans internal combustion engines; it prolongs the life of internal combustion engines. It removes excess water that would cause fuel-line freezing in cold climates. Ethanol allows gasoline to be safely used in internal combustion engines. If ethanol was not blended with gasoline for use in automobiles some other anti-knock agent would have to be used. The choices are tetraethyl lead, MTBE, or a cocktail of so-called aromatics. We already know that lead is poisonous, and after six decades it was finally banned in most instances. It was replaced by MTBEs, another poison that was banned—and replaced by ethanol. Ethanol-free gasoline is more expensive than gasoline blended with ethanol. If no anti-knock agent is used in an automobile engine it, would literally knock itself apart.

Many more vehicle manufacturers are warranting the use of E15 (as well as E10), and more vehicle manufacturers are supporting the move towards E30 as being the most optimum fuel to support new engine designs and achieve future emissions goals. I can't imagine where you got the information that fewer manufacturers are warranting ethanol-gasoline blends because it's not true. Gasoline also requires more energy than ethanol to produce, and your source is wrong on that, too.

Your entire erroneous article assumes there is something wrong with Trump trying to win the support of Iowans by appealing to an issue that is dear to them. So, what's the problem? That's what politics are about. The reason that politicians stump in an area is to attract voters to issues that are dear to them. Trump isn't doing anything any other Republican and Democrat politician hasn't done.

I resent your and the *National Review*'s insinuation that you speak for Conservatives, Republicans or American patriots. I am a Conservative Republican, a true free-market, balls-to-the-wall capitalist, entrepreneur, and American patriot. You don't know the first thing about business, and you certainly don't know anything about ethanol. The *National Review* is not doing America any favors.

I suggest the publishers of *National Review* change the name to *National Lampoon*.

Have a nice day.

Open Letter to Barry Ritholtz and *Bloomberg* About Ethanol

Originally published February 4, 2016

Hi Barry:

I read your article "Enough with Ethanol" published February 3, 2016 on *Bloombergview.com (https://www.bloomberg.com/opinion/articles/2016-02-03/enough-with-forcing-subsidized-ethanol-on-u-s)*

No matter how much criticism you receive about the article and the information you presented, you are not the biggest knucklehead disseminating incorrect information about ethanol, you are merely the latest knucklehead to do so.

The only thing you got right is that you spelled the word *ethanol* and your name correctly. You begin by complaining about ethanol subsidies. There are no ethanol subsidies. The subsidies that misinformed "experts" like you refer to expired in 2011. So, when you write "Why we still subsidize turning food into an inefficient fossil-fuel substitute is anyone's guess?" I would ask you: How is it that a person with your experience and knowledge of finance, the economy and the business world doesn't know this?

And how is it that you don't seem to know there was never an ethanol subsidy; there was a biofuel subsidy that went to the blenders (the oil industry), not to the ethanol producers?

People used to say, let's end the "ethanol" subsidy that makes ethanol-gasoline blends cheaper than ethanol-free gasoline and we'll see that ethanol can't compete in a free market. Since 2011, we've been able to see how well it does compete. Ethanol-gasoline blends are still cheaper than ethanol-free gasoline, and sometimes it's a lot cheaper, often \$2-\$3 cheaper per gallon, than ethanol-free gasoline.

Service stations selling E85 and E10 have the same general price difference they had before the subsidy ended. E85 sold for about 50 cents less than E10. The subsidy provided about 50 cents per gallon, leading many to believe that the biofuel subsidy accounted for that difference. But E85 is still about 50 cents less per gallon that E10, meaning the consumer never benefitted from the subsidy; only the oil industry benefitted and never passed the benefits to the consumer. E85 should probably have been about $1 less per gallon than E10.

There's also no "ethanol mandate." There is a biofuel mandate, not an ethanol mandate. Why don't you know why ethanol is necessary? Don't you own a car? As a teenager did you never sit around talking cars and engines with your buddies? Don't you know gasoline without an anti-knock additive will destroy a car's engine? Even if you didn't know ethanol was the first additive that could tame knock in a high-compression internal combustion engine, didn't you know that tetraethyl lead was the oil industry's answer to engine-knock (instead of ethanol)? Didn't you know that tetraethyl lead is deadly to touch or inhale?

Didn't you know the oil industry's solution to tetraethyl lead was MTBE, which is also poisonous; and how come you don't know that the oil industry's current alternative to ethanol-gasoline blends is gasoline with significant quantities of other poisons, such as benzene and toluene?

Do you know or care about the hundreds of thousands of American servicemen and women who have been killed defending the oil industry around the world? Do you not know that no American soldier, sailor or airman has ever lost his or her life defending ethanol fuel production and distribution?

You then move into a series of falsehoods and misinformation about corn crop land use, food deprivation, ethanol EROEI (energy returned on energy invested), environmental issues, and the myth of ethanol causing engine damage. I thought I was going to have to address all these issues individually. But you gave me a gift, you quoted Robert Bryce, author of such fairy tales as *Gusher of Lies.* This means that he is at the least one of your resources for bad information.

Robert Bryce is the biggest knucklehead disseminating incorrect information about ethanol. He's held this title for a number of years.

Here's the 60+ page report published in June 2013 about Mr. Bryce and his view of ethanol and alternative fuels.

You can find the report at:

https://www.theautochannel.com/news/2013/06/10/082044-gusher-lies-book-review-and-reply-to-robert-bryce.html

(The report is included later in this book.)

You also quote the Environmental Working Group as if they were an authoritative independent resource on the environment and alternative fuels. EWG is a shill for the oil industry and/or friend of the oil industry.

Be cautious when using EWG as a trusted or authoritative resource because you may find several of their opinions are in fierce conflict with opinions you and Robert Bryce have. If EWG were correct about corn crops and ethanol, they could easily be correct about all kinds of things in direct conflict with information you probably tout to your financial clients.

Read my report, consult with whomever you wish (especially Robert Bryce), and then challenge me on any of the facts. I'll publish your findings word for word, regardless of what you find out. You don't have to do it today or tomorrow; you can do it next week or next month or next summer. Prove to yourself, your clients, to *Bloomberg* that you're not a knucklehead.

Thanks for your time.

February 5, 2016

Reply from Barry Ritholtz

Sir:

I just finished reading your response to my column "Enough with Ethanol."

My only wish is that YOU had read the column and responded to what I wrote. Instead, you responded to what others have been saying about Ethanol. It appeared you misread or misunderstood what I put on paper.

First, I wondered "why we still subsidize turning food into an inefficient fossil-fuel substitute." I did not say we had tax subsidies; The word has a very precise meaning; the dictionary definition of the word *subsidize* is:

Verb... to purchase the assistance of by payment of a subsidy; to aid or promote (as a private enterprise) with public money or efforts

We have "subsidized" Corn Ethanol with direct tax subsidies through 2011; we have also subsidized ethanol by mandating the usage of ethanol through the Renewable Fuel Standard requirements.

This is not the free market at work, but rather an EPA decree that requires retailers blend almost 15 billion gallons of corn ethanol a year into the gasoline they sell to the public.

It may not be a tax-based subsidy (a phrase I never use in the article), but it is a direct support of the product. And that, sir, is a subsidy. So too was the a 54-cent per gallon import tariff, making U.S.-made ethanol that much more expensive, but that was not in the final version of the column.

That definition aside, you never respond to the direct points of my column that Ethanol is:

- Anti-Free Market
- Damaging to engines
- Not especially efficient;
- Drives up the cost of water to local Midwestern consumers
- Causes pollution via runoff to Gulf of Mexico
- Barely generates more energy than it takes to produce it

You also made a number of misstatements in your letter. You say there never was an ethanol subsidy, ignoring both the 45-cent per gallon payment to blenders, as well as the aforementioned import tariff. That subsidy, whether to farmers or blenders, still operates to lower the cost of ethanol thus distorting the market. That anyone can imagine it is not a subsidy is astonishingly silly position.

Last, I have long ago learned that people who engage in Ad hominem attacks and name-calling in debates do so because they either lack the ability to make a coherent argument, or because their position is without merit. I suspect in your case it is both.

Good day, sir.
www.Ihatecorntethanol.com

My Reply to Barry Ritholtz:
Hi Barry -

I do appreciate that you replied to my earlier letter, thank you.

You say you wish I had read your column instead of responding to what others have said about ethanol. I did read your column; and when it was evident you were parroting what others have said (most notably Robert Bryce), I thought the most expedient use of my time was not to repeat my responses to Mr. Bryce in the past, and provide a link to a lengthy report I wrote in reply to his book, *Gusher Of Lies*.

Unfortunately, it doesn't appear that YOU read your column. Both words *subsidize* and *subsidies* appear in your article. But even if the word *subsidies* didn't specifically appear in your article, you are correct that the word *subsidize* is a verb defined as "to purchase the assistance of by payment of a subsidy." Consequently, by definition, the use of the verb automatically inserts the word *subsidy* or *subsidies* into the reading of your article. I suggest that in an effort to find some fault with my criticism that you felt semantic subterfuge would suffice.

In any event, you are incorrect, especially referring to the real issue (not just the use of the words *subsidies* or *subsidize*). The Renewable Fuel Standard doesn't mandate the use of ethanol, it mandates the use of a renewable substance, of which ethanol is just one possibility, although it is the best currently available.

You say the subsidy is (or was) "...a direct support of the product" (meaning ethanol). However, the subsidy wasn't in support of ethanol, it was in support of gasoline. The subsidy went to the blenders to make up for what they lost by having to blend in what is for them an unprofitable non-petroleum ingredient. As I wrote in my earlier initial response, that subsidy should have been passed along to the consumer to make ethanol-gasoline blends more affordable than ethanol-free gasoline.

The general anti-ethanol argument that cites subsidies is preposterous because if subsidization is the issue, then the true objects of your enmity are the subsidies that go to the oil industry, which are many times greater than any subsidies of any kind that could be said to benefit ethanol.

If you dislike subsidies and think subsidies are a restriction on a free market, then you should hate all subsidies that go to the oil industry. In looking over your other published articles, I don't see any critical of the subsidies that go to the oil industry. If you don't like standing outside when it drizzles, why do you like standing outside in a downpour?

You included a link to a website called *IHateEthanol.com*. I thought it would have some information in support of your position. I was anxious to read it. The link doesn't work. Investigating a bit further, I found that it is simply a URL that you purchased a few years ago to display your dislike for ethanol and "ethanol subsidies." So, if you hate subsidies so much, do you also own the URL *IHateGasoline.com*?

Restrictions don't necessarily mean a market isn't free. It may mean that in order to ensure a free market, some regulations are required. Stop lights and stop signs at street corners don't restrict the free flow of traffic; they insure the orderly and safe free flow of traffic. Putting age restrictions on the purchase of alcoholic beverages or tobacco products are considered by most members of society as a necessity to keep the markets open, instead of being legislatively closed due to health and danger concerns. Even the most ardent libertarians understand and accept these measures.

The concern for us, as patriotic Americans, is to insure a safe and sane American free market. I'm not interested in helping foreign ethanol producers at the expense of American farmers and producers. There is no benefit in allowing someone else to dump products in America. Domestic ethanol is cheaper than gasoline and diesel; and if we want our fuels at the pump even cheaper, there should be no restrictions on how much ethanol can be sold at a fueling station.

The free-market restriction we should all be concerned about is that higher level ethanol-gasoline blends are not allowed at every single service station. Pumps should be available that allow the consumer to choose whatever blend level they desire, from E0 to E99, with all the stops along the way. Who is stopping this from happening? It's the oil industry.

What's more, it should be illegal for an oil company to impose severe financial penalties on independently owned stations for adding in independent fuel options. Let the power of the consumer's purse control the market; make all and any level blends available; remove the subsidies given to the oil industry for more than 100 years—and let the price of gasoline and diesel reflect their real cost, which might be as high as $15-$20 per gallon.

For argument's sake, let's say that all import tariffs are removed from foreign ethanol. At this point, the biggest losers wouldn't be domestic ethanol producers, it would be the oil industry. Cheaper ethanol would make ethanol-gasoline fuels even cheaper than they are now. They would be a competitive threat to the oil industry, forcing your precious oil industry to lose money or the government to impose greater at-the-pump taxes to quell the oil industry's complaints. The irony of your solution would be to create more taxes, not less.

The oil industry is the most heavily restricted industry in the world; there is nothing free about it. It's worse than diamond cartel restrictions, because not everyone needs diamonds on an every-day basis. We can get along without diamonds, but we can't get along without engine fuels. They have us over a barrel, pun intended.

To sum up the issues of subsidies and free markets: The subsidies did and do support the oil industry; import tariffs on foreign ethanol protect the oil industry much more than the ethanol industry; import tariffs on foreign ethanol protect American farmers and keep jobs, taxable revenue, and salaries in America. Having a strong domestic ethanol industry means we will never have to send troops to Brazil or any other country to defend access to ethanol.

You bring up the myths of engine damage caused by ethanol inefficiency; cost of water; pollution and EROEI. This means YOU not only didn't read your column; you didn't read my 60+ page report answering Robert Bryce. If you had, you would be embarrassed to repeat these childish points and obvious falsehoods.

Lastly, in reply to your taking umbrage to my calling you a "knucklehead," your accusation that my use of pejorative terms indicates an inability to make a coherent argument is an example of your inability to respond to facts.

Words are a wondrous thing; they are tools. And like other types of tools, the use of the right tool can accomplish a task much more efficiently than using complicated workarounds. Many words express something that might otherwise take an entire sentence or even a paragraph to explain. Given your inability to provide contrary information to the facts I've set forth regarding ethanol, along with your reluctance to read my additional writing on the subject, you have produced no evidence to show otherwise.I welcome any response you would like to make and will faithfully report every word of your response in my published reporting of our communications.

Good day to you, Barry, and have a great weekend. By the way, who are you rooting for, the Panthers or the Broncos?

February 8, 2016

Reply from Barry Ritholtz

You still have not responded to the main points I raised about ethanol.

Let's try a numbered list format.

1. Do you claim it is not damaging to engines? Can you explain why small engine makers will not warranty their engines if more than 10% of ethanol is used?

2. Do you believe ethanol as an energy source is especially efficient?

3. Do you disagree with the complaints that it drives up the cost of water to local Midwestern consumers?

4. Do you not accept the scientific assessment that it causes pollution via runoff to the Gulf of Mexico?

5. Do you disagree with the contention that after being a negative net energy producer; it now barely generates more energy than it takes to produce?

6. What makes this corn-based ethanol so special that it does not have to compete in the Anti-Free Market for customers?

Those were the key issues I raised in the article you pushed back against.

You have yet to respond to those issues, please do.

My Reply to Barry Ritholtz

Barry - So nice to hear from you again. I tried over two separate emails to suggest that you read the 60+ page report I wrote, which would have answered all your questions...and more. I provided the link, and there's no cost, so I don't know why you wouldn't have availed yourself of the opportunity since you seem to want to see how I respond to these points.

1. You use the word *damage*. What damage are you referring to? Internal combustion engines operate on thousands of small explosions per minute, regardless of the type of fuel being used. The explosions eventually "damage" the engine. Ethanol does not cause explosions that are any worse than gasoline or diesel fuel. In fact, in higher compression internal combustion engines (such as those found in cars and trucks), gasoline without ethanol or another anti-knock ingredient will knock the engine apart. That's damage.

The issue that you are probably referring to are the three complaints typically made about ethanol-gasoline fuel blends.

- The first is that the ethanol causes sludge, tar, gunk, goo, crap (or whatever you want to call it) to appear in the fuel tank, thereby clogging the filter, spark plugs, fuel injectors, carburetors, etc.
- The second complaint is that the ethanol corrodes the rubber hoses and some metal parts.
- The third is that the ethanol in the fuel-blend magically sucks water right out of the air and drops it in your fuel tank, thereby causing some type of water-related problem.

Of the three, I'd say that the first, "the sludge quagmire," is the most often heard, so I'll begin there.

To get to the bottom of this issue, basic questions must be asked:

Question #1 - Does sludge/goo/gunk/crap form in an internal combustion engine? The answer is yes.

Question #2 - Is the sludge/goo/gunk/crap caused by ethanol? The answer is no; it is caused by the gasoline and oil used in the fuel.

While ethanol is a fuel, it is also a very effective solvent. Solvents are used to clean and remove the residue and deposits left after a substance has been used, such as the residue and deposits left after gasoline and/or oil are burned.

Gasoline is also a solvent, but it's not as effective a solvent as ethanol. Gasoline and petroleum oil products are champs at leaving a varnish-like residue. If you were choosing sides to compete in a national residue-making competition, gasoline or petroleum oil would work for your team. But if you were choosing teammates for a national clean-up competition, you'd pick ethanol.

How much more residue does gasoline or a gasoline-oil combination create as compared to ethanol? Watch these videos to find out:

- *https://www.youtube.com/watch?v=WOAyoCo3xXA*
- *https://www.youtube.com/watch?v=-o5D-nic9WM*

Imagine the glass jars in the video are the insides of an internal combustion engine. Instead of burning fuel-soaked wicks for a few seconds, burn the fuel-soaked wicks for an hour or two every day for a week, a month, a year. Based on the results of the video, the residue and deposits from the gasoline and gasoline-oil fuels would be caked on.

Would the caked-on residue affect the functionality of an engine's pistons? Would the solid caked-on bits break off and affect the functionality of other parts of the engine? Of course, they would. This is one reason engines fail and mechanics have to clean out an engine. To mitigate this problem, gasoline companies add "detergents" to the gasoline to help dissolve the build-up. They add alcohol. Ethanol is that alcohol. If gasoline was a more effective solvent, it might remove the residue that burned gasoline creates—but it does not.

In an article published by *GasBuddy.com* on June 29, 2013 Gregg Laskoski wrote about automakers' efforts to encourage the development of more effective gasoline detergents. Mr. Laskoski is Senior Petroleum Analyst at GasBuddy and was once Managing Director of Public Relations at AAA Auto Club South. Here's are some excerpts from that article:

"The automakers formed a consortium, Top Tier Gasoline, that certifies retailers as meeting their standard for detergent. They say it's necessary because detergents prevent deposits of leftover material from building up in engines and exhaust systems. The deposits are like the ash that remains in a fireplace, and their presence in your car's engine can reduce fuel economy and performance."

"We strongly recommend Top Tier detergent gasoline to keep your engine clean," GM fuel specialist Bill Studzinski said. "Fuel economy, emissions and acceleration all suffer when there are deposits in an engine."

I hope you noticed that there isn't one single mention of ethanol in these two paragraphs that attribute the deposits and residue problem to ethanol. And if you have the opportunity to read the entire article, you'll see that the word ethanol never appears anywhere in the text of the story.

Let me take you in a different direction. Imagine having a fuel that creates much less residue and is a very effective solvent at the same time. In other words, something that might make a small mess, and then cleans up after itself, along with cleaning up the mess left behind by petroleum-based fuels. Isn't that better? How could it not be better?

But wait, some engines do have problems with sludge, tar, gunk, goo, crap (or whatever you want to call it) after fuel has been switched to ethanol or an ethanol blend. There, I said it; I admitted it.

If you could fault ethanol fuels for anything, it's that they are too efficient at cleaning the residue and deposits. That's like complaining your toothpaste whitens too much; that your hand-soap gets too much dirt off your hands, and that your greasy pots and pans clean up too quickly.

If you have an engine that used some type of gasoline fuel or gasoline-oil fuel, then the engine will have substantial deposits throughout, and you will have to have it serviced. You will have to change filters, spark plugs, anything and everything that comes in contact with the burned fuel deposits. Through the life of the engine and the device that the engine powers, you will have to do this, maybe multiple times. The good news is that once you have had this service completed, ethanol-gasoline fuels will keep your engine clean longer and better than any non-ethanol gasoline with detergent.

So, if petroleum oil products, like gasoline and diesel, are the cause of the build-up of all the engine "ash," or engine "plaque," why do some people and organizations claim ethanol is the cause of the problem? You could say it's a lie to blame ethanol. In fact, I will say it: It's a lie.

If you watched the videos at the links above and read the quotes from the *GasBuddy.com* article and if you have just the most rudimentary knowledge of internal combustion engines, then you know what I've written is true. If you have zero knowledge of internal combustion engines, you won't accept what I've written as true. If this is the case, you shouldn't be commenting on engine damage in the first place.

It could be said that all fuels used in any type of engine will contribute to the engine breaking down. It could be a spark-induced internal combustion engine, a compression-ignited internal combustion engine, a steam engine, a water or wind turbine, or an electric motor—they all break down, and the fuel that supplies the power helps cause it. Engines have moving parts; moving parts wear down. Internal combustion engines suffer from thousands of small explosions per minute.

When anti-ethanol people talk about a fuel damaging an engine, they make it seem as if only ethanol causes the wear and tear. They make it seem as if gasoline somehow cushions the explosions or coats the engine components to prevent deposits from forming. There is no fuel used in an internal combustion engine that is worse for the engine than gasoline. Gasoline damages engines. Small engines suffer from gasoline problems. Ethanol helps to clean the engines.

If gasoline was not harmful to engines, America would not have had thousands of engine repair shops and mechanics spread throughout the country in all the years during the age of leaded gasoline.

There are three often quoted engine problems attributed to ethanol. The above explanation only refers to the "sludge" issue. Now I'll address "corrosion."

All liquids are corrosive, including and especially water. Air, wind, solar rays are also corrosive. Gasoline is, of course, corrosive, so certain materials cannot be legally used to hold gasoline. But guys like you make it seem as if only ethanol causes corrosion. When you're faced with a corrosion-causing factor, you use materials not susceptible to corrosion. You use paint that can withstand solar rays and metals that don't rust. It's not a big deal. As corrosive as water is, we drink it and bathe in it.

During the decades in which tetraethyl lead gasoline was the only real fuel available for spark-induced internal combustion engines, engine designers used parts that were least susceptible to gasoline corrosion.

After leaded gasoline and gasoline with MTBE were banned, engine designers changed the parts susceptible to ethanol corrosion to parts that resist ethanol corrosion. The vast majority of passenger cars and lights trucks on the road today have engine parts and fuel system parts that are no more susceptible to ethanol corrosion than the previous parts were to gasoline. Small engine manufacturers have also been making the changes and more and more new equipment is approved for ethanol fuels.

Yes, older engines may require the replacement of parts, but older engines require the replacement of parts at some point anyway, so parts that are not susceptible to ethanol should be used. The replacement parts are usually readily available and cost differences, if any, are negligible.

The third engine-problem issue is the myth that alcohol (ethanol) sucks water right out of thin air and puts it in your fuel tank. This concept is a joke. Mercury Marine, the largest manufacturer of marine engines, has clearly and repeatedly stated: "ethanol does not grab water molecules out of the air." And they correctly attribute water collecting in boat fuel tanks as the result of condensation, not because the ethanol in the ethanol-gasoline blend sucked the water molecules out of the air. You can hear more about this for yourself from an extensive webinar that Mercury presented 2011. The full webinar can be found at:

https://www.youtube.com/watch?v=GbkmY7TMKMg

An abbreviated version of the results can be found at: *(http://www.boatus.com/magazine/2011/december/ethanol.asp)*

I purposely cite the Mercury Marine report because it also addresses the general nonsense about ethanol-fuels being harmful to boats. If you do avail yourself of the links just above, you'll find that Mercury says that E10 may be better for marine engines than non-ethanol gasoline.

2. Do I believe ethanol as an energy source is especially efficient? Of course, I do, as do top scientists around the world, today and for more than a century. An internal combustion engine optimized to run on ethanol will out-perform a comparable internal combustion engine optimized to run on gasoline. The early automobile inventors knew this, Henry Ford knew this, and General Motors top fuel scientists (Kettering, Midgley and Boyd) knew this. This same GM trio was responsible for inventing leaded gasoline to prevent ethanol from being used as a no-knock aid, cutting into GM profits.

In addition, USDA tests in 1906 demonstrated the efficiency of alcohol in engines and described how gasoline engines could be modified for higher power with pure alcohol fuel or for equivalent fuel consumption, depending on the need. This was testified to during the "Free Alcohol" Congressional Hearings held by the Ways and Means Committee of the 59th Congress in 1906.

The U.S. Geological Service (USGS) and the U.S. Navy performed 2,000 tests on alcohol and gasoline engines in 1907 and 1908 in Norfolk, Va. and St. Louis, Mo. They found that much higher engine compression ratios could be achieved with alcohol than with gasoline. When the compression ratios were adjusted for each fuel, fuel economy was virtually equal despite the greater BTU value of gasoline.

Elihu Thomson reported that despite a smaller heat or BTU value, "a gallon of alcohol will develop substantially the same power in an internal combustion engine as a gallon of gasoline. This is owing to the superior efficiency of operation..." (*New York Times*, Aug. 5, 1906) Other researchers confirmed the same phenomena around the same time.

In recent times, other studies have been done to prove this. For example, in 1998 the "Ethanol Vehicle Challenge" staged by Wayne State University proved that ethanol-optimized engines out-performed comparable gasoline-optimized engines.
(https://wayne.edu/newsroom/release/1998/06/04/high-performance-wayne-state-ethanol-car-wins-1998-ethanol-vehicle-challenge-1464)

In 2010, Argonne National Laboratory presented their study, "Ethanol-fueled racecar engines outpower lead-fueled engines."
(http://www.anl.gov/articles/ethanol-fueled-racecar-engines-outpower-lead-fueled-engines)

In 2011, a joint study conducted by the U.S. Dept. of Energy, Ford Motors, and AVL Powertrain Engineering presented similar positive ethanol results.
(http://www.ars.usda.gov/SP2UserFiles/Program/213/Teleseminars/Robert%20Stein%20(AVL)%203-29-11.pdf)

3. Do I disagree with the complaints that ethanol production drives up the cost of water? I don't think I've seen specific complaints that ethanol production drives up the cost of water. If you have actual quotes or statements about this, please submit them.

I am aware that the quantity of water use is cited as an issue. This was particularly true during the recent Midwest drought years. However, golf courses, residential lawn use, corporate and scholastic campuses, and ubiquitous restaurant serving of water as soon as a customer sat down at a table were also areas of concern. Therefore, to single out ethanol production is deceitful.

Is it possible that local communities raised water prices during the Midwest drought? Of course. That's what they do; that's what governments do to try and reduce the use of whatever they think is being used too much.

There are areas of the west that have been faced with rising water prices because of the drought in the west, not because the western states are growing an abundance of corn for ethanol. In California, crops include lettuce, oranges, olives, almonds, pumpkins, hops, grapes, tomatoes, cotton, walnuts, grapefruits, and flowers. Several of these crops are water intensive. Are you aware that lettuce production, for instance, has had a steady increase year after year? Maybe people should stop eating so much salad.

I checked the Internet and didn't find one source blaming lettuce growing for the increase in water prices. Following common sense, the argument about ethanol causing rising water prices is disingenuous.

What does trouble me about your asking if I disagree with the complaints that ethanol production drives up the cost of water is that this isn't even the correct water issue to wonder about. The correct issue to be concerned with is how much water is required to produce ethanol. So not only is your question irrelevant, you didn't get the question right. If we were playing Jeopardy on television you would have lost.

I'll pose the problem for you, Barry: "Hey Marc, don't you think that ethanol production uses far too much water?"

Barry, I'm glad you almost asked about this because it's one of my favorites to address and it always makes me laugh. Can you imagine the oil industry and its shills pretending that they have a concern for water issues? It is the ultimate in hypocrisy. It's double speak to the 10th power.

Are the people who brought us global fresh and saltwater disasters, who have killed untold millions of wildlife, who have destroyed untold miles of beautiful beaches raising a red flag warning about the water requirements to produce ethanol?

The issue of water use may be the single best example of the advances that have been made in ethanol production. The anti-ethanol crowd says ethanol requires huge amounts of water, not just in the actual distillation process, but for the growing the crops used to make the mash that gets fermented and then distilled. It's true. Water is required for these tasks. In his book, Robert Bryce cites various studies that reveal the amount of water required. He then uses various voodoo equations to paint a picture that would show the Earth turning into the desert planet Arrakis in Frank Herbert's *Dune*.

Mr. Bryce foresees absolutely no circumstances in which the *Dune* scenario wouldn't happen. He doesn't think desalination plants or technological advances could mitigate the problem. Nor does Bryce consider the fairly large variety of raw materials that require less water and could be used instead of corn to provide much greater yields of ethanol. Bryce, and people like him are completely wrong. Here's some facts from POET Ethanol, one the largest producers of ethanol in America:

> "The first plant that POET purchased took 17 gallons of water per gallon of ethanol in 1987. Today, POET averages less than 3 gallons of water, an 80 percent reduction from where it started. But we're not done yet. POET believes that water is a precious resource and is committed to using as little of it as possible in our ethanol production process. So, we have set a goal...of decreasing the overall annual water intake at our plants 22% by targeting an average usage rate of 2.33 gallons of water taken in per gallon of ethanol produced. This will reduce our total annual water use by one billion gallons based on POET's 2009 production capacity of 1.5 billion gallons of ethanol.
>
> "POET engineers have invented a new way for its facilities to use water which results in a decreased intake with a reasonable economic payback. Named Total Water Recovery, it ensures that our facilities don't have to choose between profit and planet. As of December 2011, eighteen POET plants had installed Total Water Recovery.
>
> "Most of the lifecycle water used in ethanol production is for growing the feedstock. According

> to the National Renewable Energy Laboratory, almost all of that is from rainfall as less than 5 percent of corn used for ethanol production is irrigated. Still, water requirements of the feedstock represent the vast majority of the total lifecycle water use for ethanol production. POET is working with our feedstock producers to understand the amount of irrigated corn coming into our facilities in order to identify opportunities to collaborate for additional water conservation."

To support POET's comments, made in 2011 by Forrest Jehlik, Research Engineer at Argonne National Laboratory:

"The amount of water used to make ethanol has declined dramatically. Today, producing one gallon of ethanol requires about 3.5 gallons of water. That's a little more than it takes to process a gallon of gasoline. Much of the criticism about ethanol's water requirements stem from the need to irrigate feedstock crops in drier climates. But most ethanol is produced from rain-fed crops grown in the Midwest.

"In addition, ethanol is not carcinogenic and doesn't poison groundwater or the ocean. Ethanol rapidly biodegrades. Concerns over ethanol spills are muted by ethanol's low toxicity. In fact, you'll find ethanol in beer, bourbon and other happy-hour beverages you've probably consumed."

Water advancements are possible, they have been made, and they will undoubtedly continue. However, the key item in Mr. Jehlik's comments is this: "Today, producing one gallon of ethanol requires about 3.5 gallons of water. That's a little more than it takes to process a gallon of gasoline." Yes, it requires lots of water to make gasoline, but you people never think of telling that to your audience.

If you'd like more information about POET and the water issue, you can listen to an audio interview I conducted with Jeff Broin, Chairman and CEO of POET on January 27, 2016:

https://www.youtube.com/watch?v=QZY7Nyun00U)

4. Do I accept the scientific assessment that it causes pollution via runoff to Gulf of Mexico? I assume you mean the "it" is ethanol. If so, the answer is no. I don't accept this scientific assessment (if there is such an assessment).

The pollution you're referring to is caused by fertilizer, not ethanol. Corn crops are not used to make fertilizer, they are used to produce food and fuel. Golf courses, residences, corporate and scholastic campuses, and other crops use fertilizer. Did you know that in the 10 states that border the Mississippi River, and the states whose minor rivers and streams feed into the Mississippi, that there are thousands of golf courses—the number may be as many as 10,000 golf courses. They all use fertilizer. That fertilizer contributes to the pollution running into the Gulf of Mexico. I am willing to bet that not one of those golf courses grows crops intended for ethanol production.

If you had a true concern for the Gulf of Mexico, you would want to ban all golf courses; all residences and campuses with lawns; and all farming, regardless of the crops grown.

Plant life requires fertilizer of one type of another. The solution is not to eliminate the use of fertilizer, it is to employ fertilizers and techniques to minimize down-river pollution. One very good concept is to harvest the algae that grows abundantly in the Gulf of Mexico because of the fertilizer and use it to produce ethanol and organic fertilizer for the farms, golf courses, campuses, etc. This be an excellent fertilizer and would diminish the amount of algae that exhausts the oxygen in the water that prohibits marine life from existing in those areas.

5. You ask if I disagree with the contention that ethanol now barely generates more energy than it takes to produce it.

I absolutely disagree with that contention, both by implication and by fact. The question implies that only ethanol requires energy to produce, or that it has the worst EROEI statistics. This is disingenuous and shows you have no understanding whatsoever of energy in general, and ethanol in particular.

All fuels require vast amounts of energy to be transformed it into a useable state and bring to market. This is true for coal, gasoline, diesel, nuclear, hydroelectric, steam, natural gas, propane, hydrogen, solar, methanol or ethanol. They all begin as energy negative.

Some fuels, such as solar, may require little energy to be captured and transformed, but its output is so restricted (with current technology), it's not very efficient or economical for broad application.

The only relevant question is, "How much energy is required to produce ethanol compared to gasoline?" The answer is that gasoline is more energy negative than ethanol. Since you cite Robert Bryce, and no other resource, I'm assuming you're referring to David Pimentel and Tad Patzek's studies. Bryce doesn't cite any other work.

If Robert Bryce is the kahuna of knuckleheads, Pimentel and Bryce are little better than village idiots. Almost as soon as their study was completed and published in 2006, their findings were rendered incorrect and irrelevant. The counter studies were conducted far and wide by different individuals and institutions, including Patzek's home facility, UC Berkeley. If you had taken the time to read my 60+ page report, you would have learned the following:

The Pimentel-Patzek studies were hatchet jobs because they calculated in extraneous materials, they under-estimated crop yields, they over-estimated other aspects of farming (either intentionally or because the information they used was out-of-date), they exaggerated possible worst case agricultural scenarios, they created deceptive algorithms to calculate and compare statistics, and they purposely ignored certain other aspects of oil refining and gasoline production that make gasoline EROEI horribly negative.

Perhaps the most shocking item that Pimentel and Patzek left out in their gasoline EROEI calculations are the hundreds of thousands of American military personnel (past and future) who were killed or wounded to protect and defend the oil industry's status quo. I'm sure they never calculated in the cost of continually treating those service personnel who sustained long-term and life-long injuries. Then there are the accidental deaths of oil industry workers (past and future). Add in the destruction of beaches, and the loss of employment for those non-oil industry workers who live near the oil disasters, and then—as in the case of the Gulf of Mexico fiasco—the destruction of millions of birds and fish.

You would have to figure in all the energy expended to make every bullet, every gun, every tank and cannon and jet and ship and pair of shoes worn by our service men and women who protect the oil industry. These are all a part of the cost of doing business oil-style.

Okay, okay, I'm not playing fair; I'm using tragically true but typically sideline aspects of oil EROEI. I don't know how you can overlook the deaths of thousands of Americans, but the oil industry does. I should be comparing apples with apples, even though Pimentel and Patzek used bad information and irrelevant factors, such as farming equipment that are unnecessary.

What Pimentel and Patzek did was to try and reinvent the wheel, so to speak, even though the wheel has been with us for thousands of years. Here's what I mean; they started with a zero-based budget as if America had no farmers. They even calculated the energy required to build some of the equipment used in farming. Not only were the tractors they selected unsuitable for the job because they were too large and cumbersome, but they ignored the fact that almost all corn farmers already own the equipment necessary to grow corn or other products. They stacked the deck against ethanol. To be fair, P&P should have calculated in every aspect that goes into what it takes to get the gasoline to the pump, and that requires adding in all the costs of war, and every single item—right down to the energy it takes to make the zippers on the pants worn by our military service personnel.

Consider results of studies that studied the Pimentel-Patzek studies. The first comes from Tad Patzek's own school, UC Berkeley. It was completed in 2006 after the Pimentel-Patzek studies and was conducted by UC Berkeley's Energy and Resources Group.

Here's some highlights of the results:

Ethanol can replace gasoline with significant energy savings, comparable impact on greenhouse gases.

The analysis, appearing in this week's issue of Science, attempts to settle the ongoing debate over whether ethanol is a good substitute for gasoline and thus can help lessen the country's reliance on foreign oil and support farmers in the bargain.

Dan Kammen and Alex Farrell of the Energy and Resources Group at UC Berkeley, with their students Rich Plevin, Brian Turner and Andy Jones along with Michael O'Hare, a professor in the Goldman School of Public Policy, deconstructed six separate high-profile studies of ethanol. They assessed the studies' assumptions and then reanalyzed each after correcting errors, inconsistencies and outdated information regarding the amount of energy used to grow corn and make ethanol, and the energy output in the form of fuel and corn byproducts.

Kammen estimates that ethanol could replace 20 to 30 percent of fuel usage in this country with little effort in just a few years. In the long term, the United States may be able to match Sweden, which recently committed to an oil-free future based on ethanol from forests and solar energy. Kammen last year published a paper, also in Science, arguing that even Africa could exploit its biomass to build a biofuel industry that could meet energy needs for the poor and develop a sustainable local fuel supply, a future much better than using fossil fuels.

The goal of the UC Berkeley analysis was to understand how six studies of fuel ethanol could come to such different conclusions about the overall energy balance in its production and use. Farrell, Kammen and their UC Berkeley colleagues dissected

> each study and recreated its analysis in a spreadsheet where they could be compared side-by-side. The team said it found numerous "errors, inconsistencies and omissions" among the studies, such as not considering the value of co-products of ethanol production—dried distillers' grains, corn gluten feed and corn oil—that boost the net energy gain from ethanol production. Other studies overestimated the energy used by farm machinery.
>
> The assumptions made by some of the authors were not based on the best data, or were just a little bit too convenient, and had a strong impact on the results," Kammen said.

Included in the UC Berkeley review were the following:

Tad W, Patzek, "Thermodynamics of the Corn-Ethanol Biofuel Cycle," *Critical Reviews in Plant Sciences* 23(6), 519-567 (2004). David Pimentel and Tad W. Patzek, "Ethanol Production Using Corn, Switchgrass, and Wood; Biodiesel Production Using Soybean and Sunflower," *Natural Resource Research,* 14(1), 65-76 (2005).

Another study, conducted by the United States Department of Agriculture, was presented in 2007 at UC Berkeley by Roger Conway, Office of Energy Policy and New Uses at the United States Department of Agriculture (USDA). The report showed huge discrepancies in the figures that Pimentel and Patzek used to arrive at their conclusions versus the figures used by USDA's efforts to conduct their own studies on ethanol vs. gasoline EROEI. The USDA studies were significantly more favorable towards ethanol production.

A Michigan State University study conducted by Bruce Dale, Professor of Chemical Engineering, found that the Pimentel-Patzek methodology is flawed. The measurements of BTU are irrelevant and that the net energy of ethanol is actually higher than gasoline (in other words, EROEI for ethanol is positive, while the EROEI of gasoline is more negative).

In 2005, Bruce Dale participated in a C-SPAN televised debate against David Pimentel and Tad Patzek. Speaking on Dale's side was John Sheehan, Senior Engineer, National Renewable Energy Laboratory. The debate was supposed to center on the issue of net energy balance. One of the most important points illuminated in the debate was that even if Pimentel-Patzek's finding that corn ethanol production results in a negative net energy equation, gasoline production is far worse and electric energy is horrendously bad compared to both ethanol and gasoline. After watching the video of the debate, it's hard to believe anyone took Pimentel-Patzek seriously, unless all other opposing information is kept from the viewer. The entire video can be watched at:

https://www.c-span.org/video/?188549-1/ethanol-energy-policy

Another report critical of Pimentel-Patzek was published in 2006 by Justus Wesseler, an agricultural economist and professor of Agricultural Economics and Rural Policy at Wageningen University in the Netherlands. He called the Pimentel-Patzek work "flawed" and "misleading." This report can be found at the Elsevier Energy Policy website.

Also in 2006, the spring edition of *The New Atlantis* (*Journal of Technology & Society*) said this about Pimentel and Patzek's studies: "Professors Pimentel and Patzek have published several studies on this subject, and these have been thoroughly and repeatedly debunked in the scientific literature, in government reports from the Department of Energy and Department of Agriculture, in congressional testimony, and elsewhere...Reputable scientists have publicly called the work of Pimentel and Patzek "shoddy," "unconvincing," and lacking in basic scientific transparency. The most recent dissection of their claims, appearing in the journal Science in January 2006, found that their results depended upon "some input data that are old and unrepresentative of current [ethanol-production] processes, or so poorly documented that their quality cannot be evaluated." The complete editorial from *The New Atlantis* can be found at:

http://www.thenewatlantis.com/publications/addicted-to-bad-data

In 2011, Forrest Jehlik, Research Engineer, Argonne National Laboratory responded to what he felt are the five most prevalent myths about ethanol. He said that ethanol does not take more energy to make than it yields, "Argonne National Laboratory research has shown that corn ethanol delivers a positive energy balance of 8.8 megajoules per liter. The energy balance from second-generation biofuels using cellulosic sources is up to six times better…"

In 2012, the U.S. Department of Energy concluded a report on LIFECYCLE ENERGY BALANCE by stating "...corn-based ethanol shows a clear benefit over gasoline." You can read this report at:

https://afdc.energy.gov/vehicles/emissions_balance.html

So, Barry, say goodbye to the issue of EROEI.

6. You ask, "What makes this corn-based ethanol so special that it does not have to compete in the Anti-Free Market for customers?"

It's not ethanol that should have to compete in the free market, it's gasoline that needs to compete without all the subsidies and allowances they have continuously gotten for more than a century. Oil companies and gasoline distributors should be charged with restraint of trade and anti-trust violations. There is nothing so special about gasoline and other oil products that deserve the protection of an anti-free market.

The concern for us, at least for me, as a patriotic American, is to insure a safe and sane American free market. I'm not interested in helping foreign ethanol producers at the expense of American farmers and producers. There is no benefit in allowing someone to dump products in America. Domestic ethanol is already cheaper than gasoline and diesel, and if we want our fuels at the pump even cheaper, there should be no restrictions on how much ethanol can be sold at a fueling station.

As I wrote in my previous reply to you, the free-market restriction we should all be concerned about is that higher level ethanol-gasoline blends are not allowed to be sold at every single service station. Pumps should be available to allow the consumer to choose whatever blend level they like, from E0 to E99, with all the stops along the way. Who is stopping this from happening? The oil industry.

What's more, it should be illegal for an oil company to impose severe financial penalties on independently owned stations for adding independent fuel options. Let the power of the consumer's purse control the market; make all and any blend level available; remove the subsidies for the oil industry and let the price of gasoline and diesel reflect their real cost, which might be as high as $15-$20 per gallon.

For argument sake, if all import tariffs are removed from foreign ethanol, the biggest losers wouldn't be domestic ethanol producers, it would be the oil industry. Cheaper ethanol would make ethanol-gasoline fuels even cheaper than they are now; they would be more of a competitive threat to the oil industry. It would force the oil industry to lose money or force the government to impose greater at-the-pump taxes to quell the oil industry's complaints. The irony of your solution would be to create more taxes, not less.

The oil industry is the most heavily restricted industry in the world; there is nothing free about it. It's worse than the diamond cartel restrictions because not everyone needs diamonds on an every-day basis. We can get along without diamonds, but we can't get along without engine fuels.

There you have it, Barry, a numbered list of responses to your numbered list of questions. I welcome any information you think contradicts, challenges and/or corrects anything I've presented.

Frankly, if *Bloomberg* doesn't terminate its relationship with you immediately, they have no credibility.

(Barry Ritholtz never replied back.)

Meet the New 'Tool' In Big Oil's Efforts to Scuttle Ethanol

An open letter to George "David" Banks

Originally published March 21, 2016

Hi David -

I just finished reading your article "Renewable Fuel Standard Continues To Devastate," published in February on the *High Plains Leader & Times* website.

(https://web.archive.orgweb/201603 26065527/http://www.leaderandtimes.com/index.php?option=com_content&view=article&id=24435:renewable-fuel-standard-continues-to-devastate&catid=29:opinion&Itemid=58.)

I immediately visited the American Council for Capital Formation (ACCF) website, the organization where you are listed as Executive Vice President. I was curious to learn your background and try to determine where your information comes from. I noticed that Richard Rahn from the CATO Institute is on your Board of Directors, so I assume you received some input from him. In fact, your article looks suspiciously like an article Mr. Rahn published on *NewsMax.com* in 2012 called "Biofuels Not Worth the Trouble or Cost."

The other thing I looked at was your biography. It appears that you never actually worked in any industry in which you would have learned about business. I'm sure you took some "business" classes in school, but it doesn't look like you ever worked in a manufacturing company or sales company; and there is definitely no indication that you ever tried to start and run your own business.

In reading press releases from ACCF, I noticed you joined this group just a few weeks before your article appeared in the High Plains Leader. Were you given the assignment to write this story as part of your initiation into ACCF? Were you told that petroleum oil groups are major contributors to ACCF, and you had to show what you can do?

I'm not criticizing anyone for taking money from the oil industry; after all, people have families and need to provide for them. Certainly, the oil industry has more money than they know what to do with, so what the heck? I only wish you would choose a way to support your families that isn't so harmful to our country.

The point of my contacting you is to make sure you know, if you don't already, that you have no idea what you are writing about. Your comments about ethanol are as ignorant as those written by Richard Rahn almost four years ago. On July 24, 2012, I wrote to Mr. Rahn and explained how he was wrong. He never replied; I guess he was embarrassed at being caught so obviously wrong. However, from a Google search of Mr. Rahn's work after that, it appears that he shied away from discussing ethanol again—so maybe my note to him was effective.

I hope you won't be so rude as to not respond to me. I'd at least like to see you give challenging the truth a try.

You start with the ridiculous "ethanol production causes food price increases" argument. This argument was specious nearly 10 years ago, when the World Bank was first putting together the paper that would blame increased ethanol production for food price increases. Two years later, the World Bank retracted the claim and placed the blame where it rightfully belonged—on petroleum oil price increases and commodity speculators. Two years after that, the World Bank reiterated that petroleum oil prices and speculators were to blame, not corn ethanol production.

I guess no one shared this information with you. I guess no one told you that even if corn ethanol caused an increase in the price of corn, the increase had virtually no affect. For example, in 2010, a $4 box of corn flakes contained about 5 cents worth of corn. So, if corn prices doubled, the corn content would only be 10 cents. The reason for a box of corn flakes to cost so much is because of distribution and marketing costs related to selling breakfast cereals.

Since 2010, corn prices have generally been low, so for you to write a current story about high corn prices causing food price increases is just plain silly. This is 2016; if there have been food price increases since 2010, it wasn't because of corn prices. There was a rise in corn prices from 2011 to 2013, but this increase coincided with crude oil prices rising to near record highs (three times more than today), and three times higher than they were in 2009 (after oil prices fell from the record setting 2008 oil prices).

It was during the 2008 record-setting high oil prices when the World Bank issued the fallacious report had to later retract. David, you need to do a lot of homework on this.

You say that America no longer depends on oil from foreign producers. But in all of 2015 and since January 2016, we imported more crude per month from Canada than at any time since 2010. Canada is a foreign country.

See: "Weekly U.S. Imports from Canada of Crude Oil" at: *https://www.eia.gov/dnav/pet/hist/LeafHandler.ashx?n=PET&s=W_EPC0_IM0_NUS-NCA_MBBLD&f=W*

Crude oil imports from Saudi Arabia for 2015, and in 2016 were approximately the same as the two years prior (2013-14). We continue to import oil from Iraq, Kuwait, Nigeria, Venezuela, Mexico, Columbia, Ecuador, and Angola. When you say that America "no longer depends on oil from foreign producers," you're wrong. Yes, import levels are the lowest they've been in a long time, but your statement is grossly incorrect.

You write "There's no evidence that the use of ethanol actually lowers oil imports." This is one of those times when having some real business experience would have been helpful to you. It's this simple: Say you had a grocery store, and you normally sell 100 gallons of milk per day from distant Dairy A; but because you are also selling milk from local Dairy B, you are only selling 60 gallons of Dairy A milk per day. Although you are still selling 100 gallons of milk every day, by selling local Dairy B milk you don't have to import as much milk from the distant Dairy A.

By simple arithmetic, if you use one gallon of ethanol in place of one gallon of gasoline, it means that you need less petroleum oil. Regardless of how much oil importation has changed, ethanol has played a role in reducing the need for foreign oil, because if we had enough domestic oil, we wouldn't have to import so much or any.

Additionally, you only attribute lower oil imports to "an explosion in shale exploration," but that is not the case. The U.S. Energy Information Administration has pointed out the following:

> There is no single explanation for the decline in U.S. oil import dependence since 2005. Rather, the trend results from a variety of factors. Chief among those is a significant contraction in consumption... and reflects factors such as changes in efficiency and consumer behavior as well as patterns of economic growth...Shifts in supply patterns, including increases in domestic biofuels production, NGL output and refinery gain, also played an important role in moderating import dependence. U.S. ethanol net inputs...help[ing] to displace traditional hydrocarbon fuels and so reducing petroleum import needs.

David, you write that between 2008 and 2014, "Domestic production increased by more than 3.5 million barrels a day. Meanwhile, ethanol production increased by a negligible 328,500 barrels daily over that same period—comparatively, a drop in the bucket."

By writing this, you intimate that ethanol only increased by such a small number because of low demand and consumer interest. This is another gross misstatement. Ethanol, because it's confined by government restriction to 10% in regular gasoline options, has had a strict limitation on its growth. If E15, E20, E30 and other ethanol-gasoline blends weren't restricted from sale at filling stations, ethanol production and use would have "explosive" growth.

But what I find peculiar is that you fail to realize that if you take into account the decline in automobile fuel consumption due to driving habits as well as the increase in engine efficiency in late model vehicles, there has still been an increase in ethanol production. Consequently, this means there has been considerable growth in ethanol acceptance. And while the increase may be small in proportion to the shift in where crude comes from, it is rather enormous compared to the drop in gasoline and diesel fuel used by passenger vehicles. In neglecting to understand and take all this into account, you prove you are a "tool" used by the oil industry.

You wrote, "To meet the growing demand for ethanol, farmers have chosen to produce corn over pork, beef, poultry, and other agricultural products. That's costing families and businesses. In 2012 alone, the average family spent an extra $2,000 on groceries thanks to the RFS. And, the RFS has cost chain restaurants an additional $3.2 billion annually—about $18,000 per restaurant."

David, it sounds like you know nothing about the farming, ranching, and restaurant industries. I guess you weren't aware that prior to the new ethanol market for corn, many of these farmers struggled to survive; and that by growing corn, they now have a profitable product that has a growing demand. I guess you don't understand that if crop farmers shifted to raising beef cattle and other agricultural products, there might be a glut on the market for those products, driving down prices, causing farmers and ranchers to suffer. This would put all farm crops and livestock meats in jeopardy. Having a vigorous external market that can absorb corn crops without destroying the meat and vegetable markets is a good thing, not a bad thing. You probably took some classes in school that covered supply and demand, but no one actually taught you what it means or how it works in the real world.

And when you wrote the part about the RFS costing chain restaurants $3.2 billion annually, you didn't consider that most of the higher costs they have been subjected to relate to the cost of petroleum oil used in transportation, marketing, and packaging.

You write that ethanol "also hurts drivers...ethanol is less efficient than regular gasoline." This is untrue. Ethanol is more efficient than gasoline. Determining "efficiency" is not just a matter of how many miles a gallon of fuel will provide; you must factor in the cost per gallon in order to arrive at a cost per mile. Generally speaking, over the last decade, the lower cost per gallon of E85 made up for or equalized any reduction in MPG. If by using E85 a consumer got 8%-10% fewer miles per gallon, the 10%-20% lower cost per gallon gave the use of E85 a net gain.

Likewise, the higher cost per gallon of ethanol-free gasoline (E0) more than eats up any increase in mileage compared to using E10. You didn't know this, did you? When you received the talking points memo from the oil industry, they didn't cover any of this, did they?

Allow me to drive the final nail in the efficiency/economy coffin for you: the reason a gasoline engine will get fewer miles per gallon when it uses ethanol (or ethanol-gasoline blends) is because the engine is optimized to run on gasoline. A comparable engine optimized to run on ethanol will deliver the same or higher miles per gallon, and the ethanol fuel is less expensive.

You basically close out your article by writing, "Some argue these costs are worth it for ethanol's environmental advantages. But research shows there's hardly anything 'green' about it. ...To get a full sense of environmental effects of ethanol production, it's necessary to take into account all emissions associated with growing corn for fuel. That includes those from farm equipment used to clear land, plant seeds, irrigate, and harvest crops. Don't forget the pollution emitted hauling ethanol from factories to refineries to gas stations. ...In short, ethanol isn't nearly as clean as its proponents suggest. Indeed, the greenhouse gases produced during a fuel cycle for ethanol are roughly the same as that produced by petroleum-based fuels...These emissions add up."

This information is so ignorant and so naive that it's hard to believe you took the time to use it. To begin with, if you want to see the difference in emissions advantage of ethanol versus gasoline, just watch either or both of these short YouTube videos:

https://www.youtube.com/watch?v=H6qKa8X6o1w
https://www.youtube.com/watch?v=mwMJ1vzrH_8

David, you say that in order to get the full sense of environmental effects, you have to calculate all the ancillary aspects of ethanol production. Did you not know and understand that gasoline and diesel fuel production also have ancillary aspects that affect the environment? Do you think that gasoline just comes straight out of the ground and goes right into the fuel pumps at filling stations all around America and the world? Didn't you know that crude oil, gasoline and diesel fuel also have to be hauled from point to point to point to point?

Didn't you know America has engaged in several wars to defend the oil industry around the world, and that these wars have cost us hundreds of thousands of lives in addition to untold environmental damage? Talk about emissions that really add up!

You may feel I have been nasty in how I have communicated this information to you. However, all I did was to match your tone. In order for you to have gone out of your way to put together so many conspicuously exaggerated statements is a terribly nasty deed. All things considered, I think I've actually been very kind to you.

Please respond. Please offer additional nonsense that you know nothing about. And please do share this letter with Richard Rahn and all your contacts in the petroleum oil industry. I would challenge you and your friends to a public debate, but I know you don't have the guts to accept.

Have a great week and Easter holiday.

(David never replied.)

Are Ethanol Opponents Sniffing Glue? Something Is Making Them Dizzy!

Originally published July 30, 2016

The U.S. Environmental Protection Agency held a public hearing today in Kansas City, Missouri. The hearing was part of the EPA process for setting/recommending the proposed rule "Renewable Fuel Standard Program: Standards for 2017 and Biomass-Based Diesel Volume for 2018."

More than 130 individuals were set to give live testimony, with additional write-in testimonies.

In preparation for this hearing and EPA process, a group called American Council for Capital Formation (an oil industry lobby) commissioned a report from Dr. Daniel De La Torre Ugarte, Research Professor, Department of Agricultural and Resource Economics University of Tennessee. A press release regarding the report can be found at:

http://accf.org/2016/06/09/10-year-review-of-renewable-fuel-standard-impacts-to-the-environment-the-economy-and-advanced-biofuels-development/

The entire report can then be accessed via the press release.

Dr. Ugarte's report pretty much follows the oil industry's standard playbook on inventing or over-exaggerating information. They wave their checkbooks and college professors or think-tank stooges magically appear. Sometimes the professors have degrees in a field related to energy and fuels; sometimes they don't. Sometimes the think-tank guys or gals have some knowledge of the industry; sometimes they don't.

Usually when I engage an oil industry shill, I am amazed at the high level of credentials he or she sports. It's David versus Goliath all over again. Lucky for me, inside my ammo bag is a very smart business partner, lots of incredible fuel/energy people to consult, and a ton of honest-to-goodness hard facts.

What always winds up happening is that the oil industry's Goliath gets left lying on the ground with a great big wound in the middle of its forehead while their college degrees scatter in the wind. It's a clear case of "the bigger they are, the harder they crash and burn." The reason they crash and burn, regardless of their credentials, is that they are simply repeating lies and exaggerations created by the oil industry. Am I overstating the results? I think my rebuttals to the oil industry shills speak for themselves.

But let me not dwell on the past, because there's a new battle to engage in today, with some new meat.

As it turns out, ACCF and its new boy, George 'David' Banks, was the subject of a reply I published this past March in response to a hit piece written by 'David.' Consequently, I already had some knowledge of the American Council for Capital Formation. I was not aware of Dr. Ugarte. He's the new meat.

I have repeatedly referred to Professors David Pimentel and Tad Patzek, and their attempts to fatally kill ethanol. These attempts have been unsuccessful in the long term. I document their figurative demise in the 60+ page report I published three years ago. You don't have to read the entire report, unless you really want to. You can go straight to the relevant chapter at:

https://www.theautochannel.com/news/2013/06/11/082057-gusher-lies-book-review-and-reply-to-robert-bryce-pt.html

Just scroll down to the section on EROEI.

I read the new ACCF press release forwarded to me by Bobby Likis (a real live expert in automobile maintenance and repair, and host of a fabulous national radio program). Then I read Dr. Ugarte's complete report, and here's what the email I sent to him and ACCF says:

> Gentlemen:
>
> I finished reading your press release, as well as the 17-page report written by Dr. Daniel De La Torre Ugarte that's referred to in the press release.

Before I dissect the report, I'd like to point out that I'm still waiting for a reply to a letter I wrote to George David Banks and your organization on March 21, 2016, subsequent to Mr. Banks' editorial "Renewable Fuel Standard Continues to Devastate."

I think it's rather irresponsible of him and your organization not to have responded to me. If you misplaced what I sent to ACCF in March, you can read the entire text in my editorial, "Meet the New 'Tool' In Big Oil's Efforts to Scuttle Ethanol."

On to the present. It seems that Dr. Ugarte has picked up the baton fumbled by Professors Pimentel and Patzek in their ill-fated attempts to present overly exaggerated, convoluted "scientific" information in place of facts and common sense. Years ago, when I was young, I worked with models. I found the glue used at that time made me dizzy. I'm wondering if the same thing happened to Dr. Ugarte with his modeling. Yes, of course, I'm being sarcastic and I'm trying to add some levity to the issue, because that's the only way to take Dr Ugarte's report.

The bottom line in the newest ACCF and Dr. Ugarte's attempt to vilify ethanol is that if their conclusions were significant and relevant, the report should have called for the immediate closure of all golf courses located in the states that border the Mississippi River, or that have rivers and streams feeding into the Mississippi River. Did you know, by the way, that there are nearly 10,000 such golf courses, and they all use fertilizer? Do we really need to have people wasting their lives with a silly game when there are hundreds of millions of starving people in the world?

These golf courses are added to all the corporate and educational campus lawns located in those states, all the residential communities, parks and lawns which use fertilizer that eventually gets washed into the rivers, streams, etc., etc.

In the same vein, I'm wondering why there is no general hue and cry to stop new vineyards and hop fields from being planted. I don't know how much you know about wine and beer, but as a wine and beer aficionado, I can tell you that we don't need any new wines or beers. We have plenty, and there's never a shortage at supermarkets and liquor stores. Perhaps the American Council for Capital Formation should address the crucial over-supply of wine and beer at a later time.

Don't forget that all these golf courses, campuses, residential communities and ever-increasing number of vineyards and hop fields require care and maintenance using tractors, lawn mowers, small engine tools, other service vehicles, in addition to all the personal internal combustion engine vehicles driven by the patrons, employees, students, home owners, etc. I can't even imagine the quantity of harmful emissions spewing out of all these machines.

Do you think Dr. Ugarte has an agricultural model that can calculate this?

Does Dr. Ugarte know that all of these activities involve soil erosion? If so, how can he in good conscience worry only about soil erosion from corn farming? Maybe if everybody just stayed home and lived off the nutrients carried in the air and vitamins from natural sunlight, we'd all be better off.

How much harmful engine emissions and soil erosion goes on around the world caused by the military efforts needed to defend oil production and shipping?

There are aircraft carriers, destroyers, tanks, jet fighters, tankers, jet bombers, jeeps, Humvees, staff cars, bombs exploding, and the production of all the material. There's also the loss of American lives, but I'm sure you can't concern yourselves over what the grand dame Leona Helmsley referred to as the little people.

But what really makes me wonder if Dr. Ugarte knows anything about the overall issue of using ethanol for fuel is his (and ACCF's) overall attack on biofuels. He and you act as if fossil fuels are not poison—as if the entire fossil fuel industry has not swallowed trillions of dollars in subsidies; as if gasoline and petroleum diesel do not damage engines; as if gasoline is a superior fuel to ethanol; as if there are no environmental and health problems caused by fossil fuel production and use, and as if ethanol (or a logical, non-poisonous anti-knock ingredient) is not needed in high-compression engines. Without ethanol we have leaded gasoline, gasoline with MTBE or gasoline loaded with toluene and benzene poison—as if the entire fossil fuel industry has not all proven poisons that have hurt many Americans and even killed them.

Why are you writing reports that make you look like you're getting paid by the petroleum oil industry to invent and disseminate lies?

I think your organization is a fraud. I think Dr. Ugarte's report is specious. Prove me wrong. Debate me. Debate my team against yours.

America's future is at stake. The free world's future is at stake. If you care enough to spend time creating and disseminating lies about this issue, then step up and put the oil industry's money where your mealy mouths are; and debate the issue in a public nationally televised forum. Let's do it before November so that the politicians will have a clear understanding of the issues.

This time I expect a reply.

As usual, I cc'd this email to several of the incredible fuel/energy people I know. Here's a reply I received from Steven Vander Griend of Urban Air Initiative and ICM Inc.

Marc - Thank you for writing these folks. I find it interesting they only look at the data they want to and totally ignore reality.

I find no mention about the protein value and just wonder how they can think cellulosic is better. We can exceed 50 percent feeding value to livestock from an acre of corn and still take all that starch to ethanol. "Last week I was talking to an agronomist here in Kansas. Technology has come a long way and even with increased yields per acre, fertilizer application is down per acre so how can these guys claim all the increase?

These guys totally ignore the fuel benefits as well and would assume they are drinking the Kool-aide of API and the manipulated vehicle studies. If they only knew the truth that many of us can blend test fuels to either make ethanol look good or have ethanol look worse. It's all about who controls the fuel blending who then also controls the outcome of the study.

They should look at the octane value, but I am sure they would just assume it is too technical—that one gallon of ethanol can displace two gallons of aromatics as it relates to octane. They just think ethanol is less efficient because ethanol has less energy per gallon; but automotive studies show that based on energy, ethanol burns three to six percent more efAciently. If they only knew what makes a better fuel and that the autos want more octane such as E25/E30 to increase efficiency.

The biggest issue in my view is the dirty little secret API and even EPA is hiding. As Carnegie Mellon and some other recent studies now point out, 40 percent of particulates by number in the eastern US is from gasoline combustion. It is the highest source of ultra-fine particulates in urban areas and the leading cause of coronary disease listed by the University of Wisconsin.

You are right Marc, we should question who is actually paying for this study. Maybe they should visit this link and start looking at petroleum's problems: *https://fixourfuel.com/*

Biofuel Blunder? Renewable Fuel Mandate Is Having A Negative Effect on Home Runs

Biofuel mandate was supposed to fix everything, but no one has hit over 60 home runs in the past 15 years

Originally published July 30, 2016

I just finished reading an anti-ethanol *Bloomberg* editorial written by Jennifer A. Dlouhy. Her story, "Biofuel blunder? Greens who once championed a renewable fuel mandate say climate benefits are coming too slowly," raises some seemingly noteworthy questions. Read it here:

https://www.bloomberg.com/news/articles/2016-07-27/as-corn-devours-u-s-prairies-greens-reconsider-biofuel-mandate

The more I thought about it, the more I realized that she may be on to something.

Not since 2001 has any professional baseball player hit more than 60 home runs in one year. What's the connection, you ask? It's obvious, 2001 is the first year of increased global biofuel production!

To make matters worse, 2001 was the first year of a decline in total home runs hit by all MLB players. In 1999 and 2000, total home runs per year topped 5,500. We have never again had numbers like that, and in some years total home runs didn't even exceed 5,000!

It's all because of biofuels, ethanol, and that dreaded "renewable fuel standard." I'm shuddering in fear and anger.

Look, if the American renewable fuel mandate can't fix man-made global warming, and it impacts on home run totals, then it can't be good for our country. It's bad enough when MPG production is called into question by blending ethanol with gasoline but messing with one of the most cherished aspects of our great national pastime is just too damn much!

That's right, I used the word "damn," because no other word so succinctly describes the outrage we must all share.

I have more startling news: Some of the top deceptively named organizations have come out to challenge the value of the renewable fuel mandate. Yes, that's right, we have PECC (Persons for the Ethical Consumption of Corn) voicing their concerns. There's AHASC, raising heck over this issue (AHASC, in case you don't know, is Americans Hanging Around Street Corners. It's very big in Brooklyn where I grew up). Naturally, LAPWEC (Let's All Pretend We're Environmentalists and Conservationists) is involved. But the most vocal of all of the groups is OPUTDT (Oil Pays Us to Do This).

Reality Check Time

Let's say man-made global climate change is real and the consequences are as bad as the most extreme predictions. How would anything we do in America reverse all the environmental damage being caused around the world? How can the renewable fuel mandate be held responsible for any of this? How would a change from corn ethanol as the renewable fuel ingredient to some poison made by the oil industry solve the problem? It wouldn't.

Let's say that man-made climate change is not real, or at least not quite as consequential as predicted. Then maybe the renewable fuel mandate is not an effective solution to the problem because the problem doesn't exist. You know, the world's greatest laundry detergent will not improve the results of plain warm water on an already clean shirt.

As for the organizations with the highfalutin' names and acronyms, remember that the Democratic People's Republic of Korea is not democratic and it's not a people's republic. It's a tightly controlled dictatorship under the control of one monster. If the dictator changed the official name of North Korea to Democratic People's Democratic Republic of Democratic Korea, it wouldn't be anything other than what it is. To paraphrase an old saying, a skunk by any other name is still a skunk.

The Center for Consumer Freedom, which changed its name

to Center for Organizational Research and Education, is a front group for the tobacco industry. The Global Climate Coalition was a front group for the petroleum oil industry. The names were intended to make it seem like they had some planet-wide environmental concerns, rather concern for their own personal business environment.

So when we get to a group like the Natural Resources Defense Council, and Environmental Working Group, I don't care how many "green" words they use in their title or if they marched down 5th Avenue with environmentally-conscious flags flying out of their rear ends. If they promote unworkable schemes to divert attention from real solutions, or they spread irrelevant and contrived information against solutions that are inexpensive and available right now (like ethanol), then they are as deceptive and damaging as any Big Oil funded "think tank." Jennifer A. Dlouhy might refer to NRDC and EWG as "respected," but I don't respect them one bit.

As it happens, I had some experience with Environmental Working Group when their Research Analyst, Emily Cassidy, published an absurd "Dirty Corn Ethanol" story on the EWG website in October 2015. She took EPA information out of context and perhaps invented some new details to suit her unwarranted attack on ethanol...and I told her so.

Ms. Dlouhy's *Bloomberg* article refers to the fertilizer runoff problem that has created a "dead zone" in the Gulf of Mexico. She does what many ethanol opponents do, she implies that the problem is caused solely by corn farmers. She conveniently fails to recognize that nearly 10,000 golf courses, thousands of other farms, hundreds of thousands of residences and commercial campuses also use fertilizer that drains into the Gulf. Animal waste is a major part of the problem, as well. Therefore, unless all farming and ranching and living is discontinued in the states with rivers that drain into the Mississippi basin, the problem will continue.

Since the end of farming and elimination of all other areas that use fertilizer is unlikely, the solution would be to do what David Blume preaches in his Permaculture Workshops: Harvest the algae, use it to make ethanol, allow oxygen back into the water, use the remnants as organic fertilizer on the farms and ranches and campuses, and help to make America energy independent.

These are my possibly noteworthy comments to Ms. Dlouhy's seemingly noteworthy questions.

The Case Against Ethanol Opponents: They Are Simply Incorrect

Originally published August 28, 2016

It seems to me the American oil industry has been ramping up its attacks on ethanol and trying to gain ground by recruiting high-powered lapdogs to press their fallacious arguments. These minions come from universities and "think tanks."

My impression is based on my own current amped-up attention to the attacks. And the reason my attention may be at a peak level is because of events I've participated in since May 2016. I've just returned to the U.S. from a very exciting and informative trip to Australia. I was there to speak at the Australian National Biofuels Symposium, and I'm honored to say I was one of the keynote speakers in a program that featured many extremely knowledgeable professionals.

Before heading to Australia, I was a speaker at a Congressional Briefing in Washington, D.C., one of four panel members at a morning session and then participated in several one-on-one meetings during the day. Prior to the Washington event, I was asked by the U.S. Department of Agriculture and the Grains Council to speak at an APEC (Asian-Pacific Economic Cooperation) Biofuels conference in Taiwan.

One of the basic errors gasoline lapdogs make is that they act as if gasoline and petroleum diesel were some type of healthful elixir. They act as if ethanol fuels are highly polluted poisons being introduced by greedy, evil farmers into a market dominated by the "perfumed, life-giving essence of fossil fuels."

The evidence lies in works of deception, such as:

- articles written by John DeCicco, Research Professor at University of Michigan Energy Institute *(https://link.springer.com/article/10.1007/s10584-016-1764-4))*
- Daniel De La Torre Ugarte, Research Professor, Department of Agricultural and Resource Economics University of Tennessee

(http://accf.org/2016/06/09/10-year-review-of-renewable-fuel-standard-impacts-to-the-environment-the-economy-and-advanced-biofuels-development/);

- and Carlisle Ford Runge, Professor of Applied Economics and Law at the University of Minnesota *(https://e360.yale.edu/features/the_case_against_ethanol_bad_for_environment)*

The reality is that gasoline and petroleum diesel fuel, along with the various additives manufactured by the oil industry, are vile, deadly substances. The use of the seemingly benign word "aromatics" to refer to poisonous additives like benzene and toluene are a cruel joke, as cruel as the Nazis telling Jews they would be taking showers in the gas chambers—and that is not a stretch. Standard Oil and Shell Oil collaborated with the Nazis during World War II.

What are the differences between gasoline/aromatics and ethanol? Here are three videos that speak volumes:

https://www.youtube.com/watch?v=H6qKa8X6o1w
https://www.youtube.com/watch?v=sg6sZq8Sefk
https://www.youtube.com/watch?v=udJSDyZUolo

The three "distinguished" professors argue that ethanol production produces more harmful gases than the production of gasoline and diesel fuel. They arrive at this spurious position by calculating the harmful emissions given off by the machines used in crop growing and ethanol delivery processes. No one explained to them that storks don't find and deliver finished fossil fuel products to filling stations.

These three gasoline apologists don't mention the CO_2 and other harmful gases produced in the process of finding, extracting, transporting and defending oil around the world. They must think planes, ships, armored vehicles, and personnel carriers protecting oil company properties in foreign countries run on non-polluting fairy dust. It seems no one explained to them that the production of weapons and ammunition to protect those sites requires energy usage. Add the energy expended to make the uniforms for all the service personnel, as well as food and housing for them. They did not include pollutants given off every time a shell or bomb explodes and every time an individual weapon is discharged. They didn't calculate all emissions produced by first-responder vehicles when called to the scene of a related terrorist attack. And they didn't add the harmful emissions caused by cleaning up oil disasters, including the bodies of all the dead animals killed as a result of those disasters.

Do you think that these three apologists for the oil industry ever considered the millions of people killed in the wars waged over oil?

Soon after John DeCicco and the University of Michigan published their story, MichBio, the biosciences industry association in Michigan, issued the following statement:

> It's unfortunate when scientists conduct research predicated on assumptions that lead to biased results, rather than conducting a truly independent analysis," said Stephen Rapundalo, PhD, President and CEO, MichBio. Moreover, the Petroleum Institute's direct funding of the research calls into question any impartiality on the part of the research team and, hence, the study's conclusions.
>
> In sum, this study is an attempt by the oil industry to smear the biofuels industry and stifle competition in the fuels market. It ignores the benefits of Michigan's homegrown biofuels industry, and if its conclusions were to be valid, stands to hinder a potential economic driver for Michigan and the U.S.

The Argonne National Laboratory stated ethanol reduces greenhouse gas (GHG) emissions by an average of 34 percent compared to gasoline, even when the highly controversial and disputed theory on Indirect Land Use Change (ILUC) is factored into the modeling. Additionally, Argonne has found that without ILUC included, ethanol reduces GHG emissions by 57 percent compared to gasoline.

When Daniel De La Torre Ugarte issued his report in May 2016, The Auto Channel and others took him to task for his oil industry paid-for conclusions.

C. Ford Runge tried to include the entire litany of oil industry-created lies about ethanol, quoting Emily Cassidy from the Environmental Working Group as an informational resource. In my opinion, Ms. Cassidy and the entire EWG could not find their way out of a paper bag with openings at both ends. It is clear Mr. Runge did nothing but restate the same old tired and well-refuted oil industry propaganda to collect his bounty from them.

I'd like to know if, C. Ford Runge is called a "distinguished professor," does that mean that he's distinguished from professors who know what they're talking about?

The oil industry's tutelage of professors makes a mockery of academia. But why should the oil industry care if they ruin the honor and prestige of the professors they're buying? After all, they are perfectly willingly to sacrifice the lives of millions of soldiers, sailors and airmen to protect their filthy empires.

Two of the most interesting things about the oil industry's machinations to defame ethanol are, first, that next generation automobile engines will require fuels with greater octane levels than E10 can provide. The engines will need E20, E30, E40, and so on. They decided that the only alternative is to increase the amount of benzene/toluene, which is costly, poisonous and dangerous. In the end, the oil industry will probably start buying up ethanol distilleries so they can keep the profits. It will be interesting to see how and if they change their tune regarding ethanol.

The second thing is that many of the same cars and boats sold in America are sold in countries like Brazil and Thailand, and they have significantly higher ethanol blend mandates then the U.S. These cars and boats all work very well. This means the warnings and warranty restrictions currently given by OEMs in America are nonsense.

It's Time to Rethink the Value of PhDs

Distrusting "Other" Doctors' Advice on Engine Fuels

Originally published September 12, 2016

About a century ago, the tobacco industry responded to growing health concerns by using medical doctors in their advertising to promote smoking. Executives at tobacco companies duped the public by relying on the near universal trust in physicians to keep the world addicted to their poison. The outcome was three decades of pseudo-medical research and MD endorsements to pretend that cigarettes were not harmful.

Tobacco advertising that promoted doctor-endorsements appeared in all consumer media and leading medical publications like *The New England Journal of Medicine* and *The Journal of the American Medical Association.*

According to Robert K. Jackler, MD, Sewall Professor and Chair, otolaryngology, chief of the head and neck surgery department at Stanford University Medical Center, "Cigarette companies also wanted doctors to smoke their brands."[1] To this end, tobacco companies participated in medical conventions by sponsoring doctors' lounges and giving away

Cigarette ads courtesy of *http://tobacco.stanford.edu*

free cigarettes. At least one cigarette manufacturer created an internal Medical Relations Division to find medical personnel to substantiate their illogical advertising claims. RJ Reynolds housed their Medical Relations Division in the offices of their advertising agency.

The petroleum oil industry, another nefarious harbinger of death, took many pages from tobacco's advertising playbook, especially the use of doctors to support their invented claims and their pseudo-scientific research. The difference is Big Oil's misuse of the "other" kind of doctor to advance their misinformation. Big Oil likes those with a Ph.D. or a J.D. after their name. When they find an individual with both a Ph.D. and J.D. after his/her name it is better than striking gold.

I'm not against people taking money from various interests to further their commercial goals; I'm against the misuse of someone's academic or non-related credentials to make didactic statements about issues they know little or nothing about. The oil industry, like the tobacco industry before them, uses the public's virtual worship of the title "doctor" to silence any criticism. Unfortunately for the health and well-being of the world, these tricks are working out well for Big Oil, even though the Ph.D. and J.D. "doctors" they're using seem devoid of any practical knowledge concerning ethanol production, internal combustion engines, or day-to-day business.

It's more than an interesting coincidence that Ph.D. is the abbreviation for "doctor of philosophy." *Philosophy* is the operative word. In other words, they may understand the notion of the subject matter, but not its hands-on application. Consider the British legal system's two types of lawyers: the solicitor and the barrister. Even in the medical field there are differences between MDs who are consulting doctors and those who are practicing surgeons. Another way to look at it is the difference between those who sit in the stands to watch a baseball game, and those on the field who can hit a 90-mph fastball into the center field stands.

What I'm getting at is that the reverence for people with Ph.D. or J.D. after their names should be tempered. Maybe a new title is in order, such as Ex.D.: Doctor of Experience. We would go to a Ph.D. to hear the philosophy of how something might work, and then rely on an Ex.D. to learn how it actually works.

In the years The Auto Channel has been actively advocating alternative fuels (particularly ethanol), we've come across a number of "degreed"[2] Ph.D.s and J.D.s who oppose alternative fuels (particularly ethanol). The opposing focus on ethanol is because ethanol is the primary competition to petroleum oil fuels. Some time ago, when compressed natural gas appeared to be the serious up-and-coming alt fuel, the oil industry made CNG a bigger target with lies and exaggerations even though CNG is one of their own derivative products.

Now that no automaker in America regularly manufactures CNG-powered vehicles, and the oil companies coerced federal and state governments to practically outlaw the conversion of gasoline engines for CNG use, the oil industry feels safe in going hog-wild in finding more natural gas and creating a fracking mess while they are at it.

On the other hand, the oil industry and their bought-and-paid-for spokespeople generally don't oppose electric vehicles because right now, they are not a real threat to the oil industry. They may become a real threat in the 22nd or 23rd century, but for the coming decades, electric passenger vehicles will remain an expensive curiosity.

I've written to and about several of the oil industry's apologists, including Richard Rahn, Barry Ritholtz, George Banks, David Pimentel, and Tim Searchinger. Of these five, only Dr. Ritholtz responded. He asked me six questions, and after I sent him my 16-page reply, I never heard back from him, though he said he would provide his own answers to the same six questions to explain his anti-ethanol position.

I recently came across a new "person of interest." Bernard L. Weinstein may be the most heavily credentialed of all oil industry spokespeople I've ever dealt with. You can do your own Google search—his academic credentials are impressive—knockdown drop-dead out-of-the-universe impressive.

The problem is, I don't think he knows what he's talking about when it comes to ethanol, alt-fuels, and energy after reading an article he wrote for *Investor's Business Daily* in

August 2016. His article is "It's Time to Rethink the Ethanol Mandate." (Weinstein's story is here: (*http://www.investors.com/politics/commentary/its-time-to-rethink-the-ethanol-mandate*).

Weinstein's article is troubling. It's troubling to me as a business expert and automotive chronicler; it's troubling to our country as we try to free ourselves from foreign oil domination and right our sinking economic ship; and it should be extremely troubling to the academic institutions he works for.

Least among the troubles are those I feel. Since I know the truth about the issues he addresses, I am only troubled because of patriotism. I can always ease my suffering by writing an exposé like this and burying my sorrow in a chilled delicious glass of Yankee-rita (a margarita made with corn alcohol instead of tequila). Our country and Mr. Weinstein's academic institutions truly have something to worry about. If our country follows the "philosophies" in Weinstein's IBD story, we will be kept chained to foreign oil, grow our national debt, and continue to allow the oil industry to force poisons down our throats—proven deadly poisons.

The only people to make out well from an article like Weinstein's IBD story are those in the oil industry. They buy a man's lifework for a measly sum and turn their collective backs on him when someone takes their apologists to task. They will abandon him, as they have abandoned others they've used. Destroying the reputation of a man like Weinstein means nothing to the oil industry, since they take no responsibility for the deaths of millions of humans around the world. What's one more professor to the devils that egged us into two world wars and several major regional armed conflicts?

Here's what he said in his IBD story:

Weinstein wrote that one of the problems with the biofuel mandate is that oil refiners and importers often have to buy credits in order to comply with blending requirements set forth in the mandate, and that the price of these credits has jumped markedly in 2016, and will depress profitability of the oil industry companies struggling with a global glut of gasoline and diesel.

The credits are necessary because of the oil industry's financial inducements to our politicians to ensure that gasoline remains our primary engine fuel. There was no practical reason for gasoline to become our primary engine fuel (as I have repeatedly shown in this book and my articles), and there is no reason for it to be our primary fuel now. All modern passenger vehicles in the United States (and many other countries) can use ethanol as primary fuel. At the most, vehicles manufactured since the mid-1990s would require nothing more than a computer software update to permit an easy and safe transition to E50 or higher ethanol-gasoline blends. Such adoption would make it possible to immediately meet all emission requirements anticipated for the next few decades.

The idea of buying "credits" is as despicable as paying outrageous scalper fees for disreputable dealer car packages, entertainment tickets, etc. Having to worry that disreputable car dealers, ticket scalpers or the like will lose profits should never concern the public. Biofuel mandate credits are a shady way to let the oil industry off the hook for poisoning us for the last century. If we are going to have a biofuel mandate for health reasons, for economic reasons, environmental reasons, or all these reasons combined, mandate requirements should be MANDATE REQUIREMENTS.

Weinstein posits the banal bromide that "research has shown that (ethanol-gasoline) blends above 10% can damage car engines and fuel systems for millions of vehicles on the road today." On the contrary, research has shown the overwhelming majority of vehicles on the road today are unharmed by ethanol-gasoline blends above 10%. Global on-road use of higher ethanol-gasoline blends prove the positive research on ethanol to be correct. What's more, as for possible damage the relatively minor number of vehicles that are not ethanol compatible can be overcome with minor parts replacements during routine maintenance. And, to top it all off, no one has suggested we remove ethanol-free gasoline from the marketplace. Those who can afford to own beautiful classic/antique cars that rarely get used can afford to find and pay for fuels they are more comfortable with.

Weinstein also hides behind the newly minted falsehoods about ethanol having limited effect on greenhouse gas emissions. This claim is easily defeated by looking-up at the skies in any large American city that used to suffer from heavy smog. That blue you see on a cloudless day is the sky. If you land at LAX Airport and look north on a cloudless day you will see something that looks like mountains. They are mountains. Before the mid-1990s you would rarely be able to see those Hollywood Hills through the smog.

The oil industries' reduced benefit claims leveled against ethanol are a concoction of irrelevant rationales. An example is a report released in mid-August by another "degreed" professor, John DeCicco. DeCicco's claims were nothing new for him and he was soundly rebutted by reports from MichBio* and the University of Illinois at Chicago:

http://www.erc.uic.edu/assets/img/documents/UIC_Review_of_DeCicco_Sept_6sub.pdf

These rationales include everything involved before, during and after production of ethanol. But if you're going to add in everything and anything related to ethanol in order to undermine its effectiveness in reducing greenhouse gases, then you have to do the same to gasoline and petroleum diesel by

---

* MichBio, the biosciences industry association in Michigan, issued the following statement in response to DeCicco:

> "It's unfortunate when scientists conduct research predicated on assumptions that lead to biased results, rather than conducting a truly independent analysis," said Stephen Rapundalo, Ph.D., President and CEO, MichBio. "Moreover, the Petroleum Institute's direct funding of the research calls into question any impartiality on the part of the research team and hence the study's conclusions.
>
> "In sum, this study is an attempt by the oil industry to smear the biofuels industry and stifle competition in the fuels market. It ignores the benefits of Michigan's homegrown biofuels industry, and if its conclusions were to be valid, stands to hinder a potential economic driver for Michigan and the U.S."

adding in anything and everything related to its production. This would have to include all the wars, all the protections of shipping, all the manufacture of war materiel, all the environmental disasters, and all the acts of terrorism.

Why does a man with so many academic accomplishments not know and understand this?

Weinstein criticizes ethanol for causing water and food prices to rise. He uses a ridiculous figure he says is the measure of water required to produce ethanol. No one bothered to inform this economics professor that the World Bank long ago retracted their original statement about food price increases being due to ethanol production. (Their initial report that begat the rising food price lies). No one informed Weinstein that most of the water for ethanol production (the growing of the base crops) comes from rain. If he's going to use a government report to diminish the greenhouse gas benefits of ethanol, he should rely on another study from the same government that reports water needed for ethanol production decreased significantly in the last 20 years, and is only slightly higher than water requirements for gasoline production.

It's one thing for Robert Bryce to cling to the outdated, debunked information he included in his book, *Gusher of Lies*, but there's no excuse for Weinstein to do so in a new August 2016 report.

The water issue naturally flows into the fertilizer issue, and what I consider the flow of sewage pouring through the mouths of the oil industry's PR firms. Yes, there is a problem with fertilizer runoff flowing in and out of the Mississippi River and the Gulf of Mexico. Most of that runoff does not come from growing corn. Much runoff comes from growing other crops, raising livestock, as well as the residential, commercial and academic properties using fertilizers for their green areas. How can Weinstein ignore all these other contributors to the fertilizer problem?

Lots of fertilizer runoff comes from the thousands of golf courses located in the states that border, or have rivers, that flow into the Mississippi River. Does America really need so many golf courses?

Why is Weinstein unaware that bothersome algae growing in the Gulf of Mexico because of fertilizer runoff can be turned into ethanol—enough ethanol that might be enough to supply the entire needs of the country? And algae do not even need rainwater!

Unless Bernard Weinstein and his related institutions are receiving hundreds of millions of dollars from the oil industry for his endorsement, there is no excuse for him to write his IBD article. As I've written previously, this kind of ignorant endorsement makes a mockery of our academic institutions.

Notes

[1]*http://www.healio.com/hematology-oncology/news/print/hemonc-today/%7B241d62a7-fe6e-4c5b-9fed-a33cc6e4bd7c%7D/cigarettes-were-once-physician-tested-approved*

[2]I'm not suggesting that these Ph.D.s and J.D.s have phony academic credentials, but I question their capability to do anything other than express general opinions.

Uncovering the Gas Roots of Contemporary Ethanol Opposition

Originally published April 3, 2017

Opposition to alternative fuels almost always stems from the petroleum oil industry—the controlling caste of the global energy world. The reason is financial. No person or entity in a position of power and wealth wants to willingly surrender his/her position, except perhaps George Washington. Going back more than two hundred years to find an exception to the rule proves the rule.

Ethanol, as the leading contender to the throne—a position it holds since the mid-19th century and the invention of the internal combustion engine—naturally gets the most enmity from oil lobbyists and their public relations machines.

The history of opposition to ethanol can be broken down chronologically.

Ancient Era

The era begins when federal taxes on ethanol (alcohol) production were levied to help fund the Civil War. This led to the rise of petroleum oil kerosene as the primary liquid fuel for lighting, heating and cooking, thus creating petroleum oil wealth. With money, the oil industry predictably paid for political tutelage to expand and protect their burgeoning industry (and its nouveau riche snake-oil salesmen).

Classic Era

This era begins with the public interest in those great new inventions, the automobile and airplane. The petroleum oil industry moves from kerosene as its cash cow to gasoline (as well as petroleum diesel fuel). Kerosene was big business, but gasoline was the mother lode. Although the oil industry was delivered a significant setback when the Free Alcohol Act was passed in 1906, removing the onerous federal taxes on ethanol production and allowing ethanol to compete in price with gasoline, the oil industry's new hero rode in on a dull black horse called Prohibition. This act killed ethanol as a fuel competitor and ushered in the age of organized crime. Prohibition was aided by the poisonous no-knock additive, tetraethyl lead, which made the world's largest oil producer, chemical company, and automaker bigger and richer than ever. (Standard Oil, DuPont, and General Motors, respectively).

Medieval Era

This era begins with the repeal of Prohibition and takes us through World War II, which includes Standard Oil's collusion with the Nazis to keep German alternative fuel technology hidden from the outside world, and to secure Standard Oil's world dominance in the event of a German victory. The era ends with the fear of an oil embargo in the 1970s and the dawning of alternative fuel consciousness.

Modern Era

In addition to the fear of a 1970's oil embargo, the era begins with the world acknowledging that tobacco products and tetraethyl lead are deadly poisons. The U.S. government places restrictions on tobacco, and issues a near-ban on tetraethyl lead in gasoline. Ethanol and methanol were promoted as suitable octane replacements for tetraethyl lead, but the oil industry was permitted to use its own concoction (another poison), MTBE. Thankfully, MTBE's use was short, and ethanol raised its sweet smell once again. Ethanol's bouquet is liked by everyone except petroleum oil industry hacks who seem to prefer the nauseating stench of death.[1]

Here I want to focus on the mid-1990s, and especially the period since 2005 and the adoption of the Renewable Fuel Standard.

The New Millennium

Faced with governmental actions detrimental to gasoline/diesel dominance, instead of the usual carte blanche pro-petroleum fuel policies, the petroleum oil industry is facing serious challenges to its primacy. For the time being, methanol has receded in importance as the solution to gasoline and petroleum diesel, and ethanol has taken center stage. Ethanol's importance is being aided by the potential of ubiquitous availability and next-generation vehicle engine development, which will require increasingly higher proportions of ethanol in every ethanol-gasoline blended fuel in order to meet performance and emissions requirements.

The petroleum oil industry has responded by opening its coffers to pay for any and all attacks on ethanol and other alternative fuels. Ironically, the oil industry originally attacked its own by-products—compressed natural gas and propane gas—as targets although they have since backed off on CNG and ignore propane as an engine fuel. The only "alt fuel" spared has been electric energy, and that's because the oil industry knows that electric-powered passenger vehicles will not be a serious contender for years to come, if ever. In the meantime, Big Oil uses electric to pretend they have a conscience.

Professional and Amateur Ethanol Haters

The bounties offered by the oil industry's coffers are substantial, so substantial that they attract highfalutin' ad agencies and PR firms, many of America's leading academic institutions, and a slew of writers and spokespeople (with and without graduate degrees). This group forms the professional corps of ethanol critics. I disagree with their messages, but I understand the financial rationale that puts people in those situations.

Using pseudo-science and wishful thinking, backed by their own academic degrees or the academic degrees held by their colleagues, these professionals put forth their views.

The non-professional corps of ethanol haters consists largely of unknown and disguised guys and gals who haunt Internet message boards and phone-in call boards of radio talk shows during discussions on fuels, energy, and government subsidies. They offer up some of the most ridiculous comments and lies you'll ever hear about any subject. They use science fiction. Why someone, who hides their name and occupation, would engage in this behavior is beyond me, since it's unlikely that they can turn in an expense sheet to the American Petroleum Institute for services rendered.

Why would a neutral reader of an absurd post believe someone who won't provide some personal identification and credentials?

One such amateur, on a YouTube video, blames all problems on ethanol because it's "acidic" and causes corrosion. Ethanol is not acidic, and he didn't stop to think about water, which also causes corrosion and is not acidic. Another amateur critic with a YouTube video claims ethanol contains particles of corn (he holds up a glass jar for everyone to see), and these particles cause blockage of the fuel system. If his ethanol-gasoline fuel has solid corn particles in it, it's because the oil company that blended and distributed the fuel did something very wrong, not because of ethanol. This ridiculous explanation defies all understanding of science and the distillation process.

Not all the amateurs are anonymous or hide behind screen names. Some do use real names when posting their own outrageous lies and exaggerations about ethanol.

Case in point: I recently came across one amateur critic. His name is Jack Flobeck, and he wrote an article published in *The Gazette*, a Pulitzer Prize-winning Colorado Springs news-paper. The story, "The Water and Corn Cost for a Gallon of Ethanol" was published on September 22, 2013. The reason I only recently found this article is because it was used by a member of our audience as an authoritative expose of ethanol. Read it here:

https://gazette.com/opinion/the-water-and-corn-cost-for-a-gallon-of-ethanol/article_30d76168-f4a2-5221-bcda-f4f753e46f79.html

Mr. Flobeck identifies himself as "the founder of Aqua Prima Center, a nonprofit-think tank for water research."

In the article, Mr. Flobeck tries to make the case that too much water is needed to make ethanol, concentrating on the amount of water is to grow corn. Right at the top of the article, Mr. Flobeck writes, "I am frequently asked, 'How much water does it take to make a gallon of ethanol?'"

I read his opening paragraph once, twice, three times. I shook my head and read it again. I couldn't figure out how his response is an answer to the question he says he is frequently asked. It appeared to be double-talk.

Mr. Flobeck uses wrong information, irrelevant information, and takes information out of context. He created a con-man shell game for unknown reasons.

He wrote:

> Most of our ethanol is made in Iowa, and ... a quick and dirty answer to our question might be the way the citizens of Burley, Idaho, analyzed the question of constructing an ethanol plant in their town simply by dividing the water used at the plant by the gallons of ethanol produced. When that number came to one-eighth of the town's daily total water supply, or three gallons of water for every one gallon of ethanol; they simply squelched the plant. However, they neglected to consider the amount of water to raise the corn used by the plant. With 50 acres needed to grow corn for a gallon of ethanol, we calculate that it would take 5,480 acre feet of water to grow the corn, which doesn't appear high, unless you realize that equals 1.7 billion gallons of water, or 75 more gallons of water per gallon of ethanol.

Here's what is wrong with Mr. Flobeck's statement. Iowa does not make most of our ethanol, only about 20 percent. Twenty percent of something is not "most." Iowa also does not grow most of the corn used to make ethanol; they grow less than two percent of the total corn grown in the U.S. Whatever the Iowa numbers are, what do they have to do with Burley, Idaho? And if a town in Idaho is reluctant to use a significant portion of their water resources to make ethanol, what does that have to do with the towns in Iowa, Illinois, Nebraska and 18 other states that make more ethanol per year than Iowa?

Mr. Flobeck is playing a shell game.

In fact, 50 acres of corn does not equal one gallon of ethanol; one acre of corn is enough to make more than 400 gallons of ethanol. He's off by a factor of 2000.

To further confuse the issue and the casual reader, Mr. Flobeck wrote about the amount of water required to grow corn to cement his double-talk testimony. To use Flobeck's own words, he "neglects to consider" that most of water required to grow corn in the top corn growing states comes from rain. And he doesn't follow through on explaining that all of the water used to grow corn (and all other crops) eventually evaporates and falls again and again and again, *ad infinitum*. It is perhaps nature's most basic cyclical process. The use and reuse of water is quite unlike the issue of petroleum oil depletion—oil taken from the ground and used in gasoline and diesel fuel will never be available for reuse.

If you look for information about Jack Flobeck and his nonprofit think tank, Aqua Prima Center, you'll find no information. The phone number provided for the organization is also used by the Woman's Club of Colorado Springs, and it seems Mr. Flobeck is in the business of selling residential water purification systems. How and why he would engage in disseminating incorrect information about ethanol is beyond reasonable conjecture.

The Professional Critics of Ethanol

On the professional ethanol opposition circuit are college professors, magazine writers, talk show hosts, auto mechanics, and energy consultants. Their names include David Pimentel, Tadeusz Patzek, Robert Bryce, Jerry Taylor, Richard Rahn, David Shepardson, Tim Searchinger, Barry Ritholtz, Jillian Kay Melchor, Jay Leno, Dennis Prager, Lauren Fix, John Stossel, Michael Economedes, Ed Wallace, Kenneth Green, and George

David Banks. In the interest of brevity, I left some names off the list. The collection of academic degrees of all of them aren't worth a nickel when it comes to knowing anything about ethanol, in my opinion.

Most of their bad scientific information comes from Pimentel and Patzek, the fount of contemporary disinformation.

In the middle of the aughts (2001-2009), Professors Pimentel and Patzek were paid to present a study claiming ethanol was energy-negative (that it takes more energy to make ethanol than it puts out). David Pimentel was at Cornell University at the time, and Tad Patzek was at University of California at Berkeley. They are the grandfathers of the energy-negative argument against ethanol (EROEI). Virtually every ethanol opponent has used information from this study at one time or another, even if they don't know where the information came from. All the opponents believe that the Pimentel-Patzek information was never disputed or rebutted, although it definitely was. If they know, they act like it didn't happen. None of these writers conducted any of their own research, they relied on Pimentel and Patzek.

The Pimentel-Patzek information was refuted and countered almost as soon as it was released by people with as many, if not more, academic credentials. For example, in 2005, Bruce Dale, Professor of Chemical Engineering at Michigan State University, found that the Pimentel-Patzek methodology is flawed. Dale stated that the measurements of BTU, used by Pimentel and Patzek, are irrelevant and that the net energy of ethanol is higher than gasoline (in other words, EROEI for ethanol is positive, while the EROEI for gasoline is more negative).

In 2005, Professor Dale, along with John Sheehan, Senior Engineer at the National Renewable Energy Laboratory, took on Pimentel and Patzek in a nationally televised debate. After watching the video of the debate, it's hard to believe anyone ever took Pimentel-Patzek seriously. The entire video can be watched at:

https://www.c-span.org/video/?188549-1/ethanol-energy-policy

In 2006, professors and students at the Resources Group at UC Berkeley (Patzek's home) published a study that contradicted the Pimentel-Patzek study. Another study, conducted by the United States Department of Agriculture, was presented in 2007 at UC Berkeley by Roger Conway, Office of Energy Policy and New Uses at the United States Department of Agriculture (USDA). The report showed huge discrepancies in the figures Pimentel and Patzek used to arrive at their conclusions versus the figures used by USDA's efforts to conduct their own studies on ethanol vs. gasoline EROEI.

The USDA studies were significantly more favorable towards ethanol production.

In 2008, Tad Patzek left UC Berkeley to join the University of Texas at Austin. Just a coincidence? Texas is the home of the petroleum oil industry in America.

Additional rebuttal studies against Pimentel and Patzek continued for years right up to the time of this writing. I document all this on *TheAutoChannel.com* at Truth About Ethanol, part 5.

https://www.theautochannel.com/news/2013/06/11/082057-gusher-lies-book-review-and-reply-to-robert-bryce-pt.html

Pimentel and Patzek's chief disciple is Robert Bryce. Nearly 10 years after he used the Pimentel-Patzek information in his book, *Gusher Of Lies*, Mr. Bryce continues to spew the same bad information without regard for any of the rebuttals, findings, and more accurate calculations related to water use, crop yields, and the like.

I had the opportunity to contact Kenneth P. Green; holder of Ph.D., B.S. and M.S. degrees. He's the author of three books on the environment and Senior Director of the Center for Natural Resource Studies at the Fraser Institute. A visitor to The Auto Channel website referred Mr. Green to me as the source for information opposing ethanol. I read the paper Mr. Green wrote in 2008 for American Enterprise Institute titled "Ethanol and the Environment" *(https://web.archive.org/web/20150313214854/https://www.aei.org/publication/ethanol-and-the-environment/)* and re-watched a video that Mr. Green did after the publication of his paper.

https://www.youtube.com/watch?v=oBv2WpYoRbQ

I wrote to Mr. Green asking if he had done any further work on this subject after 2008, and if he knew that everything he stated against ethanol was either out of date, irrelevant, or never true to begin with. He wrote back and replied "I'm quite sure that, by now, many of those citations are out of date; however, at the time, they were the most recent and best data available. You're also welcome to point out the 'never true to begin with' elements of the paper." He suggested that I go back and review the reference list located at the end of his 2008 article.

I did go back and review the reference list and found a highlighted note of appreciation to Robert Bryce. It reads, "Mr. Green would like to thank Robert Bryce, whose excellent book, *Gusher of Lies*, was an immense source of insight and who pointed the author to articles and reports cited here...." The reference list then also cited multiple specific attributions to Robert Bryce's book.

I responded to Mr. Green and congratulated him on recognizing that information is subject to correction and re-evaluation over time, and I directed him to the lengthy rebuttal I had written to Gusher of Lies in 2013, detailing many of the incorrect things Mr. Bryce wrote about ethanol. My rebuttal to *Gusher of Lies* demonstrates that a fair portion of the information Bryce relied upon at the time of its writing was already known to be wrong, out-of-date, and/or irrelevant.

The obvious intention in my response to Mr. Green is that footnoting a paper with irrelevant attributions and wrong or unfounded information does not make the paper true and authoritative. He did nothing but assemble some bad information to fit the requirements handed down to him from the oil industry and AEI.

Kenneth Green is not the first person to rely on Robert Bryce and assume he vetted the information he included in *Gusher of Lies*. To make matters worse, none of them conducted any of their own research. I doubt they ever tried running a vehicle on any ethanol-gasoline blend higher than E10.

All this reliance and reconsideration of Robert Bryce made me go back and take another look at Mr. Bryce's background. He is said to be an energy expert. In reading his articles and book, and watching videos he appeared in six, eight, ten years ago, he was introduced as an energy expert. But was he an energy expert?

Robert Bryce's bio on Wikipedia states he's written about energy issues for more than two decades. That sounds pretty good, but it turns out he was a general news reporter for *The Austin Chronicle* from about 1994 to 2006. He did some energy-related stories, and other stories, like those about Madalyn Murray O'Hair. I don't think Mr. Bryce considers himself an expert on atheism, cults, or kidnapping.

In 2006 he went to work as managing editor of *Energy Tribune* (now defunct). He doesn't seem to have been the energy expert, or one of the energy experts there; he simply managed the workflow of those who might have laid legitimate claim to being energy experts.

After about one year or so at *Energy Tribune* he wrote *Gusher of Lies*. From reading the book, one does not get the sense he conducted any of his own scientific or hands-on research. So how was he an energy expert at the time of writing *Gusher of Lies*? Mr. Bryce did spend about five months working for Robert Bradley at the Institute for Energy Research during some of the time he was working on the book, but how much could he have learned in five months? In my own personal correspondence with Robert Bradley, Mr. Bradley confessed that there are significant details related to ethanol he never knew and never included in his own published works.

Looking up an inverted funnel at how contemporary ethanol opposition is presented, we see the petroleum oil industry paying for a bogus study by Professors Pimentel and Patzek, with their information being used blindly by Robert Bryce; and then, his information is accepted by dozens or hundreds of others, as if Bryce knew what he was talking about.

It's 2017, a decade after the publication of *Gusher of Lies*. Is it possible Robert Bryce is now an energy expert? Yes, you can learn a lot in 10 years. But if he's learned a lot about the issue over the last 10 years, why is he still peddling the same old rubbish?

Notes

In writing about the history of ethanol and the oil industry, I was aided tremendously by the fantastic work done by William Kovarik (and his associates), David Blume (and his associates), Edwin Black (and his associates), Ted Chipner, Ron Lamberty, Tom Quinn, Bobby Likis, Adam Khan and Klassy Evans, Anne Korin, Josh and Rebecca Tickell, Jamie Kitman, Robert Falco, Michael Carolan, and many others.

Unmasking the Gas Roots of Contemporary Ethanol Opposition - Round 1

Breaking the Ring of Big Oil-Sponsored Deception

Originally published April 17, 2017

Two weeks before the above date, I wrote and published a story on *TheAutoChannel.com* called "Uncovering the Gas Roots of Contemporary Ethanol Opposition." It dealt with how the petroleum oil industry works to distort the truth about ethanol. This story unmasks a specific individual with a history of working to hide the truth from the public about significant health and safety issues, while using his professional and academic credentials as a front for his hostile analysis of ethanol.

I had just finished reading a new anti-ethanol commentary "How the Ethanol Mandate is Killing the American Prairie," *(https://www.investors.com/politics/commentary/how-the-ethanol-mandate-is-killing-the-american-prairie/)* written by William F. Shughart II, a professor at Utah State University and a Senior Fellow at a think tank called Independent Institute. Professor Shughart is a former economist at the Federal Trade Commission and has taught at George Mason University, Clemson University, University of Mississippi, and the University of Arizona.

Mark Perry, an economist and professor of economics and finance in the School of Management at University of Michigan-Flint, brought the article to my attention. Mr. Perry is a visiting scholar at the American Enterprise Institute and previously taught at George Mason University. Two years ago, Mr. Perry wrote and published an anti-ethanol editorial. I challenged his editorial as containing an incorrect analysis of the results of producing and using ethanol. Although he never replied to my rebuttal, he sent me an email with a link to Bill Shughart's editorial. I don't know why Mr. Perry brought this article to my attention—either he believed it would show me up, or he was hoping I would expose Mr. Shughart.

I responded directly to Mr. Shughart by email as if the claims made in his April 13, 2017 editorial were in earnest. I did this because I believe there is always great value in presenting correct information about the subject. I ended my letter with information that I believe should change everything about how America views the academics and spokespeople who come out in opposition to ethanol.

I did receive a reply from Mr. Shughart a few hours after sending the email I have included his reply and my response to that reply below.

An Open Letter to William F. Shughart II on How the Ethanol Mandate Is Killing the American Prairie April 17, 2017

Dear Bill:

I'm writing to you in regard to your commentary "How the Ethanol Mandate Is Killing the American Prairie" published in Investor's Business Daily on April 13, 2017.

Since I wasn't familiar with your name (and I'm sure you don't know me from Adam), the first thing I did after reading your commentary was do a search and see what else you've written about ethanol. The only other item I found was a story from March 2013, which was very similar to your new piece—almost identical in its wording and exaggerated concern over America's grasslands. In the older story you use the term "destruction of nation's prairies," in the new story you use "killing the American prairie."

Concern for the environment is an admirable mission. Concern for the animals who depend upon the environment is highly commendable. Frankly, I feel exactly the same about the environment and animals. This is why I can't understand how someone (you) can be concerned about "disastrous effects" related to alternative fuels, but not concerned when even worse damage is caused by the oil industry in its efforts to literally shove more poisonous gasoline down our throats. I'm sure that you must have seen photographs and videos of all the dead mammals, birds, and fish that have been killed by oil spills and explosions.

I'm also sure, Bill, that you must be aware of all the deaths caused by the oil industry to that other abundant species of mammals known as humans. Oil industry destruction of human beings occurs in several ways, most notably via respiratory diseases, cancers, and war.

Every few months I update a story I originally published in September 2014. The upshot is that no American servicemen have been killed defending the ethanol industry. By my count, it is now 15,763 consecutive days since the October 1973 oil crisis in which no American servicemen have been killed because of ethanol. Do you know how many days have gone by since an American military man has been killed defending oil? How about, oh, two days?!

If you're such a strong conservationist and humanitarian, why haven't you written any editorials about the death and destruction caused by our national dependence on petroleum oil?

You cite a report by the Organic Consumers Association as your source for the amount of grassland at risk by corn crops. Have you researched who OCA is? You are a "research fellow," aren't you? Do you know that one of their biggest claims to fame was their protest against Starbucks Coffee? Don't get me wrong, if you've ever tasted Starbuck's regular coffee, then you know there's lots to complain about, but their protests concerning the use of cow's milk in Starbuck's coffee is a bit excessive, don't you think? I understand their protest was against the use of hormones to make the cows produce more milk, but how do they square the potential health risk to humans associated with hormones against the proven health risks to humans that are associated with the use of gasoline and diesel fuel? I find their concern over America's grasslands and wildlife to be disingenuous.

Let's move past Starbuck's and their lousy regular coffee. Your editorial bemoans the loss of so much of the beautiful natural American grassland. You wrote: "When prairie is plowed under to grow corn, it becomes a barren landscape."

Actually, the bleak word picture you've created is quite wrong. When prairie is plowed under to grow corn this is the true picture:

Now, let me juxtapose the picture of what happens when prairie is plowed under to grow corn with this picture of what happens when America's natural landscape is used to lift petroleum oil out of the ground:

Without doubt, corn fields are not barren landscapes, but oil fields most definitely are. How much wildlife do you think can be supported in these oil-killed fields, Bill?

Bill, I'd also like to point out to you and the Organic Consumers Association that there are a wide number of reasons why prairie is being lost. In Indiana, they lose about 200,000 acres of farmland per year to urban growth. If you're not a fan of shopping centers and new housing communities, I guess you could call it a "barren landscape." The only way you could blame that on corn is to complain about the popcorn they sell at the new movie theaters. However, for you, or anyone serious about this subject I recommend reading "Vanishing Open Spaces" by Leon Kolankiewicz, Roy Beck and Anne Manetas.

(https://www.numbersusa.com/sites/default/files/public/assets/resources/files/vanishing-open-spaces-study.pdf)

Their study is an excellent breakdown of the loss of open space and the problem of urban sprawl and population growth. They don't seem to have an oil industry-sponsored agenda to blame it on corn. The entire study, which covers about 100 pages of text, charts, and tables, only mentions "corn" once.

The title of your article refers to an "ethanol mandate," and you imply that there a mandate to make the ethanol from corn. But there is no ethanol mandate; there is a renewable fuel mandate. It happens ethanol is the best and most affordable renewable fuel to use and, therefore, ethanol is used. There is no mandate for corn be used to make ethanol. It happens that in America the best crop to use to make ethanol is currently corn.

For many decades, the oil industry argued that there isn't enough land in the United States to grow a sufficient number of crops needed to make ethanol the primary engine fuel or for it to become the primary anti-knock gasoline additive for America. William Kovarik documents this extensively in his 1993 dissertation, *The Ethyl Controversy* which details the times and events surrounding the invention of leaded gasoline (gasoline with tetraethyl lead added to boost the octane and eliminate engine knock), and the oil industry conspiracy to keep ethanol out of the picture.

See:*https://web.archive.org/web/20170822211200/http://environmentalhistory.org/billkovarik/wp-content/uploads/2016/12/Ethyl3.pdf*

This land underestimate argument continued to be used against ethanol right up to, and through, the Pimentel-Patzek 2005 study and Robert Bryce's 2008 book, *Gusher of Lies*. The land underestimate established a land-wall position similar to the blend-wall position that has been evoked to try to keep ethanol production down. The presumption behind the blend-wall is this: Since internal combustion vehicles can only safely use E10 (10% ethanol/90% gasoline), and that national fuel usage has dropped to such low levels because of modern engine MPG efficiency, there's no reason for ethanol production levels to be increased. This spurious proposition resulted in investments in new ethanol plants being withdrawn, along with the premature demise of several operating ethanol plants.

The blend-wall was a fiction. Major studies showed the majority of passenger vehicles on the road in America could safely use E15 and E20. This effectively increased the opportunity for 50% to 100% more ethanol to be produced. Several independent experiments—including personal, long-term experiments I conducted with high-level ethanol-gasoline blends in a variety of non-flex fuel vehicles—proved that ethanol-gasoline blends ranging from E25 to E50 and higher could safely and routinely be used by the same majority of passenger vehicles on the road. These experiments effectively increased the opportunity for 200%, 300%, 400% and more ethanol to be produced.

David Blume (author of *Alcohol Can Be a Gas—http://www.permaculture.com/node/490*) and others correctly identified the true amount of American land that could be used as about 200% more than the acreage previously thought capable of growing energy crops. As David Blume and others rightfully point out, there are a number of other crops that can be easily and efficiently grown in the U.S. that permit ethanol yields that are 2X, 4X, 6X, 8X greater than the corn per acre yield. In fact, with kelp it's possible to attain up to 100 times per acre more ethanol than corn, with no irrigation or fertilizer needed.

(http://www.biomassmagazine.com/articles/2166/british-report-use-kelp-to-produce-energy)

For this to come to pass, we must drop anti-ethanol pretenses like yours, Bill, and encourage significant financial investment in next generation ethanol production.

Bill, you also wrote: "Ethanol per se is not the problem; however, Washington's lack of common sense is the problem."

I'd say that if there is a "lack-of-common-sense problem,"

there's also a "lack of truth problem," and they exist within the oil industry and those who are bought to write editorials on their behalf.

To top it all off, you and your oil-sponsored apologist colleagues obviously lack historical perspective on the issue. You exhibit no knowledge of how and why gasoline and diesel fuel became our primary engine fuels. Do you let your students get away without having a foundation when they express their opinions in written papers or oral exams? You write about the possibility of extracting oil from tar sands, but photographs show how high the environmental cost would be. When you raise the specter of environmental disasters, you automatically invoke the petroleum oil industry as a primary cause. And in case you did not know this, the actual cost at the pump for fuels made from tar sand oil would be far higher than ethanol.

You write about subsidies given to the ethanol industry, but you fail to mention subsidies have been given to the oil industry for more than 100 years and are many times greater per year than any ethanol subsidies.

Your bio identifies you as the Research Director of an Oakland, California-based think tank, the Independent Institute. How about doing some real research into the issue? How about actually doing some independent thinking before regurgitating lies told by the oil industry?

This brings me, Bill, to your history of working as a paid advocate for the Tobacco industry, not unlike what you are doing for the oil industry. You have been described as a "cash-for-comments-academic":

- on the *ScienceCorruption.com* website *(https://sciencecorruption.com/ATN183/00198.html)*
- the *SourceWatch.org* website: *(https://www.sourcewatch.org/index.php/William_F_Shughart)*
- the SourceWatch page describing the Cash-For-Comments-Economists-Network *(https://www.sourcewatch.org/index.php/Cash_for_Comments_Economists_Network)*

How little respect do you have for the academic institutions you've served and the degrees you hold that you would sell yourself to the tobacco and petroleum oil industries for a few measly dollars?

Your participation in ethanol bashing, along with the awareness of your participation in the tobacco-promotion racket, verifies what we've long believed about such academics and media spokespeople who support the oil industry: You are clueless about the subject matter and you are selling access to your professional and academic credentials without concern for the truth. You should be removed from your post at Utah State University and barred from ever teaching at another school.

REPLY #1 FROM WILLIAM SHUGHART:

Dear Mr. Rauch,

I have only two replies to your column:

First, most of the information in my recent column in *Investor's Business Daily* was taken from the peer-reviewed economics literature as well as U.S. government data sources. As far as I can tell, yours is not. Second, I reject categorically your *ad hominem* attack on my credibility and your implication that my analyses of the ethanol mandate are influenced by monies paid to me by special-interest groups. I have been a professional economist for nearly 40 years; my opinions never have been "bought!".

With best wishes,
William F. Shughart II

MY RESPONSE TO REPLY #1

Hi Bill -

I greatly appreciate your reply. I'll be adding it, my response, and any additional comments you'll make to the story.

All my writings regarding ethanol and the automobile industry are peer-reviewed. I'd be happy to provide a list of those peers who regularly review my work. Like you, I also use U.S. government sources for data, whenever it's relevant and available. If you have a particular question as to where a specific piece of information I used originates, please ask. In the instance of my referring to subsidies given to the oil industry, I am providing an abbreviated list below.

Regarding your use of peer-reviewed literature, I'm wondering if you can help identify which literature you are referring to that helped you to form your opinions, and who the peers were who reviewed it. I saw no list of any peer-reviewed literature attached to your *Investor's Business Daily* editorial. The only thing I found by doing an Internet search of your name was the anti-ethanol editorial you wrote in March 2013 for McClatchy Newspapers, which was virtually identical to the new IBD editorial (and it also lacked any reference list). I particularly ask about the peer-reviewed literature in light of the findings published on the *SourceWatch.org* website that identify you as a "major figure in the Cash For Comments Economist Network..." and that you "...played a you-scratch-my-back/peer-review game where each would generate favorable 'peer reviews' of each other's work." I'm sure you'll agree that this information is in sharp contradiction to your comment that your "opinions never have been bought."

Regards.

Notes

Abbreviated list of links related to oil/gasoline subsidies:

- *http://www.nytimes.com/2010/07/04/business/04bptax.html*
- *http://www.cato.org/pubs/pas/pa390.pdf*
- *https://www.eli.org/research-report/estimating-us-government-subsidies-energy-sources-2002-2008*
- *https://www.imf.org/external/pubs/ft/spn/2010/spn1005.pdf*
- *https://blumenauer.house.gov/energy-issues#fossil*

- *http://merkley.senate.gov/newsroom/pressrelease/?id=F68FED0D-08C0-4DC4-9035-211E28DD9B7E*
- *http://blog.ucsusa.org/jim-kliesch/oil-industry-subsidies-are-anything-but-rosy?_ga=1.161769531.686943746.1492723673*
- *https://web.archive.org/web/20110503064227/http://investmentwatchblog.com/ending-oil-subsidies/*
- *https://web.archive.org/web/20110305034655/https://www.investmentwatchblog.com/house-gop-votes-unanimously-to-protect-big-oil-subsidies/*
- *https://www.forbes.com/sites/energysource/2012/04/25/the-surprising-reason-that-oil-subsidies-persist-even-liberals-love-them/#102e12632792*
- *https://thinkprogress.org/big-oils-banner-year-higher-prices-record-profits-less-oil-d337b2dcf58c*
- *http://priceofoil.org/2012/02/07/the-big-the-bad-and-the-subsidized/*
- *http://www.foreffectivegov.org/sites/default/files/budget/Revenue/OilandGas.pdf*
- *http://www.huffingtonpost.com/2011/02/01/obama-moves-to-cut-oil-subsidies_n_816787.html)*
- *http://themoderatevoice.com/78678/unemployment-benefits-and-oil-subsidies/*
- *http://www.americanprogress.org/issues/2010/05/oil_company_subsidies.html*
- *http://energydeals.wordpress.com/2011/02/02/obama%E2%80%99s-bid-to-end-oil-subsidies-revives-debate/*
- *http://ilsr.org/wp-content/uploads/2012/05/Oil-Slickers.pdf*
- *https://www.yahoo.com/news/decades-federal-dollars-helped-fuel-141648115.html*
- *http://www.sourcewatch.org/index.php/Estimating_U.S._Government_Subsidies_to_Energy_Sources_2002-2008*

Unmasking the Gas Roots of Contemporary Ethanol Opposition - Round 2

Breaking the Ring of Big Oil Sponsored Deception, Link by Link

Originally published April 21, 2017

Hot on the heels of Economist-Professor Bill Shughart's mendacious editorial ("How The Ethanol Mandate Is Killing The American Prairie"), comes another dishonest anti-ethanol commentary by yet another "economist professor" of "destinktion" armed with his own version of subjective impressions, out-of-context references, and incorrect analysis of ethanol and the Renewable Fuels Standard.

My rebuttal to Bill Shughart corrected his statements and unmasked him for his participation in the decades-long effort by the Tobacco Industry to keep poisoning the public.

This new anti-ethanol editorial comes from Gary Wolfram, Professor in Economics and Public Policy at Hillsdale College. I emailed Mr. Wolfram to ask if he was a member of the Tobacco Cash-For-Comments Network or any similar oil industry-sponsored group of academics. Although he did respond to other points made in my email to him, he did not address this specific issue. However, regardless of whether Mr. Wolfram is a card-carrying member of a Cash-For-Comments-Oil Industry-Network or not, he performs as if he had a leading role. The coincidental timing of his editorial with Bill Shughart's editorial may mean that if they're not connected electronically, they must be using smoke signals.

Mr. Wolfram's editorial is titled "Renewable Fuel Standard Has Failed" It was published in the *Detroit Free Press* on April 15, 2017: *https://www.detroitnews.com/story/opinion/2017/04/15/wolfram-renewable-fuels/100516304/.*

As the headline advertises, Mr. Wolfram attempts to lay out the case that the RFS program has been a failure. He could have called the editorial "Traffic Stop Light Program A Dismal Failure," and filled the article with details about how many people have been killed or injured in traffic light accidents over the past century—and then punctuate the text with gory photos and videos of dead and mutilated bodies to prove the program's failure. We all know that regardless of the number of traffic light accidents, that number is infinitesimally tiny compared to the number of people who safely navigate our public thoroughfares every day, year after year. You get the idea. His story about the RFS is a renewable energy resource produced by cattle.

I'm not at all suggesting that people get killed or injured by using ethanol fuels, not even a teeny-tiny number, because they don't. But we know that in the last 150 years, hundreds of thousands—millions—of people have been injured and killed because of gasoline.

In every way imaginable the Renewable Fuel Standard has been a success. It has:

1. Given us a non-poisonous alternative to tetraethyl lead and MTBE.
2. Given us a tool to begin to wean our country away from dependence on oil (foreign and domestic).
3. Been a significant factor in cleaning the air and water.
4. Given us a fuel that cannot create the kind of environmental and animal disasters that we have seen from petroleum oil and its derivative fuels.
5. Helped stabilize the farming industry, perhaps our most important domestic industry.
6. Saved consumers on the price per gallon of fuel by helping to force oil prices down and because the ethanol component in the ethanol-gasoline blends is less expensive than the fuel or oil-industry poisonous aromatics that would be used if ethanol was not available.

Is Ethanol Dirtier Than Gasoline?

Mr. Wolfram and his anti-ethanol cronies say that ethanol is not cleaner burning than gasoline and that ethanol causes more greenhouse gases than gasoline. They say this again and again and again. They cite one oil industry sponsored study after another as proof. Well, there's one sure way to prove how much cleaner ethanol is than gasoline, watch these two videos:

https://www.youtube.com/watch?v=-o5D-nic9WM
https://www.youtube.com/watch?v=WOAyoCo3xXA

If you don't believe what is in these videos, and you think the results were accomplished by a Las Vegas magician, try the experiment yourself in your kitchen, basement or backyard—or better yet, in your neighbor's kitchen, basement or backyard.

For fun, you can read this recent study from the U.S. Department of Agriculture that says ethanol produces 43 percent less greenhouse gas emissions than gasoline:

http://blogs.usda.gov/2017/01/12/innovation-is-driving-down-greenhouse-gas-emissions-from-corn-based-ethanol

Or, you can read the report from the U.S. Energy Department ("Biofuels & Greenhouse Gas Emissions: Myths versus Facts") that unequivocally states ethanol produces far less GHG than gasoline.

https://www.energy.gov/sites/prod/files/edg/media/BiofuelsMythVFact.pdf

The U.S. Energy Dept. report is from 2007. The significance of this is that ethanol's GHG benefits have been known for years, and the newer study concludes the results have gotten even better. So, when charlatans like Messrs. Wolfram and Shughart try to pass off information that says something else, they should be rebuked in the harshest terms.

A tactic the oil industry uses to put the "ethanol-is-dirtier-than-gasoline" myth over on the public is a trick like one from a card sharp on 42nd Street playing three-card monte. Here's what they say: "The tractors used on the farms have to use diesel fuel; and the trucks to ship the ethanol have to use diesel fuel; and the blah, blah, blah to blah, blah, blah the ethanol has to use gasoline. So, all this use of diesel and gasoline creates additional harmful emissions; which, when you add it to everything else, tilts the scale in favor of fossil fuels (abiotic fuels). It seems not to occur to them that ethanol producers would fuel their vehicles with as much ethanol as they could.

What about all the petroleum fuel expended to ship oil, gasoline, and diesel? What about all the petroleum fuel expended by our military forces to send ships, planes, missiles, and servicemen and women around the world to defend the oil industry? What about all the fuel expended to manufacture all the military materiel, uniforms to clothe our military personnel, food to feed them, and medicine to treat their wounds? Have you factored in all this to your con-job rhetoric?

Tar Sands and Fracking

Wolfram and Shughart write about the potential of oil from tar sands and fracking. What a wonderful world it will be. Not. Gasoline and diesel fuel from tar sands oil and fracking is just as poisonous as gasoline produced from a Saudi oil well. If you're interested in protecting beautiful landscapes, forget tar sands mining and those fracking horizontal wells.

Regarding the environment, Gary Wolfram does what Bill Shughart tried to do, frighten the public with alarming news from a questionable environmentalist group by telling them that the American prairie is rapidly shrinking. And Gary tries to do what Bill did—blame it on ethanol—blame it ALL on ethanol.

I write to Gary Wolfram what I wrote to Bill Shughart:

> There are a wide number of reasons why prairie is being lost. In Indiana, for example, they lose about 200,000 acres of farmland per year to urban growth. If you're not a fan of shopping centers and new housing communities, I guess you could call it a 'barren landscape,' but the only way you could blame that on corn is to complain about the popcorn they sell at the new movie theaters. However, to anyone serious about this subject, I would recommend you read *Vanishing Open Spaces* by Leon Kolankiewicz, Roy Beck and

Anne Manetas. Their study provides a rather excellent break down of the loss of open space and the problem of urban sprawl and population growth—and they don't seem to have an oil industry-sponsored agenda to blame it on corn. In fact, in the entire study, which covers about 100 pages of text, charts, and tables, they only mention corn once.

Water Quality And Availability

Regarding water quality and its availability in the corn crop states, if these are really serious issues, then do something about the oily run off from parking lots and paved streets after a rainstorm. Ask the almost 10,000 golf courses in the states that feed into the Mississippi River to stop using water and fertilizers. If quality and availability of water is so important, then why waste it on golf courses?

Here's another thing, why are people in the Midwest trying to grow wine grapes and make wine? Don't they know there's a water shortage? There's no shortage of great wine from California or Australia or Brazil or Spain or Italy or France. (Well, yes, there is a shortage of good French wine. It almost doesn't exist, except in some people's imagination.) But we certainly don't need to be wasting valuable water on mediocre Midwest wine (sorry, Midwest, but I am a California wine snob).

Of course, the most ironic part of the whole 'water' issue is that it's brought up by the oil industry and its factions. How many more water disasters do we have to experience before it's completely clear to everyone that the oil industry couldn't give a rat's brass ring for what happens to our water and the animals that depend on it?

If growing corn is so harmful, why are we still selling popcorn at movie theaters and carnivals? Don't those people have chocolate bars they can snack on?

Mr. Wolfram says that 40 percent of the corn crop is being used for ethanol, thereby robbing starving people of food. This is an ignorant, out-of-context, and very old statistic. The percentage of corn being used for ethanol is about the same as it's been for decades. Yes, on a tonnage basis, more is being used for ethanol. That's because farming techniques have improved; and while more land is used, the percentage has stayed relatively static—so there's more for everybody. The more corn being used for ethanol, the more distiller's grains are available to feed livestock. This means more beef and other meats for humans to consume. In case you haven't noticed, Gary Wolfram, meat is food, too!

Then, to top it all off, if the corn industry went away tomorrow, the farmers wouldn't grow so much, so it's not like there would be an even greater surplus to be fed to people in other countries who can't pay for it. You're supposed to be a real, live professional economist; why do you need me to tell you this?

The laughable thing about Gary Wolfram and his ilk is that I get the sense they know nothing about engine fuels. I can't even find any evidence that these guys own cars or that they know how to drive. And I have never, ever come up against one of these oil industry-paid professors able to cite an actual bad experience that they had using an ethanol-gasoline blend. I get anonymous cranks who claim to be engine mechanics, but the only thing they know about automotive engines is how change oil and spark plugs.

Bottom Line: Is the RFS A Success or Failure?

I have explained why I believe Gary Wolfram is nothing more than an oil industry lackey. But what do other professorial academics make of the RFS? Here's a very new paper published by the *Utah Law Review*, written by Jay P. Kesan, University of Illinois College of Law and the Energy Biosciences Institute (EBI), University of Illinois at Urbana-Champaign; Hsiao-Shan Yang, Energy Biosciences Institute (EBI), University of Illinois at Urbana-Champaign, and Isabel F. Peres, Energy Biosciences Institute (EBI), University of Illinois at Urbana-Champaign (*http://dc.law.utah.edu/cgi/viewcontent.cgi?article=1034&context=ulr*).

This paper concludes that the RFS has been a success, but the degree of success has diminished over time because of the so-called "Blend-Wall." As I have written previously, the "Blend-Wall" is a myth. Although it's a myth, many people, especially the politicians responsible for raising or lowering the amount of ethanol used in a mandated ethanol-gasoline blend, have been hoodwinked by the oil industry. Consequently, ethanol producers are forced to keep production down. This limits all the economic, health, and national security benefits the RFS was intended to provide.

Exploding the "Blend Wall" myth is easy. There is no reason to limit non-flex fuel gasoline-engine passenger vehicles to just E10 or E15. The great preponderance of vehicles on the road today can use E20, E25, E30, E35, E40 and higher fuels. As I previously wrote to Bill Shughart, this has been proven in several studies, including a multi-year test I personally conducted. In support of this are several "Ethanol Challenges," including one that was just recently concluded by Glacial Lakes Energy and Dyno Tune Speed and Performance. *(http://ethanolproducer.com/articles/14126/passing-the-test-with-flying-colors)*

The bottom line is that the Renewable Fuel Standard should be expanded to mandate E20 for all spark-ignited internal combustion passenger cars and light trucks. Higher level blends should be available through specific pumps or blender pumps, and a bio-diesel mandate should be created. This should be done forthwith.

All academics who have brought shame to their schools for their unethical behavior should be tossed out on their ears.

SHOCKING NEWS:

Existence of Parallel Universes Explain Support of Petroleum Oil Fuels

An Explosive Review and Commentary on Meghan O'Sullivan's *Windfall*

Originally published September 18, 2017

The theory of parallel universes, the idea our universe is one of many similar universes existing on a simultaneous timeline, took a giant step forward toward acceptance with the publication of a book written by Meghan L. O'Sullivan titled *Windfall.*

In parallel universes (sometimes called "alternate universes") things are very similar to our universe, but there are some differences. For example, if you're financially poor in this universe you could be rich in one of the alternate universes; or, instead of marrying your high school sweetheart, you married his or her sister.

I'm not suggesting that Ms. O'Sullivan's *Windfall.* has anything to do with parallel universe theories; it doesn't. It deals with petroleum oil fuels and what she refers to as the "new energy abundance" – an abundance of petroleum oil and related refined fuels she believes "strengthens America's power." I'll explain the connection to parallel universes in a moment.

I learned about this book when Dennis Prager interviewed Ms. O'Sullivan on his radio show. Although I don't agree with everything that Dennis says, his show is my morning program-of-choice when I happen to be in the car.

I missed some of it, but I heard a fair portion. I did hear Dennis talk about her credentials, which includes her position as Kirkpatrick Professor of the Practice of International Affairs at Harvard University's Kennedy School, as well as several high level political/government stints. In short, Ms. O'Sullivan appears to be considered an expert in "energy, energy R&D, energy security, energy innovation policy, oil and energy prices, oil, renewable energy, and U.S. energy policy." I'm exhausted just thinking about all that.

Notwithstanding her considerable credentials, I didn't agree with Ms. O'Sullivan's comments on Dennis Prager's show. Since I didn't hear the entire interview, I thought I should read her book before cementing my position on her position(s) and publishing my comments. I hightailed it over to Barnes & Noble to pick up a copy.

Just prior to leaving to go to the bookstore, I did send Ms. O'Sullivan an email in which I inquired if she received or is receiving any remuneration from the oil industry for writing the book. While I did receive back a return-receipt email indicating that she received and (may have) read my email, she has yet to respond to the question. I consider the answer important only if the information proffered by a recipient of remuneration is wrong and/or misleading.

Circling Back to Parallel Universes

Here's why I attached the theory of parallel universes to Ms. O'Sullivan's book and appearance on Dennis Prager's radio show. *Windfall* reads as if there are no other sources of energy (or fuels) than petroleum oil-related fuels and electricity. Clearly, her book was written for a different yet similar universe.

In the universe that the book deals with, there are no alternative fuels. In 300+ pages, she never uses the words/terms "alternative energy" or "alternative fuels," nor "alt-energy" and "alt-fuels." She also makes no reference to "biofuels." The words "ethanol," "methanol," and "algae" are not to be found anywhere. Ms. O'Sullivan omits any reference to "propane," too.

In Ms. O'Sullivan's universe, her only substitutes for petroleum oil fuels are natural gas, electricity and solar. She does frequently write about "renewable energy" in the book, but the sources for natural gas and most electricity are not renewable. Additionally, most natural gas is a by-product of the oil refining process (or obtained via oil exploration) leaving solar as the only renewable source. As Ms. O'Sullivan states, electric energy is not cheap—in some cases it's very expensive. Solar electricity is much more expensive. Electricity is not yet ready to replace liquid or gaseous fuels to power personal passenger vehicles, and it may not be ready for another century.

Because the oil industry treated natural gas like an ugly stepchild for so many years, they shot themselves in the foot: lies invented and disseminated about natural gas, similar to lies told about ethanol/methanol, soured the public's perception of natural gas, despite the fact that much of the public relies on natural gas for home heating purposes. The oil industry's efforts to defeat natural gas were so pervasive, America no longer has auto manufacturers mass-producing natural gas passenger cars. Their anti-compressed natural gas (CNG) lobbying efforts were so successful that the EPA priced engine conversion kits for existing vehicles so far out of line that it's not sensible to do it. In California, our most populous state, it's not legal to do a CNG conversion to nearly all passenger vehicles, even if you wanted to spend the money. What's more, if you purchase a vehicle from another state that had an aftermarket conversion to CNG, you can't register it in California. This paucity of new or converted CNG-powered vehicles means that the largest potential use of natural gas has been wiped out.

Ford, GM, Chrysler and several other manufacturers did try mass producing CNG vehicles a decade or so ago. Ms. O'Sullivan spends considerable time extolling the new appreciation for natural gas, but this new appreciation is DOA (dead on arrival).

Consequently, in her universe, Ms. O'Sullivan only offers vanilla ice cream. I, however, live in a universe that has vanilla, chocolate, strawberry, pistachio, coffee, peach, mango, and spumoni ice creams. My universe also features additional varieties of energy fuels; some of which are better, safer, healthier, and cheaper than abiotic fuels (the so-called fossil fuels). This is not to say that there's no value to reading Windfall, because it provides some good historical information and explanations of oil price and supply cycles.

Any examination of "new energy abundance" should include a copious review of all energy sources and fuels that we have available, just as any serious discussion of ice cream must include all flavors and applicable ice cream substitutes, such as sorbets and ices. And if the new energy abundance translates to greater American strength, it should be because our rediscovery of domestically produced ethanol, for example, has allowed us to be free from dependence on any other country for our energy.

UNRAVELING *WINDFALL*

Historical information aside, Ms. O'Sullivan undermines the veracity and perspicacity of her message from the start. On the second page of the preface (page x), she writes that petroleum oil fuels brought forth "...a surge of game-changing innovations, including the internal combustion engine and the tank..." This incorrect comment echoes a false statement I heard Dennis Prager (and other petroleum oil sycophants) make on several occasions when he proclaims, "Thank God for fossil fuels" and then asserts that our high standard of living from modern inventions was made possible because of gasoline and petroleum diesel fuel. Perhaps it is why he was so generous with his appraisal of Ms. O'Sullivan's book.

Let's be clear, there have been no machines created by the invention or availability of gasoline and petroleum diesel fuel, especially not the invention of internal combustion engines. Petroleum kerosene became the dominant liquid fuel due to exorbitant taxes placed on alcohol (ethanol) production beginning in the American Civil War. Gasoline and petroleum diesel fuel did not even exist when the internal combustion engine was invented. There was nothing about petroleum-based engine fuels that made them better fuels than the alternatives, except price. But even the lower price of gasoline and petroleum diesel fuels were not responsible for any life enhancing machines. And in those industrialized regions of the world that were not affected by the excessive taxation on alcohol, ethanol was the preferred high-performance internal combustion engine fuel. If Ms. O'Sullivan can't get this basic fact right—or worse yet, chooses to ignore it—she has no business writing and lecturing about the subject.

Machines may have been later adapted to take advantage of petroleum-based fuels' lower price, but these adaptations actually resulted in significant health and societal problems, including everything from lunacy to severe chronic respiratory illnesses from leaded gas, to the increase in autism and Alzheimer's disease, as well as almost all major wars over the past 103 years.

Simply put, gasoline and petroleum diesel fuel are deadly poisonous—literally, figuratively, and metaphorically—at least in the universe that I live in. With or without the addition of ingredients like tetraethyl-lead, MTBE, benzene or toluene, petroleum-based fuels are hazardous to all animal life. A greater abundance of poison is not beneficial. Poison multiplied by more poison does not equal less poison.

Windfall argues for the status quo of petroleum oil as an internationally priced and fungible commodity, regardless of where the oil originates. Ms. O'Sullivan ridicules the idea of American energy independence on physical and philosophical levels. She posits that it's impossible to completely eliminate the need for foreign oil to meet consumer use of fuels, and that the attempt to regulate trade would create an economic disaster. She cites the current relatively low level of imported oil in an attempt to dismiss energy independence as being unnecessary. These arguments are similar to those posed by Robert Bryce in his overly quoted and thoroughly discredited book, *Gusher of Lies*. (My 60+ page review and commentary on *Gusher of Lies* is included in this book.)

I suppose that in a universe that only has petroleum oil-based fuels, it would be prudent to retain international availability and pricing. But in my universe, where it is possible to go without vanilla ice cream and not suffer from ice cream deprivation, I think Ms. O'Sullivan's vision is unfounded and far less delicious.

By standardizing U.S. engine fuel to a blend level of around E30 (30 percent ethanol and 70 percent gasoline), we would eliminate the need for imported petroleum oil. Tests have shown that E30 in any gasoline-powered passenger vehicle (flex-fuel and non-flex fuel) provides MPG performance that's equal to or better than E10 or E0 fuel. An increase to E40 or E50 would produce about the same MPG performance and would eliminate the need for new poison oil discovery and refining in America. By any measure, this would benefit American ethanol producers, and America as a whole. It would not benefit the Saudis, the Russians, the Iranians, and the Venezuelans to name just a few. I do realize that Ms. O'Sullivan may not be as graciously received in countries like Saudi Arabia (as she describes in her book) if she took a less capitulating, more America-first position. I expressed this same opinion to Robert Bryce after, he too, was magnanimously received during his visit to Saudi Arabia. When he came back, it seemed there was Bryce's near-miraculous rebirth as an opponent of American energy independence.

The bottom line to the premise in *Windfall* is that it's a blueprint for the continuation of wars, hostility between nations, and forced servitude in those countries relying on American oil. Assuming that America could become the dominant provider of petroleum oil and refined fuels to the world because of enhanced fracking and drilling techniques, why would we want it? The era of colonization is over. We shouldn't want to be or need to be the police force and nursemaid to the world. And if petroleum-based fuels can't be safely produced and rendered harmless to living animals, why would we want to defend the misery associated with the production and distribution of those deadly substances?

America's financial success should be built upon innovative products and technologies, and the exportation of those things to other countries. In some instances, we would license entities in the other countries to build or employ local versions of the innovations/technologies. If our primary contribution to the world is controlling the fuels, we become a one trick pony, like Saudi Arabia. Then, we have to continuously fight to keep our fuel as the sole supply source—reminiscent of England's and France's wars with Asian countries to force them to continue to allow the selling of opiates to their respective publics—or we wither away when a cheaper or better fuel is introduced and we are not the ones producing it.

Fuel in America should be abundantly available at a fair price. It should be made in America by Americans and never be used to dictate personal taste or purchasing decisions for any type of vehicle. If someone wants and needs a large vehicle, it should be the price of the vehicle that determines a consumer's purchase, not the price of the fuel required to run it. Nor should they be bullied by the false impression that they are being wasteful of fuel. There's no reason to be rewarding Saudi Arabia or Russia or Venezuela or Iran by "stabilizing" the price of oil. We don't need them. We should be seeking ways to bankrupt them, or at the least to neutralize their impact on the world scene.

(https://www.eia.gov/energyexplained/index.cfm?page=electricity_in_the_united_states)

Three Degrees Can't Make Ethanol Slam Stick
A Response to Peter Z. Grossman and the *Washington Examiner*

Originally published December 8, 2017

A new op/ed article disparaging ethanol appeared in the *Washington Examiner*. This one titled "The Washington Swamp Is Filled with Ethanol" *(https://www.washingtonexaminer.com/the-washington-swamp-is-filled-with-ethanol),* was written by Peter Z. Grossman. Peter is the Efroymson Professor of Economics at Butler University (Indianapolis), and is the author of *U.S. Energy Policy and the Pursuit of Failure* (2013). He holds three university degrees: Washington University (Economics) Ph.D. 1992; Washington University (Economics) M.A. 1990; and Columbia University (Philosophy) B.A. 1970.

In referring to the RFS, Mr. Grossman writes that it's "better known as the ethanol mandate." The funny thing is that the only people who refer to the Renewable Fuel Standard as "the ethanol mandate" are those people who want to mock the RFS and paint it as an insignificant act designed to benefit a few blue-collar farmers sprinkled around the middle of the country. It's sort of like referring to Lincoln's "Gettysburg Address" as irrelevant "silly remarks."

(http://blog.pennlive.com/gettysburg-150/2013/11/lincoln_gettysburg_address_har.html)

There is no "ethanol mandate." There is a "biofuel mandate." Ethanol emerged as the primary substance to use because it is the best, least expensive, and safest substance for the job. This is not an insignificant point.

What job is a biofuel mandate intended to do? Peter Grossman says it was written because Congress was besieged by the public for a solution to rising gasoline prices. Two of Grossman's advanced degrees are in economics, and he's an economics professor with a specialty in economic history. For him to dumb-down the RFS and say it was a way to lower gasoline prices ignores three important events: The 1973 Oil Crisis, rising fuel and food prices over decades, and the deadly health consequences of gasoline additives (tetraethyl lead and MTBE). Ethanol replaces gasoline.

Grossman might as well argue that if we do away with breathing, we wouldn't have to worry about the health problems caused by leaded gasoline.

Using gasoline in a high-compression internal combustion engine requires an additive to mitigate engine knock and to keep the price of gasoline relatively affordable. With tetraethyl lead and MTBE out of consideration, the only options are increased aromatics or ethanol. Aromatics (benzene, toluene, xylene) are far more expensive than ethanol and they are poisonous. If we only consider the cost, and forget about the health issue, we are left with ethanol-gasoline blends that cost $2 to $3 per gallon, or ethanol-free gasoline with increased aromatics that cost $4 to $8 per gallon. You don't have to be a Ph.D. in economics to appreciate the difference. But if you are Ph.D. in economics and you don't understand the difference, you should be stripped of your degree.

The "1973 Oil Crisis" happened. Yes, it was a manufactured crisis, in that it did not occur because of any true oil shortage or worldwide depletion of resources But the effect of the crisis (the control of our economy and lifestyle) were real, and it's precisely why we want fuel independence. If the consequence of oil independence results in having ethanol fuels or ethanol-gasoline blended fuels that are less expensive and healthier than ethanol-free gasoline, why is this a problem? A professor of economics should know this. A six-year old child should know this.

Grossman cites a *Wall Street Journal* comment that claimed the use of a biofuel in our gasoline would cause "a gigantic transfer of wealth," and this would be a negative effect. But the transfer of wealth that WSJ is talking about is taking money from foreign dictators and terrorists and giving it to American farmers and consumers. Which precept in economics makes this a bad thing?

The good professor should have concluded his misguided missive with the cost issue. But for some strange reason he decided to push on and get into the corn canards. He trotted them all out: the one about America not having enough corn to supply all of our ethanol needs; that it would take 100 percent of the American corn crop to do so.

So, what if it was necessary to use all corn grown in the U.S. to make enough ethanol to replace gasoline? What deprivation would that cause? Won't there be enough popcorn for movie theaters? Will the tortilla chips Jay Leno assured us would always be there, survive? If not, we'll eat more potato chips; movie goers will buy Raisinets and licorice; and kids can learn to love green beans instead of corn-on-the-cob.

The response to the stupidity behind the corn issue is that corn isn't the only crop capable of making ethanol, nor that it's the best source to use (from a yield perspective). I'm not even referring to cellulosic raw materials. I'm talking about sorghum, sugar cane, sugar beets, cattails, Jerusalem artichokes, buffalo gourds, and algae. The reason we don't use all these other crops is that we don't yet need to, we have plenty of corn for our current needs, and we could grow enough corn if we switched from E10 to E85 as our basic engine fuel. Even then, we could probably still grow enough corn to keep popcorn eaters rotund and happy.

For more info about available land to grow ethanol crops go to:

http://www.theautochannel.com/news/2013/06/11/082056-gusher-lies-book-review-and-reply-to-robert-bryce-pt.html

Professor, didn't you do any research when you accepted this writing gig from the oil industry? Who did you speak to about any of this? Did you speak to anyone about any of this?

Grossman then ventures into environmental concerns. If you want to see proof positive that gasoline is more hazardous to the environment, watch this video. It compares burning ethanol to burning gasoline.

https://www.youtube.com/watch?v=H6qKa8X6o1w

If this isn't enough for you, go to Los Angeles, or San Francisco, or New York, and look up. If you notice that the sky is far cleaner than it was when we were growing up, it's due in large part to the use of ethanol-gasoline blends. Fracking for more petroleum oil doesn't solve environmental issues, it creates more.

Fox News' Revolting Take on Ethanol

Originally published December 12, 2017

Steve Hilton's "The Next Revolution" *Fox News* show has a regular segment called "Swamp Watch." On December 10, 2017 "Swamp Watch" addressed what Mr. Hilton called "The Corrupt Ethanol Industry."

(http://video.foxnews.com/v/5674983957001/?#sp=show-clips)

The story took me by surprise because I'd never heard of Steve Hilton, and I don't believe I've ever watched him before. I watch a lot of *Fox News;* it's the news channel of choice in my home. (To be fair, Mr. Hilton would probably say that although he's driven cars for decades, he's never heard of The Auto Channel.) I was more surprised because almost everything he said about ethanol was incorrect—profoundly incorrect.

He presented the kind of fake and wrong information that I have come to expect from other news channels. I let other news channels and outlets slide with dopey information because that's all you can expect from them. Because they generally take the political positions I'm rooted in—I get very upset when *Fox* doesn't get a story right. I think when they get facts wrong, it reflects badly on the rest of the conservative, patriotic American perspective.

Mr. Hilton took off on ethanol as if it were a dangerous poison, as dangerous as gasoline. He went after large ethanol producers as if they were corrupt, greedy, monopolistic corporate monsters, worse than corrupt, greedy, monopolistic monster petroleum oil producers.

He went after government-sponsored biofuel legislation designed to clean the air, to free us from foreign dependence on oil, and to help American farmers as if it was an ill-considered, restrictive, crime-producing Constitutional amendment like the Volstead Act (known as "The Prohibition" of the manufacture and sale of ethanol, that was made possible by the generous bribes paid by John D. Rockefeller to politicians).

Mr. Hilton began by saying the federal government forces energy companies to produce a certain amount of ethanol every year. The federal government in the United States of America doesn't force anyone to make a certain amount of ethanol every year. The growing of corn is a voluntary business decision. Distilling that corn into ethanol is a voluntary business decision. And the blending of ethanol with gasoline instead of any other biofuel or petroleum by-product is a voluntary business decision. As it happens, ethanol is the best, least expensive, and safest substance to blend with gasoline, so they use it.

In America we grow corn instead of another crop to meet ethanol needs because we have lots of farmers who know how to grow corn and because the climate in the agricultural states that need a money-making crop is more suited to corn than to sugar cane.

He said the Renewable Fuel Standard (RFS) is really the "Renewable Fuel Racket," and says that it "artificially boosts the ethanol market, that in turn boosts the demand for corn," and he called it a "massive boondoggle."

Not knowing what the Renewable Fuel Standard is about explains why he calls it a "Renewable Fuel Racket" and a "massive boondoggle." On top of not understanding the RFS, Hilton clearly has no knowledge of history or automotive technology. This is also surprising given Hilton's previous position as director of strategy for David Cameron, the former Prime Minister of Great Britain.

High-compression internal combustion engines used in most motorcars for the past 100 years require a fuel that does not cause engine knock. Engine knock creates a rough running vehicle and can literally shake an engine apart. In order to quell the knock, an additive has to be combined with gasoline (gasoline, also called "petrol" in the UK, is the primary fuel used in America and certain other countries for these past 100 years). For many years the anti-knock additive was tetraethyl lead. Tetra-ethyl lead (TEL) is highly poisonous, even more poisonous than gasoline. It was finally banned by Congress after they and the public were lied to by the petroleum oil industry for more than 70 years. For a few years, the replacement for TEL was MTBE, another poison. It was also banned.

So here we were, the highly industrialized world with millions and millions of on-the-road vehicles in need of a fuel that could power all these vehicles without engine knock. The options are to use more poisonous "aromatics" such as benzene, toluene, and xylene or non-poisonous ethanol. Or to put it another way, the option is expensive poisons like benzene, toluene, and xylene or an inexpensive, safe substance like ethanol.

The "big government boondoggle," as Steve Hilton calls it, would be to bypass ethanol and use the expensive and poisonous aromatics. He needs to read a book about fuels and how they work...or don't.

I laughed when Steve Hilton said that the need to replace tetraethyllead (abbreviated TEL) and MTBE, "artificially boost the ethanol market, and in turn the market for corn." There's nothing artificial about it. When a need arises, and the solution requires some new element or raw material, there is nothing artificial about a burgeoning demand or market for that ingredient. This is basic business economics. When it was decided to use rubber on the wheels of the new-fangled machine called the automobile, the market for rubber wasn't "artificially" boosted because of the demand, there was a true need for rubber, so the rubber market bloomed. Maybe you could also use a good book about marketing for Christmas (or Chanukah), Steve.

What I have written so far covers the first one minute and fifteen seconds of the almost seven-minute segment of "Swamp Watch."

Hilton said the Manhattan Institute estimated that ethanol costs us $10 billion per year in extra fuel costs. He neglected to mention that the Manhattan Institute is home to such ethanol bashers as Robert Bryce and Mark Mills, who I refuted in my 60+ page rebuttal of Bryce's poor excuse for a book, *Gusher of Lies.*

Do you know what costs the American public trillions—not billions—of dollars? All the wars we've fought over the past 103 years to protect and defend someone else's petroleum oil. That does not include the loss of American lives to preserve access to petroleum oil for Great Britain, France, Germany, Scandinavia, Canada, Japan, Korea, Australia, New Zealand, etc., etc. What's the cost of a million American lives? What's the cost of tens of millions of injuries sustained during those wars? How do you put a price tag on this?

Next, Steve Hilton brought up the rising cost of grocery food due to the use of corn for ethanol. Hey, Steve, the World Bank called, they want their myth back. In fact, they want it back so badly, they've thrice retracted the information in the original report because it was WRONG—that's three times in the last ten years they said they were wrong. Please don't use that information. But Hilton used it, just like the rest of the apologists do. Ethanol made from corn does not cause food prices to increase. The increasing cost of petroleum oil raises the cost of transportation, as do plastic wrapping and printing inks. That's what leads to the biggest food price increases.

In referring to ethanol's environmental benefits over gasoline and diesel fuel, Hilton said, "And here's the crazy part...a report from the National Academy of Sciences found that the production, farming, and transportation of corn-based ethanol actually creates more emissions than the production and use of regular gasoline." And, he said that ethanol contributes to pollution in the Gulf of Mexico because of all the fertilizer used to grow corn and soybeans.

Here's the real crazy part of what Steve Hilton calls the "crazy part." He's quoting from a study by the National Academy of Sciences (NAS), an organization that he vehemently criticizes in his opposition to man-made climate change activism. Hilton has stated that climate change activists are the biggest threat to our environment. Well, the NAS is a climate change activist and they rather vehemently state the position you disagree with. How about some consistency, Steve, not just adherence to whatever single position or benefactor can do the most for you?

This is the problem. *Fox News* needs to be consistent. It's okay to have different opinions on different issues, but don't hold out something as a trusted or important resource when you otherwise consider its opinion as being wrong. If the NAS position on man-made climate change is wrong, and that same position relates to a judgment that they made about ethanol, then maybe the ethanol comment is also wrong. But if they're correct, then Hilton's position is wrong; and if Steve Hilton can be wrong about this, then he can easily be wrong about ethanol.

The truth about pollution in the Gulf of Mexico is that if it's such a bad situation, and it's a situation that requires immediate and drastic attention, then why not call for the elimination of all corn crops, not just corn for ethanol? As I've written in the past, we don't need popcorn in movie theaters, there's plenty of other candy to buy. We don't need corn-on-the-cob at picnics, there's macaroni salad. And if corn and soybean crops account for about half of the pollution in the Gulf of Mexico, why not also take aim at other gross users of fertilizers that drain into the Gulf? For example, golf courses use a tremendous amount of fertilizer to keep their fairways and greens looking so green. There are thousands of golf courses located in the states that are connected to the Mississippi River. Why not abolish golf; it's just a game? And why not abolish other agricultural uses that require fertilizer—like growing grapes and hops in states that feed into the Mississippi. We don't need wine grapes or beer hops from Indiana or South Dakota, we have plenty of vineyards in California, Oregon and Washington.

Do not get me started on flower farms. All that water and fertilizer just for some piece of decoration that will be dead in a few days. Just buy silk flowers, they last for years and require no fertilizer and no water. Oh, and they're allergy-free!

At around the 2:10 mark in the video, Mr. Hilton poses the following question: "If the ethanol industry hurts consumers, and the environment, why on Earth is it propped up by the Federal Government?" He answers his own question by saying that, "As usual, in Washington, it's not just lobbying; it's straight forward out-and-out corruption—big business buying off politicians to get what they want."

The question is absurd because it incorrectly presupposes that something about the ethanol industry is hurting Americans and the environment, and wrongly implies that the alternative to ethanol does not hurt Americans and the environment. His answer is ludicrous because he treats the ethanol industry as if it were the petroleum oil industry.

If you're going to talk about big business, you can't in the same breath mention the petroleum oil industry and the ethanol industry. The largest petroleum oil company in America can make a hundred times more net profit in a single calendar quarter than America's largest agricultural company does in an entire year. It's comparing Macy's in Manhattan to a 1,500 square foot dry goods store in Topeka.

And if we're talking about the out-and-out corruption of buying off politicians to get what you want, how can you compare anything to 150 years of petroleum oil political corruption to keep ethanol, the cleanest safest fuel from having a level playing field? How do you explain the oil industry's collusion with foreign enemy governments and companies (Standard Oil working with the Nazis), followed by decades of lying about tetraethyl lead? Why is it more important to poison your customers than give them clean air?

The question Steve Hilton should be posing on "Swamp Watch" is, "Given the facts that petroleum-oil fuels are poison, that the environment has been permanently fouled because of it, that it has been responsible for killing untold millions of animals in the wild, that petroleum-oil is the root cause of so many wars that have killed and wounded so many millions of people, and so many millions of others have died or been disabled because of gasoline emissions, how is it that everyone involved in the petroleum oil industry hasn't been arrested and thrown in jail?"

There's still more than four minutes of his video to go, and my computer keyboard still has lots of life in it. If you stay with me, I'm delighted; if you leave now because you've reached your laugh quota for the day, I understand.

Marching Forward

After Hilton asked his own senseless question and answered it with an absurd answer, he goes personal. He attacks Senators

Grassley and Ernst for being on committees that deal with issues related to their constituents.

He attacks Grassley and Ernst for money contributed to their re-election campaigns by the ethanol industry. He attacks the Renewable Fuels Association (RFA) as "the ring master of this swampy circus" that has spent nearly $8 million dollars on lobbying since 2010 (about $1 million per year).

Hilton, in his big boy voice, asked, "How dare Senator Ernst use hard working family farmers in Iowa to cover up her corruption?" And he said that, "Giant corporations (Cargill and Monsanto) are shoveling money into her pocket..." and that, "they are destroying family farms in Iowa, and Joni Ernst is helping them do it."

This portion of the video was a combination of the most ignorant and most disingenuous part of his entire diatribe. I thought perhaps the entire episode was just a joke. But he never breaks a smile, he never gives a wink to indicate that he's joking. I even fast forwarded to the end of the video clip to see if he broke character. No, he was serious. He is that ignorant and unconscious about life in the world we live in.

Stevie rails against RFA for spending about $1 million per year on lobbying. The major oil companies spend more than $115 million per year on lobbying. The American Petroleum Institute (API), ring master of the gasoline swampy circus, contributes the largest part of the $115 million per year, they spend $65 million per year. That's 65 times more per year than RFA.

Lobbying is not an illegal activity, but if spending $1 million per year on lobbying is bad, how atrocious is spending $65 million per year on lobbying?

Hilton "exposes" U.S. Senators, Grassley and Ernst, for taking money from the ethanol lobby. There are 100 U.S. Senators. The website OpenSecrets.org lists 95 U.S. Senators who accepted lobbying money from the oil industry in 2016. If two Senators accepting lobbyist money is bad, how disgusting is 95 Senators accepting money from Big Oil lobbyists? So, if you want a textbook example of how a fake story makes it to broadcast or publication on news media, this is it.

Steve still had about 90 seconds of junk information to peddle.

He bemoans the fates of poor small oil refiners who have to go out of their way to buy and blend ethanol into their poison gasoline blends. I agree it was probably more profitable when all they had to do was use tetraethyl lead. So what if millions of people developed respiratory illnesses, and more and more children are developing autism? Breathing is overrated anyway. Plus look at all the carbon dioxide emissions from breathing. If we can get everyone to exhale only every other breath, we could make a serious dent in CO_2 pollution.

Steve Hilton closed out the video by calling Joni Ernst a pig and a stooge. He didn't use that description for the 95 Senators who accepted oil industry money, and he doesn't consider those 95 to be stooges for the oil industry.

I'm glad Steve Hilton worked in the circus tableau because he unwittingly gave us insight into the job he should really have—the company clown.

If you want to drain the swamp, you have to get rid of the cause of the swamp. The swamp is caused by the purveyors of that gooey, smelly, war-inciting gunk called petroleum oil.

January 7, 2018 Update

Steve Hilton took off on the petroleum oil industry. Perhaps my harsh criticism paid off. Or perhaps he thought he should do some research and found out he was wrong and grossly over-exaggerated his anti-ethanol comments.

In this "Swamp Watch" report, Steve acknowledged the huge amount of money spent by the oil industry to lobby politicians, and then asks why the oil industry has to resort to spending so much money and using such high-handed tactics in Washington. The answer is simple: petroleum oil fuels are poisonous. Petroleum oil fuels cause wars, serious respiratory problems, hazardous pollution and other environmental disasters. The way to overcome these problems is by spending lots of money on politicians.

Now, it would be nice if Steve acknowledged that he knows nothing about ethanol and that he apologizes for his previous remarks.

The Immorality of Arguing There's a Moral Case for Fossil Fuels

Originally published February 19, 2018

In 2014, a young man named Alex Epstein, described as a "philosopher, energy theorist, and industrial policy pundit," wrote *The Moral Case for Fossil Fuels.*

(http://www.moralcaseforfossilfuels.com/)

The book was Alex's springboard to a career arguing the merits of his case on a website, in magazine articles, in YouTube videos and live debates. His website says Alex will "accept a debate any time, any place."

In 2014, Kathleen Hartnett White wrote a short book, *Fossil Fuels: The Moral Case.*

(https://files.texaspolicy.com/uploads/2018/08/16101046/2014-06-RR03-FossilFuelsTheMoralCase-KathleenWhite-ACEE.pdf)

She held a number of senior positions with government agencies and industry groups as an environment and energy expert and followed up in 2016 with another book, *Fueling Freedom: Exposing the Mad War on Energy*, co-authored with Stephen Moore. Mr. Moore is also described as an "economics and energy expert."

Hartnett White rode these two books and her previous positions to a presidential nomination for White House senior advisor on environmental policy. (Her name was withdrawn from consideration in early February).

The 2014 books written by Alex Epstein and Kathleen Hartnett White appear to be unrelated though they share similar titles, themes, and the same year of publication. Hartnett White's book was published in June 2014; Epstein's came out in November 2014. It's reasonable to conclude that two or more independent people can simultaneously come up with a bad idea; just as two or more independent people can simultaneously come up with a good idea.

Alex and Kathleen do not cite the other in their respective books, but both do cite Robert L. Bradley and Robert Bryce. Those who have read my previous editorials and papers on the subject of alternative energy and fuels may recall that I have written about Messrs. Bradley and Bryce.

(My exchanges with Rob Bradley resulted in his asking me to write a thesis for his use in ongoing presentations and for his website. My exchange with Robert Bryce resulted in my 60+ page rebuke of his book *Gusher of Lies*, which has elicited only stony silence from him).

The premise of the Epstein/White books is that there is a moral reason for the use and continued use of so-called fossil fuels. I write "so-called," because as my friend Tom Quinn (chairman and CEO of E-Fuel Corporation) always reminds me, the correct term is "abiotic fuels." The term fossil fuels evoke images from the old animated Chevron and Sinclair gasoline television commercials suggesting that dinosaur carcasses turned into pools of crude oil.

(https://www.youtube.com/watch?v=h0zjoPVL__Q)
(https://www.youtube.com/watch?v=gei3gpG2IYU)

The reality is that fossil fuels (abiotic fuels) come from a time prior to the existence of dinosaurs and is derived from compressed non-biological minerals and water—not giant lizards. My use of the term "fossil fuels" hereafter is done for simplicity.

The central point both authors make is that America and the world would not be enjoying the high quality of life we do if not for fossil fuels. To aid in their effort to make this point, they connect the use of coal with liquid petroleum oil fuels and natural gas. This is basically sleight-of-hand.

Joining coal with the other abiotic fuels (fossil fuels) is the "sleight-of-hand" in the opening paragraphs of Kathleen's book, where she includes a brief tale of wilderness deprivation in the year 1800 from *The Rational Optimist*, a book by Matt Ridley. She then weaves Ridley's tale into the following statement.

> Use of the energy in fossil fuels unleashed economic productivity on a scale previously unimaginable. When innovative minds developed a steam engine which could convert the stored heat energy in coal into mechanical energy, the economic limits under which all human societies had formerly existed were blown apart. A life of back-breaking drudgery was no longer the inescapable condition of the overwhelming majority of mankind.

Her premise is incorrect. I'm not saying that coal isn't related to crude oil and natural gas in its original derivation, but I am saying that you can no more credit coal with our high standard of living then you can credit gasoline, petroleum diesel fuel, and natural gas for our high quality of life.

In the first place, any accolades for the high quality of life should be given to the inventions that use various fuels...no matter what those fuels are, not to the fuels themselves, and certainly not to coal or gasoline or petroleum diesel or natural gas. The inventions were all created without consideration to any specific fossil fuel. Internal combustion engines were created before the invention of either gasoline or diesel petroleum fuel. And they ran on alcohol.

Alex also references Matt Ridley and quotes him on page 81 as follows: "A modern combine harvester, driven by a single man, can reap enough wheat in a single day to make half a million loaves."

Based on this quote, Alex writes: "A single man, made into an agricultural Superman by the power of oil," and follows it with, "Another example: Oil-based transportation causes a dramatic increase in the amount of farm products that can be brought to market."

This single man isn't transformed into an agricultural Superman by oil, but by the machine, a machine that could just as easily and efficiently run on an alcohol-based fuel. The same dramatic increase in the amount of farm products to be brought to market can be ethanol-based transportation—which was always cleaner and safer than fossil fuels.

Coal was used for heat and light for thousands of years before the invention of the practical steam engine. You can't applaud coal for making life better via the steam engine, because of all those centuries prior to the invention of the steam engine were hardly easy. If coal was the salient ingredient to a high quality of life, surely the great scientific geniuses of earlier times would have invented usable steam-powered devices to propel trains, boats, and stationary machinery. They knew that steam existed, and how to produce it; and they had the ability to make and mold metals.

An interesting side-note to the general understanding of how we are enslaved and chained to a dependence on fossil fuels is that coal mining in the early days of the industrial revolution was literally legal slavery.

(https://en.wikipedia.org/wiki/History_of_coal_mining)

If you want to credit anything with the success of the steam engine to break the back of mankind's inescapable drudgery, the credit should go to water, not coal, wood, natural gas or any petroleum-based liquid. Without water there is no steam to power the engines. Without water there is no life, period! The mere burning of wood, coal, natural gas or oil-based liquid is incapable of powering any device.

Moving forward—technologically—to the invention of the internal combustion engines, coal is irrelevant, gasoline is irrelevant, and petroleum diesel fuel is irrelevant. These fuels did not make the invention of the internal combustion engine possible. Samuel Morey, Nicholas Otto, Karl Benz, and Henry Ford never said to themselves, "Oh good, now that there's gasoline and diesel fuel, I can invent the car."

Once again, internal combustion engines weren't invented as a result of the availability of gasoline and diesel fuel. You can run an engine without them. Through a series of political events and circumstances, petroleum oil fuels gained a foothold in the market. Then, via political corruption and tutelage, the oil industry bought their position as the dominant engine fuel, bringing with it wars, disease, and environmental disasters. On page 71 of his book, Alex references the dominance of petroleum oil fuels as THE transportation fuel, but he never explains how the oil industry bought that position. Instead, he dances around the wrong notion that gasoline and petroleum diesel earned the position because they are superior to other engine fuels.

Alex errs again and again in lauding fossil fuels for the societal benefits of the devices that use these fuels; devices that could be better powered by non-fossil fuels. And by "better" I mean safer, cleaner, cheaper, healthier—and with increased performance. He does this so often, that I might be able to page through the entire book and find at least one on each page. Here are a few other examples:

On page 13, Alex writes:

> This book is about morality, about right and wrong. To me, the question of what to do about fossil fuels and any other moral issue comes down to: What will promote human life? What will promote human flourishing—realizing the full potential of life?

On pages 29-30, Alex writes:

> Ultimately, when thinking about fossil fuels, we are trying to figure out the right thing to do, the right choices to make. But what exactly do we mean by right and wrong, good and bad? What is our standard of value? By what standard or measure are we saying something is good or bad, great or catastrophic, right or wrong, moral or immoral?
>
> I hold human life as the standard of value, and you can see that in my earlier arguments: I think that our fossil fuel use so far has been a moral choice because it has enabled billions of people to live longer and more fulfilling lives....

On page 84, Alex writes:

> Without the energy industry, the agricultural industry would not exist; the world could not support a population of 7 billion or 3.6 billion and perhaps

> not even 1 billion. To starve our machines of energy would be to starve ourselves.

In the chapter, "The Greatest Energy Technology of All Time," under the heading "More Fossil Fuels, More Ability," Alex claims that only petroleum oil fuels could power the machines of energy (so cheaply and so reliably) to feed the country.

I've heard this claim made by others. It goes something like this: "To grow the crops to produce ethanol, farm machinery must run on gasoline or petroleum diesel."

That's wrong. The internal combustion engine machinery they refer to can run on ethanol or biodiesel made from ethanol—ethanol that can be produced on the very farm where the crop is grown. That was part of Henry Ford's vision for building the Model T, as well as for farm equipment powered by virtually the same engine as the one in a passenger Model T. Instead of needing fuel sourced from oil wells hundreds or thousands of miles away, then shipped/trucked/piped to a refinery another few hundred or thousands of miles away, and finally shipped/trucked/piped to a local distribution point—which costs lots of money—the farmer can tap his home-made supply of fuel on his very own property.

In Chapter 7, page 151, "Reducing Risks and Side Effects," Alex writes about many of the risks and side effects of fossil fuels we all know all too well. He writes that there have been improvements and technological advances. I agree, there have been improvements, no thanks to the decades of oil industry lies and obstruction. However, gasoline, petroleum diesel and coal are still dirty, and their use creates terrible health risks. In addition, there's a very major risk/side effect of petroleum oil fuels that neither Alex Epstein nor Kathleen Hartnett White point out: WAR!

Fossil fuels have been the cause of wars, disease, and ecological and environmental disasters. Every significant war in the past 104 years has been caused by or fought over petroleum oil. Hundreds of millions of people have been killed in these wars. To the war dead, we have to add the people who died as a result of the illnesses caused by the use of petroleum oil fuels. Then there are the life-long injuries and disabilities suffered by untold millions more. There's nothing moral about any of this.

Even if it was correct to give some credit to fossil fuels for advances in mechanical devices (which it is not), the death and destruction caused by fossil fuels makes the argument for using them immoral.

And if you're the type of person who considers other animal life to be essential, how can the literal slaughter of birds, fish, dolphins, turtles and other animals from the wars and environmental disasters be anything less than wicked and depraved?

To be fair, the books written by Epstein and White are really meant to fight against the belief in man-made climate change (global warming, cooling, etc.), not to denigrate alternative fuels. Therefore, they make their arguments to show how the benefits of using coal, gasoline, diesel fuels, and natural gas outweigh the concerns of (possible) climate change.

Had Epstein and White only made their arguments a debate on the legitimacy of man-made climate change, I probably would not have gotten involved. I've already published my own position on man-made climate change, and so there is little I could add or disagree with in their respective books. However, in their literary enthusiasm—or their attempt to curry favor with the oil industry, they chose to include criticisms of ethanol and other alternative fuels or energy sources. Hence, I entered the fray. If they broke new ground in arguing against ethanol and were able to provide proof of their complaints against ethanol, it would have justified their inclusion in the debate about climate change, and I might have learned something new.

Instead, they both rely on lies and exaggerations about ethanol to make all the same old sophomoric arguments related to national economy, food, subsidies, and comparative energy content. This shows that they either know nothing about the workings of an internal combustion engine, or subsidies, or business and economics; or they simply decided to put their knowledge aside in the hope that anyone reading their books would also not know anything.

In the first chapter of his book, Epstein recounts a chance meeting he had with a college-age Greenpeace volunteer on the streets of Irvine, California. The young woman calls out to him, "Do you want to help us end our addiction to dirty fossil fuels and use clean, renewable energy instead?"

To which Epstein replied, "Actually, I study energy for a living—and I think it's good that we use a lot of fossil fuels. I think the world would be a much better place if people used a lot more."

Epstein writes that he expected a series of retorts from the young lady, and how he would have responded to them. Some of the things he hoped she would say were:

- That fossil fuels cause climate change.
- That fossil fuels cause pollution.
- That fossil fuels are not renewable.
- That solar and wind can replace fossil fuels.

Epstein's hypothetical responses would have basically concluded with details about how ingenuity and technology can (or eventually will) clean up the pollution, and how solar and wind is intermittent and unreliable. His closing comment would be that fossil fuels "...is the only source of energy that has been able to provide cheap, plentiful, reliable energy for the billions of people whose lives depend on it."

However, Epstein writes, the young woman didn't respond in the manner he expected. I would say that he was left in a state of unfulfilled satisfaction, and therefore created this straw man argument.

Had I been there to overhear the exchange between Alex and Miss Greenpeace, and the conversation took place along the lines Alex thought it might, I would have said, "Imagine that if instead of using petroleum oil fuels that fill our skies with deadly pollutants, we used clean fuels that can accomplish even better performance results with fewer wars. How much better off we would be? Think of all the people who wouldn't have died or become disabled."

I would have added that neither a solar source of energy nor a wind source of energy is unreliable and intermittent. Indeed, should the day(s) ever come in which the darkness of night doesn't turn to a lightened day, then our entire existence on planet Earth would come to an end. And since wind currents are a direct result of a revolving planet and the same solar source, wind is a totally reliable source of energy. The only requirement is that the solar collectors be sufficiently sensitive to cloud-covered days, and that the wind generators be positioned in places where there is always wind. But, these requirements are virtually no different than the requirement for fossil fuel electric generating plants to be located within proximity to users.

To the point that only fossil fuels can provide cheap and plentiful energy, I would have informed him that ethanol could have, should have, and would have been the fuel to provide cheap and plentiful energy had the oil industry not bought their way into the dominant position.

Epstein used this young woman to create a straw man argument to demean her and her street-corner demonstration. Alex often criticizes attacks against his position as being straw man arguments. He raises this flag as if such an argument was comparable to a "low blow" in a boxing match. Instead of countering criticism with a substantive rebuttal, Alex uses the phrase to disqualify criticism from having any value. He does this in some videos on the Internet. After a rebuttal I made to Epstein's "fossil fuel is moral" position, I was accused by Epstein sycophants who condemned my criticism as a straw-man argument against him. I had one person online ask me, "Who made you God?" I guess he felt that only God was competent enough to challenge Epstein's ridiculous ideas.

Just because we know fossil fuels are far more abundant than we ever thought possible, doesn't mean we should be using them. Gasoline and petroleum diesel fuels are poisonous. They have always been poisonous, and they will always be poisonous. What's more, the additives required to make gasoline and petroleum diesel fuels non-corrosive and anti-knock are even more poisonous than gasoline and diesel fuel. Further, there is no such thing as clean coal. It's possible to reduce the harmful effects of mining/processing/burning coal, but not to entirely eliminate the terrible effects. And as the population continues to grow, and more coal energy plants are used, the harmful effects continue to grow in the aggregate, even if individual plants spew fewer emissions.

No matter what one's position is on man-made climate change, ethanol is the solution. Whether ethanol should be a long-term permanent replacement for petroleum oil fuels, or just a mid-range solution until electric and electric-power generation from a harmless and low-cost renewable source can dominate, both sides of the climate change argument should rally around ethanol. Only the mistaken belief that there is something wrong with ethanol—a view usually funded by the oil industry—stops people on both sides from agreeing on ethanol's benefits.

If catastrophic man-made climate change is real, and it is not too late to prevent the cataclysmic results, then ethanol should be immediately adopted as the world's primary engine fuel because it would mitigate any further damage. If catastrophic man-made climate change is a myth, we still have pollution, diseases, war, and a struggling national economy to worry about. The switch to domestically produced ethanol would immediately end and erase all of these problems.

Epstein and White frame their entire position on the argument that "mankind's use of fossil fuels is supremely virtuous— because human life is the standard of value." Using their own measuring stick, it must be concluded that using fossil fuels is immoral, and that the proposition that fossil fuels are moral is immoral, period.

Alex Epstein's website states that he "will accept a debate any time, any place." On January 19, 2018 exactly one month before I wrote this article, I challenged him to debate the issue. I received a very cordial and quick reply from him. Epstein wrote, "I have at least three debates lined up already this year, so whether I take additional ones will depend on how prominent I can expect the debate to be..." There were a couple of other minor requirements, like how much would he get paid, and he asked if I could refer him to any of my work on the subject in order to ascertain how prominent an opponent I might be. I sent him several links (included below), and we have touched base once since January 19, 2018. Apparently, he's still processing the material I sent because I haven't heard from him.

I don't expect Epstein will accept my challenge. Undoubtedly, he'll find some reason to discount me as not being worthy of his effort. But the real reason he won't accept the debate is because he'd lose, and then the career he built arguing that fossil fuels are moral would be destroyed. I mention this, hoping that it stirs him into action, but I still don't expect him to follow through. I will be sending Epstein and White a link to this story, and if they choose to respond I will add their response below.

P.S. It is now seven months since I challenged Epstein to a debate and corresponded with him. I have still not heard back from him and have never heard anything from Kathleen Hartnett White. It's probably safe to say that they feel they are incapable of defending their (ridiculous) opinions. A third party—someone who had worked with Epstein on another project—tried to organize a debate and meet his financial requirements but was unable to coax him onto a stage with me. However, I'm not closing the doors to a debate, I'd still love to do it.

Notes

NOTE: The page numbers referenced above relate to the Kindle e-book versions of Alex's and Kathleen's books. It is possible that other e-book versions or hard copy versions have slightly different page numbering.)

LINKS SENT TO ALEX EPSTEIN: (all of these are included in this book)

Life as We Might Have Known It: What If Ethanol Was Our Primary Engine Fuel *(https://www.theautochannel.com/news/2016/03/07/205279-life-as-we-might-have-known-it-imagining-bio-fuel.html)*

TRUTH ABOUT ETHANOL - 60+ page Reply to Robert Bryce's *Gusher of Lies (https://www.theautochannel.com/news/*

2013/06/10/082044-gusher-lies-book-review-and-reply-to-robert-bryce.html)

Existence of Parallel Universes Explain Support of Petroleum Oil Fuels *(https://www.theautochannel.com/news/ 2017/ 09/18/435077-shocking-news-existence-parallel-universes-explain-support-petroleum-oil-fuels.html)*

Open Letter to Barry Ritholtz and Bloomberg About Ethanol *(https://www.theautochannel.com/news/2016/02/04/191905-open-letter-to-barry-ritholtz-and-bloomberg-about-ethanol.html)*

American Petroleum Institute Goes Bonkers: Issues Fatuous Lies Against E15 Ethanol Blends

Originally published April 12, 2018

A pathetic apologist for the American Petroleum Institute made a series of preposterous claims against E15 that prove neither he nor anyone currently employed by API know anything about motor fuels or internal combustion engines. The apologist I'm referring to is API Downstream Group Director Frank Macchiarola.

In a press release that seemed to be hastily prepared by nitwits, halfwits, and truth-challenged people at Big Oil's public relations firm to counteract reports that President Trump favors year-round sales of E15.

(http://energy.agwired.com/2018/04/12/president-trump-supports-year-round-e15-sales/)

Mr. Macchiarola signed off on the document saying that:

- Forcing higher ethanol fuel blends into the marketplace could have negative consequences for U.S. consumers.
- The RFS is broken and we continue to believe the best solution is comprehensive legislation.
- Testing by the auto and oil industries found that higher ethanol blends, such as E-15, can harm engines and fuel systems.
- E15 potentially forces drivers to pay for unexpected and costly repairs themselves.
- Nearly three out of four vehicles in the U.S. fleet were not built for E15 and this fuel is not compatible with motorcycles, boats, lawn equipment, and ATVs.
- And many automakers say that using E15 could potentially void new car warranties.

The truth is that:

• The only negative consequences to using higher ethanol blends is to oil industry profits—there are no negative consequences to consumers who own internal combustion engines (ICE).
• If there's anything wrong with the RFS, it's that it should mandate much higher levels of biofuels to be used. The legislation that's needed is to ban the manufacture of any new gasoline-powered ICE engines for use in America.
• Any testing conducted by the auto and oil industries that found E15 can harm engines and fuel systems were fraudulently performed and/or the analysis of any problems were incorrectly blamed on ethanol.
• If E15 was the cause of a problem that required unexpected and costly repairs, then it might "force" drivers to pay for the repairs. However, E15 does not cause engine and fuel system problems. Gasoline causes the problems. And the problems that are not caused by gasoline are problems that are attributable to any multi-part working mechanism...in other words, parts normally breakdown from wear and tear.
• All vehicles in the U.S. fleet were built for E15 (and higher blends). How do I know? For one thing, the oil industry told us so, for nearly 100 years...but the people working for API weren't told this. See: "The Hypocrisy of Big Oil" *(https://www.theautochannel.com/news/2017/12/20/478776-hypocrisy-big-oil.html).*
• As for claiming that E15 is not compatible with motorcycles, boats, and ATVs, that's hogwash. A good many of the very same bikes, boats, and buggies sold in America are sold in Brazil, where the standard fuel is E27. And if, by some strange reason, someone has a lawn mower that doesn't like E15, then they shouldn't use it; they should use E10 or E0. No one is forcing them to use E15.

• If automakers could get away with it, they would say that starting your engine voids new car warranties. Congress should pass legislation that immediately requires all automakers to extend warranties for any new ICE cars and light trucks built since the year 2010* to include up to the use of E40. Additional legislation should be passed that warranties for all ICE cars and light trucks built after 2020 extend to the use of E50.

Soon after this, the American Coalition of Ethanol (ACE) released a video that exposed the problems caused by the use of E15. The video is titled "Retailers Highlight E15 Liability Issues." The morons at API and their PR firms must have looked only at the title and not the video. The video itself is a fabulous tongue-in-cheek response to the same kind of irresponsible yellow journalism featured in API's fake news press release.

Here it is: *https://www.youtube.com/watch?v=s6Bi1-IdUTM*

If the folks at API and its PR firms don't like being called truth-challenged, they could come out and publicly debate me.

(The reason I selected 2010 as the date is because of the enormous debris caused by gasoline. Any engine used for an extended period should be given a thorough cleaning before using much higher ethanol blends. Ethanol cleans engines; but when you have too much garbage, even the largest drains need clearing.)

Here's the insidious press release from API:

> WASHINGTON, April 12, 2018 /PRNewswire-USNewswire/ -- Today the American Petroleum Institute said forcing higher ethanol fuel blends into the marketplace could have negative consequences for the American consumer.
>
> "We are concerned about news reports suggesting administrative attempts to grant an E-15 summertime waiver," said API Downstream Group Director Frank Macchiarola. "Forcing higher ethanol fuel blends into the marketplace could have negative consequences for U.S. consumers."

> EPA has previously stated that it does not have the legal authority to grant the E-15 waiver, and we agree with that assessment. The industry plans to consider all options to prevent such a waiver. The RFS is broken and we continue to believe the best solution is comprehensive legislation. API urges Congress to take action to address this important issue.

Testing by the auto and oil industries found that higher ethanol blends, such as E-15, can harm engines and fuel systems and—potentially force drivers to pay for unexpected and costly repairs themselves. Nearly three out of four vehicles in the U.S. fleet were not built for E-15 and this fuel is not compatible with motorcycles, boats, lawn equipment, and ATVs. And many automakers say that using E-15 could potentially void new car warranties.

Big Oil Uses New Dopes to Spread Old Dope

API is getting really worried by Trump's comments about year-round E15 use

Originally published April 20, 2018

On Friday, April 20, 2018, a new hit piece against ethanol was published on *TheNewAmerican.com* website, by Michael Tennant*. The piece, titled "Federal Biofuel Mandates Are an Expensive Flop," features Philip Rossetti from an organization called American Action Forum (AAF). American Action Forum appears to me like another API sponsored group.

(https://www.thenewamerican.com/economy/sectors/item/28795-federal-biofuel-mandates-are-an-expensive-flop)

The piece written by Mr. Tennant starts off with this paragraph:

"Federal biofuel mandates have failed to accomplish their stated objectives while forcing consumers to pay higher prices for food and fuel, a new American Action Forum (AAF) report concludes."

A NEW report? The "higher prices for food and fuel" drivel is at least a decade old, and has been roundly and soundly refuted at least three times by The World Bank, the group that started the lies about food priced and ethanol production in the first place. (See my 60+ page rebuttal to Robert Bryce's *Gusher of Lies* in this book.)

Rossetti had put together numbers supposedly showing there are some ancillary issues that make ethanol look bad. But it's all wretched refuse. He says there has been a wealth transfer

* "Michael Tennant" may not even exist, because I wanted to send him a copy of my comments, but I can't find any information or an email for him.

from consumers to farmers. Why is he so unhappy when American farmers have a profitable crop to grow? American farmers are Americans. Why does he think wealth transfer from consumers to dictators and terrorist regimes on the other side of the world makes more sense?

Ethanol cost less than gasoline or diesel. Ethanol-gasoline blends save consumers money.

The people behind The New American website should know there is nothing patriotic about pushing deadly petroleum oil fuels, especially when America imports them from other countries. They are traitors, not patriots.

Ethanol Criticism Polluted by Oiliness

Originally published May 17, 2018

In 2016, I wrote "It's Time to Rethink the Value of Ph.D.," an editorial questioning the credentials and motivation of certain academics (university professors and highly degreed individuals) who appear to go out of their way to attack ethanol, alternative fuels, and the Renewable Fuel Standard (RFS).

In 2017, I followed up on that with a series of three editorials that explored the roots of contemporary ethanol opposition as well as identifying two specific academics who typify those people tasked (asked/paid) by the petroleum oil industry to attack ethanol and the RFS. These three editorials were:

- Uncovering the Gas Roots of Contemporary Ethanol Opposition
- Unmasking the Gas Roots of Contemporary Ethanol Opposition
- Unmasking the Gas Roots of Contemporary Ethanol Opposition - Round 2

The "Unmasking" editorials compared tactics used by the tobacco industry tactics to lie to the public about the dangers of smoking with tactics the oil industry uses to lie to the public about the dangers of petroleum oil fuels.

There is no shortage of professionals coming forward to take Big Oil's money in exchange for whatever influence their names carry.

The latest such professor is Veronique de Rugy, adjunct scholar at the CATO Institute and senior research fellow at the Mercatus Center. She is a mix of seemingly impressive academic credentials with what appears to be a complete and total lack of knowledge about engine fuels and the issues surrounding their use.

Her editorial "Is Ethanol Cronyism on the Ropes?" (*https://reason.com/2018/05/03/is-ethanol-cronyism-on-the-ropes/*)

was published on May 3, 2018 and appeared on a variety of websites and blogs. The entities carrying the piece are, I'm sorry to say, largely those that espouse conservative American values while supporting American dependence on foreign petroleum oil. This gross inconsistency is hypocrisy, but a much better word to describe it is "oiliness." Nothing says slimy like oiliness.

Basic Facts I Suspect Ms. De Rugy Doesn't Know

Internal combustion engines (ICE) run on liquid or gaseous ignitable fuels. They can run on gasoline, petroleum diesel, compressed natural gas, propane gas, ethanol, methanol, biodiesel, and a few other liquid or gaseous substances. The only true issues of concern, or questions, about vehicle fuels should be:

1. Engine performance
2. Cost at the pump
3. Availability
4. Health and public safety related to contact with, or inhalation of, the substances
5. Environmental damage

All other issues are irrelevant. When the internal combustion engine was invented, alcohol mixed with wood turpentine was the fuel of choice. When the diesel engine was invented, a fuel produced from peanuts was used. Neither device required fuel produced from petroleum. There is/was nothing about petroleum fuels that enhances engine performance. Everything about petroleum oil fuels reeks of health dangers and environmental damage. When I say it "reeks of health dangers and environmental damage" I mean it figuratively and literally.

So, right away, issues #1, #4, and #5 make petroleum oil fuels a bad choice.

In America, because of excessive taxes on alcohol production, the nascent petroleum oil industry took advantage of the situation and made great strides to become a primary fuel source as ICE and vehicle technology advanced. Elsewhere, in other leading industrial countries, ethanol (alcohol) fuels were competitive with the new petrol fuel gasoline. Almost everywhere, ethanol was the fuel of choice for motor sports, regardless of cost, because it provides better engine performance.

When the extreme taxes on ethanol in the U.S. were lifted in 1907—making ethanol as inexpensive as gasoline, ethanol was set to become the prominent "fuel of the future" because high compression engines required the higher octane provided by ethanol (as compared to conventional gasoline that caused engine knock). To prevent that from happening, General Motors' scientists created leaded gasoline as the "fuel of the future." See "Exemplars and Case Studies—Tetraethyllead: Kettering and Midgley."

https://chem.libretexts.org/Bookshelves/Ancillary_Materials/Exemplars_and_Case_Studies/Case_Studies/Tetraethyllead/01%3A_Tetraethyllead%3A_Kettering_and_Midgley

However, upon discovering that their discovery (leaded gasoline) could be patented and be eventually worth billions of dollars, they changed direction and pushed their home brew over ethanol and ethanol-gasoline blends.

See "The Rise & Fall of General Motors and the Subjugation of the Industrialized World" (located in this book).

In order to destroy the ethanol market, John D. Rockefeller of Standard Oil, donated funds to politicians who worked to ratify Prohibition, the 18th Amendment to the U.S. Constitution. This Rockefeller-sponsored law made alcohol production and distribution illegal. Nothing kills your competition better than making it illegal. It's a sure-fire marketing tactic. Think how valuable it would be to Pepsi-Cola if they could bribe enough politicians to make Coca-Cola illegal!

Fortunately, Prohibition was repealed about 13 years later. But great damage was done. One result was that disorganized crime became organized crime. The other problem was that the ethanol (alcohol) industry had to start from scratch and compete against an industry that owned America, had the financial war chest to stay in control, and had inundated the American public with false information about ethanol.

That is how issues #2 and #3 were created by external forces, outside any legitimate qualitative consideration of ethanol as a fuel and made ethanol too costly and unavailable.

Originally, gasoline with tetraethyl lead (leaded gasoline) wasn't globally accepted because tetraethyl lead (TEL) is a highly dangerous poison, and it wasn't needed in countries to mitigate engine knock. As incredible as it may seem, the same oil companies selling leaded gasoline in America and bad-mouthed ethanol here, were actively engaged in selling ethanol-gasoline blends elsewhere (and touted "power alcohol" as being cleaner, more powerful, safer, and providing better mileage than ethanol-free gasoline). See: "The Hypocrisy of Big Oil" (located in this book)

Leaded gasoline wasn't only highly poisonous because of tetraethyl lead. It also required the use of another poison (ethylene bromide) to make the TEL less corrosive. Why wonder which is more poisonous—TEL or ethylene bromide? Ethylene bromide did not magically make leaded gasoline less poisonous. The use of ethylene bromide to lessen the severe effects of TEL corrosion is little known by the general public, many of whom believe that leaded gasoline was not itself corrosive. Ironically these purveyors of poison often describe ethanol as being highly corrosive to engine and fuel system parts. It is not.

The Second World War brought many drastic changes to the world. I'm not referring to the political and geographic changes from the war. I'm referring to the economic and social changes that came with America becoming the dominant country of the world. America's economic might and generosity was much needed. The price of America's generosity was to accept or be "receptive" to American technology and our way of life. One such example was the television transmission system known as National Television System Committee (NTSC). Another was the near global acceptance of leaded gasoline, which was still under the exclusive patents owned by Standard Oil, General Motors, and DuPont Chemical, all ostensibly "American owned" and the world's largest companies in their respective industries.

Despite Standard's and General Motors' collaboration with our enemies in the great struggle (some call it treasonous because they did collaborate with the Axis countries), the world chose to ignore or forgive what they did. Consciously or unconsciously, the world rewarded the leaded-gasoline patent holders with billions and trillions of dollars, pounds, marks, yen, pesos, lire, francs, gilders, and rupees by buying leaded gasoline.

Coming full circle in a world that generally doesn't impose prohibition on the manufacturing and distribution of ethanol, and without outrageous excessive taxation, we can see that all five of the issues listed above make ethanol the smarter, less expensive, safer, healthier, and superior fuel to use in internal combustion engines.

My business partner (Bob Gordon) and I believe that anyone with brains, with a modicum of common sense and thoughtfulness, should easily understand the significance of what I've laid out. We should not have to explain things to highly degreed individuals like Ms. du Rugy. They should know the history and understand the implications of the historical facts. Unless they each are being paid many millions of dollars—enough money to never have to work again in their lives—it is unconscionable for them to take the anti-ethanol positions they do and write anti-ethanol drivel.

Veronique du Rugy's article "Is Ethanol Cronyism on the Ropes?" is preposterous.

If any industry is guilty of cronyism it's the petroleum oil industry. If any individuals are guilty of cronyism, they are the people who work at the behest of the petroleum oil industry. They stink. They reek the way New Jersey used to smell when you crossed the Goethals Bridge from Staten Island in New York. The Elizabeth area in New Jersey stank because petrochemical plants were located there. How petroleum oil industries have conducted themselves with those in political power defines the word cronyism. If other industries and non-oil individuals do unsavory lobbying, it is because the oil industry has set the bar. You know the term "grease the wheels?" "Oiliness" is required.

Du Rugy's subtitle is "It's time to end government handouts for corn farmers." Does she know that less than half of all corn grown in America is used to produce ethanol? Does she know that a good portion of corn waste from the production of ethanol is also used to feed livestock we eat? Does she realize that corn used to make ethanol is also used to make corn oil and other products? Does she know that farms in America are American farms located right here in America? Does she know that the Renewable Fuel Standard does not mandate corn ethanol be used? Does she know ethanol is used for the RFS mandate because it is the best, safest, and least expensive ingredient that can be used? Does she know that no American military man or woman has ever been killed or wounded while protecting ethanol production and distribution?

I believe she knows none of these things. If she does, maybe she doesn't give a hoot because she is too busy wallowing in money the oil industry must have paid her for sullying her academic virtue.

She refers to a 2016 University of Minnesota study (I presume she means the study done by John DeCicco, although she doesn't specify) claiming an "unintended consequence of the biofuel mandate is that it actually increases net greenhouse gas emissions" because the lower prices of ethanol-gasoline blends encourages drivers to drive more. Well, excuse us for living! Excuse the American people for wanting to go where they want to go, when they want to go! Her interpretation of DeCicco's study is the dumbest, shallowest, most idiotic criticism against ethanol I have ever read. And keep in mind that the crappy study conducted by John DeCicco was rebuked, repudiated, and laughed at from the moment that it was published.

• API-Funded Study Obscures Lifecycle Emission Benefits of Biofuels

(https://ethanolrfa.org/2016/08/api-funded-study-obscures-lifecycle-emission-benefits-of-biofuels/)

• MichBio Responds to Flawed University of Michigan Research Study on Biofuels and Greenhouse Gas Emissions

(https://www.theautochannel.com/news/2016/08/26/275399-michbio-responds-to-flawed-university-michigan-research-study-biofuels-and.html)

• The Case Against Ethanol Opponents: They Are Simply Incorrect

(https://www.theautochannel.com/news/2016/08/28/275763-case-against-ethanol-opponents-they-are-simply-incorrect-video.html)

She then concludes her comments with this: "The RFS program has failed to achieve its stated policy objectives of improving the environment and promoting energy independence. Rather, it primarily exists today as a handout for corn farmers. This is made clear by the fact that reform proposals are evaluated primarily by their impact on farmers and that the most strident defenders of the status quo in Congress come from agricultural states. A president who is serious about draining the swamp wouldn't succumb to their demands but would instead push for the permanent reforms needed to reverse an ill-conceived market intervention."

The RFS program has not failed; hundreds of millions of Americans use ethanol in their vehicles every single day, and there are no examples of ethanol causing any engine or fuel system problems that are not (or cannot be) blamed on ethanol-free fuel. Our air quality is far better today than it was prior to the regular use of E10 and E85, even though there are more cars on the road today. American farmers who grow corn have the opportunity to grow a product for which there is a robust profitable market. And every single gallon of ethanol used means that we are using one gallon of poison gasoline less. This is called success.

Permanent reforms are, however, necessary. It's time to stop selling any new vehicles powered by gasoline or petroleum diesel engines. Ten years ago, The Auto Channel called for an end to new gasoline-powered vehicles by 2014. We gave them five years to make it happen. The grace period is over; we're taking off the gloves!

Open Letter to Washington Examiner and John Siciliano About Ethanol

Originally published June 7, 2018

Hi John -

I just finished reading your article "EPA's New Super Fuel Could Spell Trouble for Ethanol" *(https://www.washingtonexaminer.com/policy/energy/epas-new-super-fuel-could-spell-trouble-for-ethanol)* published June 5, 2018 in the *Washington Examiner*.

I noticed you are the *Examiner*'s Energy and Environment Reporter, and you have been in the position for about three years now.

I'm wondering when you might begin to acquire some knowledge about energy and the environment? I'm not saying that you can't become super knowledgeable about a subject in three years. I'm just wondering how long you think it might take for you to become knowledgeable about the subject.

You reported, "The boating industry has identified biobutanol as a suitable and safe alternative to ethanol, which can damage engines, including those that power recreational boats." When they and you say that biobutanol is a safe alternative to ethanol, what do you mean by safe? Are you suggesting you can rub biobutanol over your hands? Are you suggesting you can drink biobutanol? In case you don't know, you can't do those things—because butanol (regardless of the raw material used to create it) is highly toxic. How is it a safe alternative? You can drink ethanol. You can rub ethanol on your body. You can use ethanol to clean an open skin wound. Sure, ethanol can be made poisonous by denaturizing it with gasoline; but that's because of the gasoline, not the ethanol.

By making the statement, "The boating industry has identified biobutanol as a suitable and safe alternative to ethanol...," didn't you know that the boating industry has already stated that ethanol-gasoline blends are better than ethanol-free gasoline, and that they can be used in marine engines? Are you unaware of the statements made by Mercury Marine, the world's largest manufacturer of marine engines, that E10 will not cause any engine damage? Did you not know that Mercury Marine and many other marine engine manufacturers sell their engines and boats in Brazil, where standard fuel is E27? Are you in possession of any documents from Brazil that show E27 to be specifically damaging to marine or automobile engines?

When you report "damage," did you ask anyone what they mean by damage? You use the words "damage," "damaging," "corrosive," and "harmful effects." Like what? You quote John McKnight, of the National Marine Manufacturers Association, as saying that they tested E15 and that they "helped conduct a number of tests on the effects of 15-percent ethanol blends on the engines in collaboration with the Department of Energy's Argonne National Laboratory, and we have pictures. They just blew up...I mean, they could not run through a normal durability cycle on E15."

What tests is he talking about? In what year were those tests carried out? Where is the report on those tests? Where are the pictures? What does he mean that they just blew up? What blew up? Did something actually explode?

I'm not saying that butanol is a bad fuel for internal combustion engines, but I have trouble understanding why America and the world would want to use another poisonous fuel made by the oil industry when we have the opportunity to finally get away from that. Why do we need more neurotoxins in the air we breathe? Weren't six decades of tetraethyl lead bad enough? Don't we have enough respiratory illnesses? Don't we have enough children with autism?

And while I'm on the subject, why would anyone in their right mind want to listen to anything BP has to say? They've proven they have no interest in humans and other living things.

You mention ethanol's "energy density" as a problem that makes cars less fuel efficient when added to gasoline. Energy density (energy content) is irrelevant in an internal combustion engine. Mechanical engine optimization is the key. An engine mechanically optimized to run on ethanol will provide comparable or better MPG than a gasoline optimized engine running on gasoline, and tests have shown that certain blend levels (i.e., E30) will deliver more MPG in a non-flex fuel vehicle than E0. See:

- "The Irrelevance of BTU Rating - Big Oil's Gimmick to Hoodwink the Public" *(https://www.theautochannel.com/news/2015/10/19/144405-irrelevance-btu-rating.html)*
- "Fuel Economy and Power Generation of 30% Ethanol (E30)" *(https://web.archive.org/web/20170821103837/http://www.sdfu.org/assets/docs/uploads/gle-e30-challenge-white-paper-1-19-17final.pdf Effects of High-Octane Ethanol Blends. https://bioenergykdf.net/system/files/1/Thomas_Effects%20of%20High-Octane%20Ethanol%20Blends%20on%20Four%20Legacy%20Flex-Fuel%20Vehicles%20and%20a%20Turbocharged%20GDI%20Vehicle.pdf)*
- "Study Finds Certain Ethanol Blends Can Provide Better Fuel Economy Than Gasoline" *(https://www.green-carcongress.com/2007/12/study-finds-cer.html)*
- "The Effect of Ethanol-Gasoline Blends—University of Nottingham" *(https://eprints.nottingham.ac.uk/29749/1/546471.pdf)*
- "The Effect of Ethanol-Gasoline Blends" *(http://www.stonis-world.com/docs/SuTCAF.pdf)*
- "High performance Wayne State ethanol car wins 1998 Ethanol Vehicle Challenge" *(https://today.wayne.edu/news/1998/06/04/high-performance-wayne-state-ethanol-car-wins-1998-ethanol-vehicle-challenge-1464)*
- "Ethanol Vehicle Challenge" *(https://avtcseries.org/competitions/ethanol-vehicle-challenge/?section=innovations-highlights)*

John, why does the *Washington Examiner* continue to publish absurd, incorrect stories about ethanol? Intentionally publishing incorrect information about one subject negatively affects your credibility on all other subjects.

We Need A Lasting Solution to The Lies Told by Big Oil and API

A Response to Jack Gerard and the American Petroleum Institute

Originally published June 15, 2018

On June 14, 2018, "my very good friend," the president and CEO of the American Petroleum Institute (API), sent me a letter headlined "We Need a Lasting Solution to the Broken Renewable Fuel Standard."

I say "my very good friend" sent me a letter because the email was addressed to me specifically and the letter itself starts with "Dear Marc."

The first paragraph reads, "Without reform, the Renewable Fuel Standard (RFS) is headed down a path that could limit consumer choices, raise their fuel costs and could even damage their vehicles."

I thought to myself, "Hallelujah, Jack Gerard and I finally agree on something!"

Not only was it Flag Day, but a special "ENERGY INDEPENDENCE FLAG DAY!!!"

If the American flag outside my house wasn't already flying high and proud, I would have run out and run it up the flagpole.

We need to reform the RFS, to extend consumer choice, lower our fuel costs, and stop using the poison gasoline shoved down our throats for more than 100 years. Let me say it again: "HALLELUJAH!"

But then I read the rest of what he wrote:

> Based on decade-old fuel consumption projections that have missed the mark, the RFS forces increasing

> amounts of ethanol into the fuel supply each year, regardless of market realities...
>
> Unless ethanol volume requirements are adjusted to reflect actual fuel consumption trends, we could end up breaching the blend wall...
>
> The problem: Almost three out of four of vehicles on the road today are not manufacturer-approved to use higher ethanol blends like E15 (15 percent ethanol fuel) ...
>
> Extensive testing by the Coordinating Research Council (CRC)—the gold standard in vehicle research—has determined E15 could damage engines and fuel systems...
>
> When the RFS mandates were developed more than a decade ago, the mandated use of corn-based ethanol seemed to some like a viable option to lower fuel costs and imports while reducing emissions. Now that the United States is the world's largest producer and refiner of oil and natural gas, we're in a new era. We've transitioned from a net importer of refined petroleum products to a net exporter, and U.S. carbon emissions have reached 25-year lows due primarily to greater use of clean natural gas.

Oy Vey!

It dawned on me that this wasn't a personal message from one friend to another, it was a form letter that probably went out to API's entire list of people, regardless of their names.

Jack was just trying to butter me up before launching into the same old despicable ignorant lies the oil industry and API have been telling us for decade after decade.

The RFS doesn't "force increasing amounts of ethanol into the fuel supply each year (regardless of market realities)." The Renewable Fuel Standard (RFS) called for the use of a biofuel to be added to gasoline for non-flex fuel vehicles of about 10 percent. Ethanol didn't have to be the biofuel; it just happened to be the best, the safest, and the least expensive to use. Between the time that the RFS was enacted and today, the mandated percentage has not changed. Therefore, the words "force increasing amounts of ethanol into the fuel supply each year" is a lie. And since ethanol did not have to be the biofuel used, it's an exaggerated lie. And since market realities could have impacted the use of 10% ethanol in the primary internal combustion engine fuel supply, it's an exaggerated lie with a rotten cherry on top.

Jack is worried about "breaching the blend wall," but the blend wall is a scam. It doesn't exist: It's a scam wrapped in deceit, inside a fraud—created and perpetuated by the apologists who work for the petroleum oil industry and API.

Every single gasoline-powered passenger vehicle on the road today can safely and reliably use E15, E20, E25, E30, E35, E40, E45 or E50 or higher. The only engine and fuel system issue that might occur is the Check-Engine light might illuminate. Or, the ethanol does such a good job at cleaning the gunk-goo-crap off the engine walls (formed from years of gasoline use) that a filter might have to be changed. So, if all gasoline-powered passenger vehicles can use blends between E15 and E50 (or higher), then the pretend E10 blend wall isn't a figment of someone's imagination; it's a delusional nightmare propagated to keep us addicted to the poison known as gasoline.

(See "The Figurative Ethanol Blend Wall is a Fictional Ethanol Blend Wall" in this book.)

Jack claims the "extensive testing" done by the Coordinating Research Council (CRC) is "the gold standard in vehicle research." It's fool's gold. He says the CRC determined E15 could damage engines and fuel systems. You know what can and does cause damage to every single gasoline-powered engine? Gasoline. Making a broad statement like "E15 could cause damage..." is as pointless as saying that "If you go outside without an umbrella and it rains, you might get wet." On the other hand, there is no equivocating over my statement that every single gasoline-powered engine has been damaged by gasoline; it's a fact!

(See: "Every Spark-Ignited Internal Combustion Engine Ever Produced Has Been Damaged By Gasoline.")

https://www.theautochannel.com/news/2016/06/12/248417-every-spark-ignited-internal-combustion-engine-ever-produced-has-been.html)

I guess no one ever told Jack the extensive testing of ethanol by true 24kt gold labs and research teams not connected to the oil industry have proven again, and again, and again that ethanol is beneficial to engines and does not harm engines and fuel systems (certainly not to any more than the damage caused by gasoline or petroleum-diesel in diesel engines). I know this for a fact because the reports from these organizations tell me so and the oil industry itself has said so: See: "The Hypocrisy of Big Oil" in this book.

Jack writes about why the RFS mandates were developed more than ten years ago and suggests two extraordinarily erroneous notions: First, that the biofuel mandates aren't needed because the U.S. is the world's largest producer and refiner of oil and natural gas; and second, U.S. carbon emissions have reached 25-year lows due primarily to greater use of clean natural gas.

What good does it do us being the world's largest producer and refiner of oil and natural gas? Gasoline prices are still high, we have to import a sizable amount of foreign oil, and we embroil ourselves in the affairs of other countries because of their petroleum oil resources. The only thing that can really be said for our production and refining capabilities is that we may be the world's largest provider of poison to the world. "Hey world, looking for a little more respiratory illnesses, autism, and wildlife disasters? Look no further than the good ol' U.S. of A." This distinction is nothing to brag about.

Claiming U.S. carbon emissions have reached a 25-year low due to greater use of clean natural gas is a joke. The overwhelming and largest contributor to carbon emissions is the burning of fossil fuels (abiotic fuels) that include gasoline, petroleum diesel, natural gas, and coal. While it is true that natural gas produces less carbon than coal and the switch from using natural gas to generate electricity is helpful, the worst cause of air pollution in cities like Los Angeles was/is the result of the transportation use of gasoline and diesel. The reduction of harmful vehicle emissions is not due to any increased use of natural gas, but to the use of catalytic converters, improved MPG efficiency, and ethanol blended with gasoline.

Natural gas, in the form of compressed natural gas (CNG) could have played a larger role in reducing harmful vehicle emissions, but API's denigration of CNG as a vehicle fuel was so successful, engine conversions from gasoline to CNG was (for all intents and purposes) made illegal and uneconomical. All regular production of CNG-powered passenger vehicles for the United States ceased several years ago.

In the meantime, the number of vehicles on the road using gasoline or petro-diesel continues to increase, meaning that more and more harmful emissions are being produced. To continue to reduce emissions requires significantly lowering emissions from vehicles currently on the road within a sensible time frame. This can only happen by increasing the percentage of ethanol used in the fuel powering those vehicles. The move to E15 and beyond is the only realistic, quick, and workable solution for the next several decades, at least.

Consequently, this puts Jack and me back to agreeing on something. A change in the RFS is needed. The question is, should the change be to pour more petroleum poison down our throats and into the air, which is what Jack and the API want, or do we make a change that gives us what any sane person would want—a healthier, cleaner, and safer environment?

UK's Department for Transport Uses Boogeyman Allusions to Sidetrack E10 Adoption

Originally published September 2, 2018

Here's a good example of how bad information about ethanol gets passed along and how it propagates insidiously bad information.

Henny Hemmes, The Auto Channel's Senior European Editor, tipped me off to a story on the website for Auto Express magazine. Auto Express is a well-known British publication about the same age as The Auto Channel. The article is titled "New E10 Petrol Plans: 800,000 Cars Would Be Incompatible." (*https://www.autoexpress.co.uk/car-news/104126/new-e10-petrol-plans-800000-cars-would-be-incompatible*)

Petrol is the British word for gasoline. Henny Hemmes is one of the most successful Dutch race car drivers from the seventies to the present. She was the first female driver to win a Dutch FIA Championship, and has been a test driver and automotive writer for more than 30 years.

The Auto Express story talks about the UK's plans to raise standard gasoline from E5 to E10, and how the increase from about 5% ethanol-gasoline blend to about 10% ethanol-gasoline blend would be incompatible with around 800,000 cars currently on the road throughout the UK—according to the UK government.

The Auto Express article lists five cars they claimed are incompatible with E10. These examples from what the story calls "fresh research" from the RAC Foundation. The RAC Foundation is part of Royal Automobile Club Motoring Services, equivalent to America's American Automobile Association (AAA). The five examples are:

Volkswagen Golf
MG MGB
Mazda MX-5
Nissan Micra
Morris Minor

The article doesn't give the model years for these "incompatible" vehicles, so I assume it's a blanket statement of incompatibility.

Well, the list made me laugh. Three of the vehicles are sold in America and/or Brazil: the Golf, the MX-5 (also known as the Miata), and the Micra. In the U.S., our standard gasoline is E10. In Brazil, their standard gasoline is E27. If the Golf and the MX-5 run on E10 in America, and all three run on E27 in Brazil, why can't they run on E10 in the UK?

Answer: They're perfectly capable of running on E10 in England, Scotland, Wales, and Northern Ireland. What the article should say, but doesn't, is that it's not about the vehicles being "unable" to run on E10, but one in which the manufacturers have chosen to limit their warranty liability by claiming that no fuel above E5 should be used. This is a far cry from vehicles being able to run and not being able to run.

The other two vehicles are older vehicles, classics of a sort. Morris Minors were built between 1948 and 1971, and MG manufactured MGBs from 1962 to about 1980. From the 1920s to the 1970s, at least two gasoline companies in Great Britain marketed ethanol-gasoline blends of E10 to E30. One of the gasoline companies was owned by Standard Oil (Esso), the other by Cities Service. These were and are major oil companies. Consequently, both the Morris Minor and the MG MGB were manufactured to run on "power alcohol" fuels (power alcohol is what they call ethanol-gasoline blends).

I tried doing an Internet search for literature from those early days to see if the respective manufacturers restricted their vehicles from using power alcohol fuels but couldn't find any. (If any reader has such documentation, please send me a copy). The big question, is: if the Minor and the MGB were able to run on E10 to E30 when they were built, why can't they run on E10 now?

There are very few Morris Minors and MGBs in America today, but my guess is that many are running on our standard E10 fuel. I found an ad selling an original 1967 Morris Minor woody wagon in the Los Angeles area and spoke with the car's owner. The right-hand drive car looks to be in very good condition and was imported from England 12 years ago. The engine has original parts. I asked the gentleman what type of fuel the car uses. He said, "Regular gas." I asked if he meant regular E10 gasoline, and he said, "Yes, the regular gas you find anywhere." If a 1967 Morris Minor is able to run on E10 in Los Angeles and gets about 40 miles to the gallon (according to the owner), why can't it run on E10 in Manchester or Liverpool or Cornwall or Edinburgh?

I couldn't find any Morris Minors for sale in Brazil, but a quick search netted two MG MGBs for sale in Brazil. One is a 1973, the other was a 1974. The regular fuel in Brazil is E27. If an MGB can run on E27 in Sao Paulo and Bahia, why is it unable to run on E10 in London, Bath or Glasgow?

The Auto Express article states, "The higher bioethanol content of E10 can dislodge deposits in older engines and fuel systems, causing blockages. It can also cause some seals, gaskets, metals and plastics to corrode in unsuitable vehicles."

Deposits? Yes. The built-up gunk that forms from burned gasoline! The deposits don't come from ethanol, because ethanol burns very cleanly. They come from gasoline—petrol. In other words, the ethanol acts like a cleaner and gets rid of deposits caused by gasoline. Why is that a bad thing?

It is true that if someone used ethanol-free gasoline in their vehicle for many years, and then suddenly introduced a high blend level of ethanol-gasoline to their car's engine, some of those cruddy, gooey deposits could come loose and gum-up the works. And at that point some maintenance and new parts might be necessary. But if that's the case, then the vehicle would require the same work at some point, even if they continued using ethanol-free gasoline, which has been the case for the entire history of the automobile. Engine repairmen and facilities are not new; they didn't suddenly crop-up when the Renewable Fuel Standard was enacted.

If the vehicles have been using E5 for some time, there's a likelihood that the small amount of ethanol in the E5 petrol was already slowly cleaning the engine. By making the small move from E5 to E10, there should be no drastic avalanche of gunk to create a blockage. Instead, the car owner is getting what amounts to a free engine cleaning and crud prevention service.

When Brazil was contemplating the move from mandated E25 to E27, the naysayers came out with the same fear-mongering rhetoric about the jump from 25% ethanol to 27% ethanol would somehow be the tipping point at which automobile engines might disintegrate. Tests found nothing to conclude that E27 would be any different than E25—or for that matter, from E20, E15, or E10.

The accusation also says ethanol is the sole or primary cause of corrosion to some seals, gaskets, metals and plastics. Which seals, gaskets, metals and plastics? In my review of compatibility websites that compare different substances and their compatibility with various materials, it's glaringly obvious that ethanol is far more compatible with materials likely to be used in automobiles than either gasoline or the aromatics often added to gasoline as a replacement for ethanol as an oxygenate (such as toluene, benzene, and Xylene). If you're keen to see the comparisons for yourself, check out these web resources:

https://www.customadvanced.com/printable-chemical-resistance-chart.html

http://www.graco.com/content/dam/graco/ipd/literature/misc/chemical-compatibility-guide/Graco_ChemCompGuideEN-B.pdf

https://www.calpaclab.com/aluminum-chemical-compatibility-chart/

http://www.plasticsintl.com/plastics_chemical_resistence_chart.html

https://www.calpaclab.com/chemical-compatibility-charts/

I wrote an email to Hugo Griffiths, author of the *Auto Express* story (and copied a couple of the editors at the same publication); and in my usual charming bedside manner, I told him that his story was nonsense. Of course, after I sent the email, I was immediately remorseful...not because of my harsh condemnation, but because I should have used the word "rubbish" instead of "nonsense." Rubbish seems to be a better British way to say nonsense according to all the wonderful British TV shows I binge-watch.

In any event, I received back a short but very nice and polite email from Hugo. He wrote,

> "Hi Marc - Thanks for getting in touch.
>
> "I appreciate your points, and this is clearly an issue on which opinions are divided.
>
> "It's not just the RAC Foundation that considers a large number of cars are incompatible with E10, though. The UK Government estimates there are around million cars on the road 'for which the manufacturer has not approved the vehicle for use with E10'."

Hugo then included a website link to a recent report issued by the UK Department for Transport. That report can be found here: *https://assets.publishing.service.gov.uk/government/uploads/system/uploads/attachment_data/file/727547/e10-petrol-consumer-protection-fuel-pump-labelling.pdf*

I read the report.

The report doesn't state that ethanol is damaging, or how it can cause damage, or which vehicles it is incompatible with. It makes reference to concerns voiced about ethanol. The report alludes to an ethanol-boogeyman in the same way adults use an allegorical boogeyman to frighten children into good behavior.

This was my reply to Hugo.

> "I understand there is great division of thought on this issue. Oil industry money has created this division and kept it alive.
>
> "However, the facts are that no gasoline-powered vehicles are damaged because of ethanol-gasoline blends, not in the 60 years that ethanol-gasoline blends were extensively marketed in Great Britain between

the 1920s to the 1970s, not in France or Germany or Sweden or any other European country, not in Brazil for the past 50+ years, and not even in Brazil in the years since they've mandated E25 and then E27.

"The RAC Foundation, like the AAA, has probably received significant remuneration from the oil industry and so they are happy to go along with the rubbish.

"Ethanol does not damage engines, it cleans engines. This isn't to say that internal combustion engines that run on ethanol-gasoline blends don't experience deterioration over time owing to wear and tear and the general nature of the fact that there are moving parts, and that fuel is exploding and combusting within its walls. But to blame ethanol for this, as if gasoline (petrol) is not the primary cause of the problems, is ludicrous.

"Ethanol has been blamed for corrosion to rubber, plastic and aluminum; however, it is gasoline that is corrosive to those substances. It is gasoline that required the invention of a new rubber (Viton) many years ago, to stop corrosion. Not ethanol. And any metallic substance that can experience rust problems because of ethanol will experience the same results from gasoline.

"I read through the UK Department for Transport document via the link you sent. It is incorrect and it ignores the history of Great Britain's use of ethanol-gasoline blends from Cleveland Discol and Cities Service for many decades. The Department for Transport's document should have started off by reminding any reader that ethanol-gasoline blends have a long and successful history of use in Great Britain. And the report should conclude with a statement like: 'E10 to E30 blends provided us with a cleaner, safer, more powerful, and less expensive fuel for more than half a century. There's no reason to think it won't do the same now.'

> "I'll be sending an email to the Department of Transport about this..."

The *Auto Express* article is a good example of how bad information gets passed along about ethanol, and how it can propagate more insidiously bad information. At the bottom of the webpage hosting the AE story are comments from readers. Other than my comments about the story, there are several comments from readers who use the article to tee-off on ethanol with negative comments about ethanol, such as:

> There are hundreds of thousands of classic cars from all eras that would suffer from this fuel; you only have to read the comments on any number of Facebook group pages to read the havoc it's causing in the USA to know it'll do huge harm to classic cars.

Although there does seem to be "any number of Facebook group pages" that claim ethanol is causing problems, the havoc is caused not by ethanol itself, but by the routine empty, ignorant lies and exaggerations about the ethanol-boogeyman invented and promulgated by the petroleum oil industry long ago.

The final result is that we're left with an *Auto Express* story that only alludes to problems caused by ethanol, and instead of specific examples to prove the problems exist, they rely on "fresh research" provided by RAC Foundation that offers no specific examples of how or why damage from ethanol is caused.

Auto Express then backs up all the rumors of ethanol damage with a UK government report that simply refers to unspecified concerns about ethanol causing problems. And we get to read a vacuous comment by an unknown reader about some Facebook pages in the U.S. that claim ethanol is harmful to classic cars.

Meanwhile, because of this chain of bad information, you know that there will be people running around blabbering that hundreds of thousands of otherwise perfectly good cars will just not operate on any gasoline that contains 10% or more ethanol.

The *Auto Express* story is nonsense. The information supplied by RAC Foundation is rubbish. The report by the UK Department for Transport is worthless. If you think my take on all this is wrong then read my editorial "The Hypocrisy of Big Oil and API" contained within this book, and try to explain why, for almost 60 years, British petrol companies and automobile publications stated emphatically that ethanol-gasoline blends were cleaner, safer, more powerful, produced better mileage, and were less expensive than ethanol-free gasoline.

Notes

SEE ALSO:

- "AAA Blunder on Ethanol Sets Off Firestorm of Criticism," in this book
- "Why Does AAA and Big Oil Feel They Must Lie to America?" in this book
- "AAA Gets It Wrong on Ethanol Again," in this book
- "Every Spark-Ignited Internal Combustion Engine Ever Produced Has Been Damaged by Gasoline" *(https://www.theautochannel.com/news/2016/06/12/248417-every-spark-ignited-internal-combustion-engine-ever-produced-has-been.html)*

Truth About Ethanol

The Book Within This Book

Review and Reply to Robert Bryce's *Gusher of Lies*, A Book Aptly Named for What It Is

Originally published June 10, 2013

Part 1

In 2008, Robert Bryce wrote *Gusher of Lies: The Dangerous Delusions of Energy Independence*. The book attacks America's quest for energy independence and Bryce attempts to discredit "green" alternative solutions as the answers to our dependence on oil—foreign and domestic.

Judging from all the complimentary and congratulatory testimonials in the first few pages of the book, as well as those used online for promotion, it's obvious that Mr. Bryce has friends in high places who know how to write endorsements.

I did find it amusing that the very first testimonial at the beginning of the book was written by "R-Squared Energy Blog." R-Squared Energy is Robert Rapier, who is identified, as is Robert Bryce, as an "energy expert."*

Robert (Rapier) has used some of my writing about CNG on his blog for the past few years to explain why there is such a high cost to convert existing gasoline-powered engines to CNG in the United States. I've gotten to know Robert Rapier better in the past year because The Auto Channel has been featuring some of his videos on our broadcast TV network. In my personal conversations with Robert, he's told me quite a bit about his strong support for ethanol and how he feels his advocacy has been misunderstood by some because he worked for several years in the oil industry.

When he sent me his latest video on ethanol, he included the following comment: "If Midwestern governors would follow my suggestions, the entire Midwest could be energy independent." The suggestions he makes in the video pertain to ethanol, and the energy independence he refers to is become independent from oil (gasoline). He makes it clear that he supports the Renewable Fuel Standard (RFS) and the use of American farmers to grow crops from which ethanol can be produced.

Robert Rapier's position, as you will learn, is patently different than Robert Bryce's abhorrence to ethanol and other alternative fuel solutions as set forth in *Gusher of Lies*. You can therefore understand my amusement over the use of the R-Squared Energy Blog testimonial in the book. Watch Robert Rapier's video here: *https://www.youtube.com/watch?v=fIOkbRS7yGs*

Why I Wrote This Book Review

I've come across Robert Bryce's name over the years I've researched aspects of the oil industry and alternative fuels. I read a couple of his articles and heard him interviewed on radio. It wasn't until a radio interview with Mark Mills, a colleague of Mr. Bryce from the Manhattan Institute of Policy, that I took notice of what Manhattan Institute and its "experts" were saying about oil and alternative fuels.

I've had considerable disagreement with some of the folks at the CATO Institute, another organization that professes right wing-ish, conservative, open market, libertarian perspectives, like those generally espoused by Manhattan Institute. As I've chronicled on The Auto Channel website, I've found CATO's Jerry Taylor and Richard Rahn support an energy policy that is as far from being open market and libertarian as Earth is from the Andromeda galaxy. I wanted to see how Manhattan Institute's policies compare with CATO's.

I was unaware of *Gusher of Lies* until I received a reply from Mr. Bryce to this email I sent him.

> I'm wondering if you've ever had the chance to actually research what you say about ethanol and E15 or if you are merely repeating the information points sent to you by one of the oil industry public relations groups. I realize that it might appear as if I am simply being confrontational, but I'm not trying to be. I'm seriously asking how you have come to the opinions that you have.
>
> In my own experience with covering the automotive world for nearly 30 years I didn't start with any agenda regarding fuels. But in my personal research of ethanol, which includes purchasing a non-flex fuel vehicle so that I could experiment with various blend levels of ethanol-gasoline, I can tell you that the problems you write about are false.
>
> In addition, your apparent recitation of issues related to using corn for the production of ethanol lacks basic business acumen. Perhaps business has not been a focus for you; therefore, the principles of production have not been apparent. But if you're going to attack the use of corn you should make an attempt at understanding the business side.

If you've ever read any of my editorials or news articles you might recognize some recurring themes. For example, I put great stock in personal hands-on experience. Studying someone else's work is fine, but there comes a time when you have to get out there and get your hands dirty. You've got to kick some tires, wipe sweat from your brow, and spill some of your own blood.

I also have this thing about understanding business, but not the kind of understanding that comes from just reading textbooks and listening to lectures at college. I'm talking about hands-on production and management experience in real world situations where your personal efforts and decisions are the difference between whether you and your co-workers will have the carfare to get home that night and back to work the next morning. So, when I read certain accounts and hear people talk about the business aspect of some issue; if I don't detect any evidence that they have experience of "working in the field," I become very suspect and critical of their position.

In the brief research I did on Robert Bryce before writing my email to him, I didn't come across any information that indicated he had any experience getting his hands dirty as a "field tester" or in day-to-day product production and distribution. So, I asked Mr. Bryce the questions I did, in the manner I did.

Happily, I received a rather quick response to my email; sometimes I never get any responses to my emails. Robert Bryce wrote:

> I am fundamentally opposed to use of food to make motor fuel. Further, my opposition to biofuels is based on simple math and basic physics. The power density of biofuels is too low to have it make any sense from a public policy standpoint. I have never been given any "information points" by PR firms on ethanol. My work on ethanol is my own."
>
> ... Further, I suggest you read my third book, *Gusher of Lies,* in which about 20 percent is devoted to the ethanol scam.

So, I read his third book, *Gusher of Lies,* I plunked my ten bucks down on Amazon for the Kindle version, and this project got under way.

There's a lot that Robert Bryce and I seem to share. His personal bio indicates that he leans towards libertarianism; I consider myself to be a conservative libertarian. He doesn't believe in the man-made global warming alarms, and neither do I. And he "believes wholeheartedly in the U.S. Constitution and the Bill of Rights," just like me.

But then we take divergent paths. In the "Author's Note" section of the book after the "Praise Pages," Bryce talks about his change in perspective from thinking there should be major government involvement in energy to now believing that the government should "quit meddling in the energy market."

As a conservative libertarian, I would normally eschew all government involvement in any enterprise. Calling off government involvement and its built-in political attachments after all the government and political tutelage that made the petroleum oil industry so dominant and petroleum oil-based fuels our primary engine fuels, is terribly naïve or horribly disingenuous. Gasoline didn't become so prominent because it was the best fuel; it became the primary fuel because of intentional (or unintentional-that-became-intentional) government intrusion to sideline alcohol with onerous taxation, followed by protecting the oil industry from alcohol with fourteen years of Prohibition. These fourteen years were perhaps the most critical years in the development and growth of the automobile industry, which became and remains the largest, most important industry in the world.

It is comically dishonest for Bryce, or anyone, arguing the anti-alternative fuel position, to declare the government must not help domestically produced alternative fuel to compete in the government-created, lopsided market that has favored the oil industry for more than 100 years. It's also unrealistic. Does Bryce really expect politicians, who have watered at Big Oil's money trough for so many years, will willingly go cold turkey? It's unlikely.

There's an often-used analogy for why certain government controls and intervention are necessary, and it's an analogy that even the most strident libertarians have trouble arguing with. The analogy is the necessity for traffic lights. It goes like this. If it weren't for the orderly control of any thoroughfare, there would be mass confusion; and the free flow of traffic would come to a grinding halt because of accidents blocking the road. Only by applying certain controls can we can ensure a generally freer traveling experience.

America can certainly benefit from some governmental controls and oversights in the energy arena.

For the discussion of gasoline versus alternative fuels, I like to use a different analogy. This analogy is horse racing. In order to make a horse race fair, there are strict regulations concerning the breed of horse running, the weight the horses must carry, and the implements and substances that can be used to affect the horses' performances.

The oil and gasoline industry received enormous benefits from government intervention over the past 150 years. For example, there was the $2-plus per gallon tax imposed on alcohol production during the Civil War, which was never rescinded until more than forty years later. There was the imposition of Prohibition that made all alcohol production illegal for fourteen years. And there have been the billions and billions of dollars in continuous subsidies, allotments, allowances, and tax incentives, that continue to this very day, even though the oil industry is hugely profitable. The Big Oil lobby likes to complain about and overdramatize the subsidies given to various alternative energy and fuel sources; then they deny or ignore the fact that they are and always have been the biggest recipients of government subsidies.

Another incredible financial break the oil industry gets from the American people and our government is free security service from the U.S. military. We have spent trillions of dollars to protect the oil industry.

For alternative fuels to enter the horse race and gain consumer acceptance, great adjustments must be made to make the race fair, or just fairer. I've come across many people who consciously or unconsciously support Big Oil by repeating what they've heard some oil industry wag say about subsidies to ethanol and CNG: "The government should just let the free market take over and remove the alternate fuel subsidies. Then we'll see how competitive ethanol (or CNG) is."

Paradoxically, they are correct; the government should remove all subsidies, but it should be the other way around. Remove all the petroleum-oil and gasoline subsidies, make the oil industry foot the bill for our military, and then let's see how uncompetitive gasoline and diesel really are. If all oil industry related subsidies were removed and they had to pay for the military protection they get, gasoline in America could cost as much as $15 per gallon.

Some of the oil industry apologists, like Jerry Taylor, claim America's service stations are primarily selling gasoline and diesel because that's "what the public wants." Taylor likes to say that most filling stations are now independently owned and operated, and that the oil companies have no control over what they sell. Taylor cites this as an example of the free and open market that we have. This is entirely untrue.

Yes, most filling stations are independently owned and operated, but most are under contractual obligations to their gasoline providers to not sell any type of fuel product that the supplier doesn't produce or distribute. If a particular oil distributor doesn't offer an ethanol blend, such as E85 or E15, then in most instances, the independent owner-operator cannot accept another distributor's ethanol product without paying a burdensome fee to do so. That is not a free and open marketplace.

Intervention to ensure fair competition is necessary. As a libertarian, I agree it's unfortunate that intervention has to come from the government, but who else is capable of applying and enforcing competitive regulations designed to be fair?

Bryce also thanks various oil-related Arab people for a trip he took to Saudi Arabia and the United Arab Emirates, which helped him "...understand the Arab world and the on-going trend of globalization..." of the energy sector.

While I don't want to make too much of the "help" that Bryce might have received from the Arabs, I also don't want to make too little of it.

In my email to Bryce, I asked if he is "...merely repeating the information points sent to you by one of the oil industry public relations groups." And Bryce emphatically replied, "I have never been given any information points by PR firms on ethanol." Well, let me draw this connection, Saudi Arabia's public relations firm in the United States is Edelman Public Relations. Edelman is also the PR firm for the American Petroleum Institute (API), which is the primary mouthpiece for the oil industry in our country. It was just weeks before I read *Gusher of Lies*, that I received an email from Edelman PR setting forth "informational points" about ethanol:

> As travelers plan to journey an average of 690 miles (on the Memorial Day holiday weekend), those filling up on high ethanol-gasoline blends, like E85 (up to 85% ethanol), or mid-level ethanol blends, such as E15 (15% ethanol), will notice that they'll need to stop at the pump more often. Ethanol contains 33% less energy than traditional gasoline (and is more expensive when adjusted for this), meaning consumers will get fewer miles per gallon. Unfortunately, 95% of consumers surveyed by AAA were not familiar with E15, meaning that travelers could unintentionally fill up on the fuel and end up spending more on gas and potentially wrecking their engines.
>
> More and more ethanol is being forced into U.S. gasoline every year as a result of the Renewable Fuel Standard (RFS), a policy that is diverting 40% of all U.S. corn to use in fuel instead of food. The RFS is also increasing prices for meat, poultry and dairy products, and will make Memorial Day cookouts more expensive. Since the RFS was expanded in 2007, prices for cereal and bakery products have risen 77%; prices for meat, poultry, fish and eggs have increased 78%; and prices for vegetable oil and fats (e.g., butter) are up by 444%.

The email from Edelman PR offered to arrange for me to speak with representatives from various Big Oil advocacy groups about these points.

I challenged Edelman PR on these points in an editorial that The Auto Channel published on May 23, 2013; "Why Does AAA and Big Oil Feel They Must Lie to America?" See earlier in this book.

Bryce's anti-ethanol information points are essentially the same anti-ethanol information points Edelman PR and other oil lobbying organizations have used for many years.

These anti-ethanol points are the basic anti-ethanol points used in *Gusher of Lies*, Yes, Bryce does add subsidy issues and the issue of "power density." But these issues are not new or unique anti-ethanol points discovered by Bryce; they just were not included in that particular Edelman message. The ethanol subsidy issue and the power density issue are typically included in oil industry-sponsored anti-ethanol messages and reports created by others, such as David Pimentel, Tim Searchinger, and Jerry Taylor.

Is it coincidence that Bryce uses the same informational points in his arguments against ethanol, or is it a concerted effort by Big Oil to make all anti-ethanol arguments consistent?

I acknowledge that my pro-ethanol arguments follow a consistent pattern; a pattern that is not just consistent in how I respond, but in how other ethanol proponents make their case for ethanol. The single most important reason I purchased a dedicated CNG-powered vehicle and then a non-flex fuel gasoline-powered vehicle was so I could experience firsthand the benefits or problems, if any, of using CNG or ethanol.

I find it crucially important to have personal hands-on experience. I was never comfortable just repeating what someone else said about an alternative energy solution. I felt I must live it. I've done it and continue doing it with various high-level ethanol-gasoline blends in a wide variety of vehicles, and in particular, in my own dedicated, non-flex-fuel vehicle over a long term. I wondered if Bryce would respond to my email by saying he arrived at his conclusions because he tried using various ethanol-gasoline blends in his personal vehicle(s) and that he "wrecked" the engines, just as the oil industry warned he would. But he offered no insight to any personal experimentation one way or the other in his email response, and none in the book.

The email I received from Edelman PR was not sent to me exclusively, nor is it the first time they and others in the oil lobby have sent such messages to journalists and media personalities. It is a normal part of their propaganda effort. As Robert Rapier confirmed to me in a phone conversation, he used to receive informational messages like this from Big Oil; and if he would write an article containing the information and get it published, there would be some financial remuneration. I believe there is a great likelihood Bryce received anti-ethanol information points from the oil lobby, even though he says he hasn't.

Moving Forward

Bryce's next couple of chapters treat us to a sermon called "Why We Think We Want Energy Independence."

He could have used, "Why Struggle with Freedom When Slavery Has so Much to Offer?"

Bryce's discourse reminded me of two things: First, the slave owner who says to his slaves, "What's the big deal about freedom? If I free you, you'll just have to go out and struggle to find a job, then a house, then clothes to wear, and food to eat. If you get sick, you have to pay the doctor's bills. Why not stay here and I'll take care of those things for you? What's a little whipping every so often..."

The second thing concerns those Russians who were unhappy when the Soviet Union collapsed and were faced with the prospect of independence and self-reliance. Between communism and life under the Czars, Russians had never experienced freedom—I'm talking hundreds and hundreds of years. The newly free Russians were lost; they had no direction; they didn't know how to adjust to a life where they could make a decision. Those in the black market had no problem with entrepreneurship, but they were in the great minority.

Bryce says, "Americans are fixated" on the issue. There's truth to that—but we aren't fixated enough on the issue. He makes the picture look cheerful by telling us that we're paying less for gasoline now than we did in 1913, based on adjusted numbers for inflation, improvements in lifestyle, Gross Domestic Production figures and other arcane comparisons. He attempts to alleviate our concerns when he writes that oil isn't the only foreign product we depend on. And that's true.

I think it's distressing to know or feel that we are incapable of producing steel nuts and bolts to build a new San Francisco Bay Area Bridge, and that we need clothing factories in Bangladesh to manufacture $100 jeans for our malls to sell, and that we need China to build solar panels that don't work effectively. It's sad, but I'm still not getting the connection between our production inabilities and why it's okay to be reliant on foreign sources for our engine fuels.

Bryce also reminds us in *Gusher of Lies* that we're not the only country enslaved to foreign oil; those two great allies in World War II, the Germans and the Japanese, are also locked in oil chains.

Wait, did I just write, "the Germans and the Japanese?" Oh, my gosh, weren't they our enemies in WWII? Didn't we win the war and they lost it? Didn't we spend billions of dollars to rebuild their societies? Didn't we get pulled into that war specifically because Japan went on a campaign of aggression to conquer oil producing countries in the Far East and we objected to it? Did we do all that just so we can bring ourselves down to the level of defeated nations?

Hey, everybody it's okay that we're enslaved to oil that comes from people who hate us because it's the cool thing to do! Slavery is easy; it's simple; it doesn't require troublesome decision making. If only there weren't those occasional lashings....

I'm telepathically hearing you ask, "What occasional lashings are you talking about?"

I'm talking about getting American service men and women blown up on the other side of the world, for starters; having a bomb blow up at the finish line of a marathon; having a fully loaded jet fly into one of our office buildings; having an American doctor overseas trying to help poor people get kidnapped and killed; owing questionably-friendly nations trillions of dollars; you know—those little everyday nuisances.

Review and Reply to Robert Bryce's *Gusher of Lies*, A Book Aptly Named for What It Is

Originally published June 10, 2013

Part 2

How Does Belief in The Constitution Allow for Foreign Oil Dependency?

I'm still having trouble wrapping my mind around Bryce saying he believes in the U.S. Constitution and Bill of Rights; that he is very thankful to the Saudis and Emiratis; and that he philosophically moved from believing there should be major government involvement in energy to believing that the government should "quit meddling in the energy market."

I got the picture when he took a fabulous trip to Saudi Arabia and UAE. I'm sure he was very impressed by what he saw. I'm sure the accomplishments of what trillions of dollars in oil revenue can do for desert nations that had no money less than 70 years ago is absolutely breathtaking. Good for them, but what does that have to do with us?

When I travel and see fabulous things, I am inspired to want all of that for America. Don't get me wrong, we have a tremendous amount of fabulous things already, but my love for the Constitution and the Bill of Rights, and other things American, doesn't make me want to sacrifice what we have to make life better for people who are diametrically opposite to us in so many ways. If they can be as successful as us without it infringing on our ability to be successful, I'm okay with that. I'm not okay with turning America into a second-rate nation to feed their gluttonous appetites. The images of modern Dubai are indeed wondrous; wouldn't it be great if we could make Detroit and the South Bronx look that good?

While writing this section I found a video on YouTube I thought brought great clarity on this point. Finding it when I did was almost like an epiphany. Watch it here:

https://www.youtube.com/watch?v=oOdaCpkEqWE

If we could find the national determination to end our dependence on foreign oil, we would keep hundreds of billions of dollars in America every year. This would be great for our economy. If we could find the grit to do that, then we might also find the resolve to make some other things happen, such as requiring that every vehicle sold in America be made in America by American citizens; every refrigerator, every TV, every computer, every dishwasher, every washer and dryer sold in America should be manufactured in America, by American citizens. The manufacturing companies can be foreign-owned, and I hope they would be wildly profitable to encourage more investment. However, all jobs must be filled by American citizens.

What this would do is take the approximately $400 billion per year that we now send out of the country and recycle it several times by buying American goods that make more money available to Americans to buy more American-produced items that are built by Americans, etc., etc.., etc. It would give us an annual stimulus of a few trillion dollars. If the $800 billion Troubled Asset Relief Program (TARP) stimulus program in 2008 was supposed to be good for our economy, just imagine what an annual $3 trillion stimulus would do every year for 5 years, 10 years or 25 years.

Instead of all the time Bryce devotes his book to arcane figures and charts to show us how gasoline is cheaper today than it was 100 years ago, I think energy independence is a far better direction for America. I think that if you are an American libertarian who believes in the United States Constitution and the United States Bill of Rights, that this is the only way to go.

I didn't even figure in how much we would be saving by not using our military to defend the world's oil. Yes, I did say "the world's oil," and not just "American oil." We're even protecting oil for countries that hate us.

It's disappointing that Robert Bryce and other patriotic Constitution-loving Americans like him don't understand our approach and haven't jumped on the bandwagon.

I'm disappointed, but not surprised.

I am not surprised because this goes back to the other "thing" I have, the thing about the need to understand business—to really understand business. I don't think Bryce ever had any real business experience. I don't think that Pimentel had any business experience. I don't think that Taylor or Searchinger or Rahn or David Fridley from Lawrence-Livermore Labs ever had any business experience. They may have studied business for a time in school. Clearly Bryce consulted "business experts" from some universities and/or his Manhattan Institute, and they gave him some graphs and formulas to explain what they read about in school but never put into practice.

How do I think this affects Robert Bryce's business observations? I think it means he relied on someone else's interpretation without their ever engaging in it.

I still haven't included the cost of our military defending the world's oil, and I haven't factored in the incalculable cost of the lost and wounded lives of our military service personnel and American civilians.

Insight into Bryce's Thoughts

In the chapter, "The Emotional Appeal of Energy Independence," Bryce does a pretty good job detailing the history of oil-related terrorism and American military involvement because of our reliance on foreign oil, although he does omit the First World War. I was very glad to see that he didn't ignore this issue or try to mitigate the impact of our involvement the way Jerry Taylor from the CATO Institute did. He didn't shillyshally about why we were "there," wherever the "there" was. We were "there" for the oil.

But he concludes by writing:

> The point is clear: Oil related terror attacks will continue...But even if the U.S. were somehow free

> of imported oil, American consumers would still be buying oil at the global price and would, therefore, not be immune from the price spikes caused by... (terrorist) attacks.
>
> Why? Well, because the price of oil is set globally, American oil traders are not going to sell their domestically produced crude in the U.S. if they can get a substantially higher price (elsewhere)...Thus the U.S. cannot isolate itself from the rest of the global oil market.

In other words, Bryce is telling us to just lie back and continue to accept things the way they've been, because they will only continue.

The Nitty Gritty

While I still have much to say about *Gusher of Lies*, we have now come to the crux of the matter, the bottom line.

What most people don't understand, and what the oil lobby desperately doesn't want you to know, is that regardless of how much more drilling for oil is done in America, regardless of how much fracking goes on, regardless of how many pipelines are built from Canada to the Gulf of Mexico, America will never be free from dependence on foreign oil. Control of our primary engine fuel, the fuel that has been dictating our buying and driving decisions, is in the control of foreign regimes. For as long as it continues, we will never have gasoline prices that are truly cheap; the best we will ever be able to say is, "Well, it's not so bad, remember last year when it was over $X a gallon."

In Bryce's own words, based upon his trip to Dubai and his intentional ignorance of fossil fuel alternatives, we will never be free from oil-sponsored terrorism. We will never have energy independence, ever.

I agree with Bryce that we will never, ever have energy independence; our military will never abandon the Middle East for as long as we use gasoline as our primary engine fuel. The solution is to change the paradigm. The solution is to make a non-fossil fuel our primary fuel, and that fuel is ethanol.

My business partner, Bob Gordon, recently wrote an editorial titled "When All American Drivers Use Non-Exportable Domestic Fuel, MPG Hysteria Will Go Away." *(https://www.theautochannel.com/news/2013/05/20/076908-when-every-passenger-vehicle-america-s-roads-uses-non-exportable.html)*

On this very point, Bob writes, "There are no our American oil companies; there are no American companies pricing our domestic oil and its products at a hometown price." He's right. Ethanol can give every American community a home-town price, like eggs and milk. As Bob points out, Venezuela has a hometown price on its oil, as does Saudi Arabia and Iran on their respective oil. Why not us?

Bryce goes into long discourses on the mechanics of oil trading and how it's a marvelous method by which, he says, we arrive at "a fair price". It costs the Saudis perhaps a dollar or two to lift a barrel of oil out of the ground, and the Saudis think that $75 per barrel is fair. As a consumer I wouldn't call that fair. I'm a for-profit kind of business guy, but I would say that 7,000% profit is a tad excessive, by about 6,700%.

Reading the machinations of oil trading are like listening to someone try to explain and justify derivative investing. Hey, Robert, oil is a commodity, like an apple. The world doesn't need to be betting on the size of the seeds inside the apple; we just want an apple that tastes good.

Bryce says that our fixation on oil dependency is also based on our fear and/or loathing of Islam. He writes that things are changing in the Muslim world: elections are sometimes being held; women are being allowed to attend school; banks are becoming more liberal. Good for them; welcome to the 18th century. Oh wait, this is the 21st century.

Bryce complains about editorial writers, like Thomas Friedman of the *New York Times*, who think we need to change Muslim societies. Bryce uses this to return to the concept that slavery isn't all bad, that we can learn to live with it (the way the Saudis do). I thought Robert was a libertarian. If he is, then know this: Us real libertarians don't want to nation-build at all, whether it's Friedman's vision or Bryce's vision of letting them run amuck. I may not like what I've seen from Islam, but I personally have no interest in who or what they want to worship, as long as they don't chop my head off because I choose not to worship their way.

Energy Independence Is Possible

Now, my dear readers, this is where I will finally address what Robert Bryce refers to as *Gusher of Lies*, and why I sent him that email.

But first, these words:

Bryce spends a fair amount of time discussing "Peak Oil" and "Climate Change." Other than a brief mention of "Electric Vehicles," which is almost not even worth mentioning, he doesn't write about them. But I'm going to throw that into the mix because it has great significance in any discussion of the overall energy picture.

I have no interest in discussing the subjects of "Peak Oil" and "Climate Change." I feel they are irrelevant to energy independence. That may be disappointing to some, but here's why. First, if we had a viable solution(s) to fossil-fuel fuels, it doesn't matter if or when the world runs out of oil or if production capability is maxed-out. Second, I never believed that the world would run out of oil. I didn't believe it in the 1970s when it was very fashionable to worry about it, and I don't believe it now; certainly not after learning about huge fossil fuel deposits discovered off the coast of Brazil, in the Gulf Mexico, in the Mediterranean off the coast of Israel and Lebanon, and so on.

Ironically, the fear of eventual global oil depletion goes back more than 100 years (Bryce references this fact, too). There were many responsible people who were alarmed at various times about the possibility, but the answer was always rather simple: invent new methods to detect and get at the deposits. They did and they did.

Climate change, or man-made global warming, is completely immaterial to how I view the entire energy issue. To begin with, I don't believe in "Al Gore science." I also didn't believe in the man-made global cooling theories that were rampant a generation ago. More importantly, I already support a rapid end to the use of gasoline and diesel. I know that ethanol is a cleaner fuel. I'm doing everything I can to shut off the flow of the "gusher of lies" that emanates not from the ethanol industry, but from Big Oil's propaganda machine.

If man-made global warming is real, fine. As I said, I'm already doing what I can to get everyone to reduce those harmful emissions. If catastrophic man-made global warming is a fraud, it doesn't change my opinion about ending our dependence on oil. Vehicles that run on gasoline and diesel definitely cause pollution—whether the result will be the end of mankind or not. I am a great lover of clean skies and clean water, so lowering the amount of pollution that spews from our internal combustion engines is a worthy goal.

I have always disagreed with combining the issues of energy independence and global warming, and I have counseled global warming believers to do the same. Ending our dependence on foreign oil has sufficient justification without having to rely on something that even the most ardent global warming alarmists are not quite certain about. If, and I believe when, the theory of global warming is finally disproven, it will shut down whatever alternative energy efforts were begun simply because of the alarm. That would be very unfortunate for alternate fuels. The battle for the use of ethanol as a viable alternative to oil should be fought on its own merits; and as you will read, it is fully armed to do so.

Electric vehicles are a wonderful promise; and when the time comes that technology and cost of production make it possible for the car manufacturers to profitably produce them in sufficient quantities, it will be a boon to mankind. However, the technologies are not yet here, and the cost is way too high. So maybe in 20, 30, 40 years, or 80 years electric vehicles will have a great impact on our reliance on any liquid fuel. But for the foreseeable future, they are what my business partner, Bob Gordon, long ago dubbed as "a diversion, not a solution."

Bob has written extensively about electric cars as a diversion sponsored and used by the oil industry as a divide-and-conquer tactic. He believes for as long as the public can be dazzled and bamboozled by the glitz of electric wizardry, our minds will forget about deployment of existing non-fossil fuel alternatives, and I agree with him.

Every time we learn of another electric vehicle company closing its doors, our position on electric vehicles is reinforced.

I'm not done yet. Robert Bryce helps prove the point that the oil industry and its apologists are backing electric vehicles as a divide and conquer scheme. About a year after Bryce finished the final edits on the first edition of *Gusher of Lies,* he added an "Epilogue to The Paperback Edition."

Throughout the original book and the epilogue, Bryce has nothing good to say about any non-fossil fuel or alternative energy solutions, which includes wind energy and solar. Needless to say, he is against all subsidies given to alternative solutions. He dislikes everything and rips them to shreds to the best of his ability. It doesn't matter if the data he relies on is right or wrong; he has nothing good to say. Except that, in the middle of the epilogue, he slips in two positive mentions about electric cars. His first is, "While the continued calls for energy independence and the increased use of corn ethanol are distressing, 2008 also brought some positive news on the energy front. Several automakers announced plans to develop and sell electric cars in the U.S. market by 2010."

A few pages later, he writes, "Meanwhile, numerous battery companies, both in the U.S. and abroad, are continuing to develop better batteries. All of those developments leave me optimistic about our potential to meet the energy challenges that face the U.S. and the rest of the world."

Whether you've read *Gusher of Lies* or not, you'll see Bryce is "Mr. I Hate Subsidies." He never mentioned subsidies to electric vehicle manufacturers in these two quips. Doesn't that seem odd? The entire modern electric car movement is about government subsidies, unproven technologies, insufficient production capabilities, unrealistic purchase prices, everything Bryce rails against when considering non-fossil fuels.

When it comes to electric cars, he is optimistic. All the subsidies the government gave this particular non-fossil fuel alternative are good; he now understands why a government subsidy can encourage innovation and provide a little bit of leveling of the playing field. I guess he was swayed either by GM's promise that they would produce 7,000 Volt electric cars for 2011, or one of the handlers at BigOil-R-Us gave him the secret sign it was okay.

About 12 million new cars and trucks were sold in the U.S. in 2011, so he was optimistic about sales that represented less than 0.7 percent of the new vehicle market. If that made him happy, he would probably be delirious to know a few 2011's are still available, and he can even choose the color.

The Gushing Truth About Alternative Fuels

In 2008, around the time Robert Bryce published *Gusher of Lies,* I wrote another long article published on The Auto Channel website, "No New Gasoline-Powered Vehicles in the U.S. By 2014—Can It Be Done?" earlier in this book.

The article suggested a plan whereby the U.S. would become free from foreign oil in a rather short period of time. I didn't fantasize flipping a switch at midnight and expect a complete and total change by dawn. It called for a gradual, aggressive switch to alternative fuels. It didn't call for the prohibition of all domestic oil production—unlike the 18th Amendment to the Constitution, which stopped all legal distillation of alcohol.

My plan understood the need to keep existing domestic oil and gasoline jobs fully functioning, the need for petroleum-based fuels for older vehicles, as well as for myriad products requiring petroleum-based chemicals. I reviewed the various possibilities, from compressed air to algae to propane and CNG to ethanol, and made my comments about them.

I understood then, and I understand now, that technologies change, improve and evolve. It happened with powered flight, with the telephone, with television, with refrigerators, etc. If in 2008 there had been a national imperative to make an alternative fuel our primary fuel—and if Barack Obama was serious when he gave his August 4th Energy Policy speech—I have no doubt that the technology used today, just five short years later, would have shown considerable improvement over the technology that was used in 2008 to produce whichever of the alternatives was selected.

In 1994, when *TheAutoChannel.com* was in its infancy, we knew the new media, called the Internet, would one day converge with television. That was our vision. To make that vision a reality, we sought technological solutions. I visited a teleconferencing exposition in Anaheim, California and approached one particular video teleconference vendor. I told him what we had in mind and asked if his service could be used by us. The vendor became angry and yelled at me. People in the surrounding exhibit booths stopped what they were doing and turned their attention to the commotion. The vendor shouted, "What are you stupid? It will never be possible to send video over the Internet; the fastest data will ever travel over twisted-pair copper wires is 28k." (This was in the days when 9600 baud modems were considered high-tech and only a relative handful of geeks had 14.4k connectivity.)

I answered the irate exhibitor:

> The fact that you are saying right now that there will never be a way to transmit video over the Internet means that there are people in garages and basements at this very moment figuring out ways to make transmission faster and to compress the data smaller.

A month later we found a company that had a streaming video solution, and three months later we became to first website in the world to deploy streaming video. Less than one year after that, we were streaming live video and broadcasting the first live motor sports event on the Internet.

In *Gusher of Lies*, Robert Bryce never anticipated technological improvement in the production of alternative non-oil-based fuels—except to expressly discount its possibility. He locked himself into a bygone era. He repeats and presents data that was based on information that was out-of-date probably from the day it was created. To make matters worse, some of the information was never correct even when it was created, and the passage of time didn't magically make it accurate.

Some readers will take me to task for these comments, saying I purposely chose an easy target to go after. The book was originally written five long years ago. But I didn't choose the book, Robert Bryce did. If he had something more up-to-date or pertinent, I think he would have suggested the more current item. Immediately before sending my email to him, I read an article he wrote for *The Daily Beast* on "the myth of energy independence." As far as I could tell, the article contained no new information compared to the book.

Before I delve into what Bryce labels as ethanol lies, I think he greatly tipped his hand as being a premier apologist for the oil industry by not including dedicated non-fossil fuel proponents when he did his research for *Gusher of Lies*. The people he thanks in the acknowledgements seem to be mouthpieces for Big Oil. I sent a follow up email asking Bryce if he had spoken with any specific ethanol fans. He did not reply.

The five primary falsehoods told about ethanol relate to:

1. Energy content and MPG of gasoline vs. ethanol.
2. Ethanol causes engine damage.
3. Corn ethanol deprives starving people of food and causes food prices to increase.
4. Producing ethanol requires more energy to make than it puts out (EROEI).
5. Government Subsidies

There are other lies as well; most are variations on a theme, but a couple are stand-alone lies. Bryce uses them all, and I'll try to dispose of them in order. Keep in mind, these are not Bryce's fabrications. These falsities were conceived and honed over the years by people working in or for the oil industry. Bryce merely repackaged them in his writing style.

As he begins his attack on ethanol, Bryce warns the reader, "the discussion contains quite a few numbers... and that getting a grip on basic production numbers is an essential part of understanding why biofuels and alternative fuels cannot replace fossil fuels." This warning reminds me of the shell game of conmen plying their trade around Times Square in New

York, pulling in the suckers. I have to laugh because Bryce is the victim of the scam; he's the one who never got a grip on what the numbers mean.

Energy Content

British Thermal Units (BTUs) measure energy output, and they are at the heart of the first lie. Gasoline has a rating of about 116,000 BTUs, and ethanol has a rating of about 76,000 BTUs, which means ethanol has about 33 percent fewer BTUs than gasoline. Another way to look at it is that the energy content of ethanol is about a third (33 percent) less than gasoline.

Bryce uses this difference in energy content in three ways. First, he says one gallon of ethanol equals only 0.66 gallons of gasoline. Second, he says that because of the lower energy content, ethanol will deliver fewer miles per gallon (MPG) than gasoline. And third, he uses the lower BTUs to compute how many bushels of corn will be required to produce the ethanol needed to replace gasoline. When combined with incorrect data about available land, water requirements, CO2 emissions, and overall energy needed to grow crops, he assumes his sleight-of-hand is undetectable.

Bryce gets good mileage (pun intended) out of the energy content issue as he charts what the comparative numbers look like, and then writes paragraph after paragraph about how this principle, almost by itself, negates any sensible consideration of ethanol as a replacement for gasoline. Indeed, all ethanol opponents widely use this simple formula to create flashy algebraic equations and other graphic displays of what they think is irrefutable proof of ethanol's inferiority to gasoline based upon the law of physics.

But there is one major problem. The so-called energy content of ethanol and gasoline is irrelevant when it comes to internal combustion engines. The law of physics, in this case, is out of its jurisdiction.

BTU measurement is used to explain and calculate how much energy is required to heat water. If we were discussing a steam powered engine, heating water is a significant issue. But internal combustion engines in automobiles are not heating water.

Using ethanol in a typical gasoline-powered engine will produce fewer miles than gasoline because the engine is optimized to run on gasoline, not because there is a difference in BTUs. Using ethanol in the same engine that has been optimized to run on ethanol will deliver at least the same mileage and greater horsepower. This principle has been known and understood for more than a century.

In 1906, during the hearings before Congress on the Free Alcohol Bill, the legislation that finally removed the $2-plus per gallon tax on alcohol production, this was testified to several times by different witnesses.

Also, in 1906, the U.S. Department of Agriculture conducted experiments with ethanol and gasoline, and concluded the power from ethanol is materially higher than gasoline.

In 1907 and 1908, the U.S. Navy and the U.S. Geological Service jointly conducted 2,000 tests on ethanol and gasoline and concluded: "While gasoline carries about one-third more energy than an equal amount of alcohol, the higher engine compression ratios allowed by the latter made the two fuels highly comparable."

In 1906, the Edison Electric Testing Laboratory and Columbia University began testing ethanol. Elihu Thomson, who worked with Thomas Edison and was a co-founder of General Electric, reported that despite a smaller heat or BTU value, "a gallon of alcohol will develop substantially the same power in an internal combustion engine as a gallon of gasoline. This is owing to the superior efficiency of operation." Thompson became acting president of MIT in the early 1920s.

Studies and trials comparing gasoline-optimized engines to ethanol-optimized engines continue to take place during current times, such as the Advanced Vehicle Technology Competitions conducted by Argonne National Laboratories. In 1998, the U.S. Department of Energy, General Motors, and Natural Resources Canada sponsored the 1998 Ethanol Vehicle Challenge, a technology competition involving engineering students from 14 American and Canada universities and colleges. The goal was to convert Chevrolet Malibu sedans into optimized vehicles fueled solely by E85. Results of the competition showed the optimized Malibus were 13%-15% more efficient than stock Malibus.

The 1999 and 2000 versions of the Ethanol Vehicle Challenge used Chevrolet Silverado trucks instead of Malibu sedans. Ethanol-optimized vehicles again were more efficient than stock gasoline-optimized Silverados.

Another way to demonstrate engine optimization and why BTU rating is irrelevant, is to compare gasoline with diesel fuel. Diesel, which is also produced from petroleum oil, is rated at about 129,000 BTUs (higher than gasoline's 116,000 BTUs). Using Bryce's "basic physics," it would stand to reason that if diesel fuel was used in a gasoline-optimized engine, it should produce about 10 percent more miles per gallon. But if you fill your gasoline vehicle's tank with diesel, you get NO miles per gallon because the engine won't run. Conversely, if you fill a diesel-powered vehicle with gasoline, you don't get 10 percent fewer miles, you'll probably get zero miles, if the vehicle even starts. So regardless of BTU rating, engine optimization is the key to better mileage.

There are three things different between a gasoline-optimized internal combustion engine (ICE) and an ethanol-optimized internal combustion engine. They are spark timing, distance of piston stroke, and the quantity of fuel dispersed by the fuel injectors. Building internal combustion engines optimized to run on ethanol would require no additional assembly time, and there would be no difference in costs. (There can be a slight difference in cost to produce current flex-fuel engines, but that's largely because of the low volume of flex-fuel engine production.)

In any event, the actual difference in MPG when ethanol is used in a gasoline-optimized internal combustion engine is nowhere near as great as the theoretical 33 percent figure based on BTU comparison.

The differences in MPG are typically offset by the lower cost of ethanol, even with the pump subsidies removed since Bryce wrote *Gusher of Lies*. Consequently, there is a net gain when ethanol (E85 typically) is used in place of regular gasoline (E10 or gasoline without any ethanol, if you can find it). As I've reported many times, I have field-tested a wide variety of vehicles with many different ethanol-gasoline blend levels. I purchased a 2002 non-flex fuel Ford Taurus for the express purpose of gaining long term hands-on experience and knowledge about using ethanol. I have never experienced an MPG loss that wasn't offset by the savings in the price of E85 compared to regular gasoline.

A long time ago, someone concocted the BTU gasoline vs. ethanol story, and this fake workhorse has been ridden hard—too hard. It's time to put the entire issue to pasture, except it is an excellent example of the horse manure the oil industry and its avid disciples have used to lie about ethanol. It boggles the mind to think of all the trumped-up BTU comparisons and graphs and diagrams used thousands of times to prove the unprovable and to hoodwink the public.

I think it's unconscionable for anyone, especially an "energy expert," to use the BTU argument to denigrate ethanol. In all the research Bryce claims to have conducted, in all the books and articles he's written, and with all the public appearances he has made, has he never worked this issue through to arrive at the logical conclusion—a conclusion recognized since the invention of gasoline? Ethanol is better.

Update: On March 25, 2016, I sent an email to the Department of Energy (DOE) and suggested they make a change to their Alternative Fuels Data Center webpage. I showed how the information related to ethanol's energy content is being misused to incorrectly state MPG results in passenger vehicles, and that it should be changed.

The basis of my argument to the DOE is that BTU rating has no relationship to performance characteristics of an internal combustion engine, and that using a "blind" mathematical formula using BTU ratings to calculate mileage is as deceptive as an automobile manufacturer who invents MPG figures.

The DOE wrote back and agreed to make a change to the website. Effective in April 2016, they no longer used the irrelevant 33 percent less energy content calculation. Instead, they state MPG is dependent upon engine-optimization of the fuel used, and that engines optimized to run on ethanol can produce equal results to engines optimized to run on gasoline. This means the end of the BTU/gasoline argument, although I acknowledge that it will take a long time before the ethanol opponents willfully acknowledge it.

For another take on the BTU issue, consider these two "scientific" references:

In 1936, William J. Hale, Ph.D., published a book titled *Prosperity Beckons: Dawn of the Alcohol Era.* Hale was a leading chemical engineer of his day. He was Director of the Dow Chemical Organic Research Laboratory, Chairman of Division Chemistry and Chemical Technology, National Research Council, Washington, D.C., consultant to the Chemical Warfare Service, visiting Professor of Chemurgy at Connecticut and President of the National Agrol Company.

In his book, Dr. Hale writes that the inefficiency of burning gasoline in a gasoline-optimized internal combustion engine loses 25 percent of it's energy value, whereas alcohol (ethanol) "enters into complete combustion" in the engine. Therefore, instead of comparing gasoline's 116,000 BTUs to ethanol's 76,000 BTUs (and advancing the idea that ethanol has 33% less energy than gasoline, as the oil industry lays out their story), you should deduct one-fourth of the gasoline BTUs. This the makes the comparison 87,000 BTUs for gasoline versus 76,000 for ethanol, presumably giving ethanol 12 percent less BTUs, not 33 percent less. "Thus," Hale writes, "the custom of comparing fuels on their potentially available British Thermal Units becomes at once obsolete."

Coincidentally, this difference of about 12 percent fewer miles is much closer to real world MPG results when an ethanol-gasoline blend is used instead of ethanol-free gasoline.

The significance to William Hale's position is that even if you want to discard my explanation for why ethanol might deliver fewer miles per gallon of fuel in a gasoline optimized engine (that is, that the difference is related entirely to mechanical adjustments and not inherent energy content), now you have an expert-chemist's explanation of why the stated BTU ratings for the two substances is irrelevant when discussing their use as a fuel in an internal combustion engine.

In 2007, Bruce Dale, Professor of Chemical Engineering and Chair of the Department of Chemical Engineering and Materials Science at Michigan State University, put together a Power Point presentation that he used to explain the misuse of "net energy" and to rebut the bogus claims made by David Pimentel and Tad Patzek. This presentation can be viewed at *https://www.c-span.org/video/?188549-1/ethanol-energy-policy.* It provides the same substantially lower energy-content information that William Hale wrote about.

Review and Reply to Robert Bryce's *Gusher of Lies*, A Book Aptly Named for What It Is

Originally published June 10, 2013

Part 3

Laws of Thermodynamics

In continuing to discredit and denigrate alternative fuels, Bryce states:

> Precious few Americans understand the rules of the energy game, and that lack of scientific knowledge allows them to be easily gulled by (outside groups) …, who want to advance a no-pain agenda that sounds appealing but has no basis in reality.

I'd say a much more accurate comment would be:

> Far too many so-called energy experts who purport to understand the rules of the energy game lack the scientific knowledge to do so and have been easily gulled by the lure of oil industry compensation to help advance an agenda that has no basis in fact.

Bryce goes through a diatribe on two laws of thermodynamics to convince us that Big Oil has not launched a scandalous plot against alternative energy. It's a summation of what can be found in greater detail almost anywhere on the Internet. He explains it so that a four-year old child can understand it. I asked my wife to go out and find a four-year old child, because I couldn't make heads or tails of what he was talking about!

I understood what he's talking about. The second law of thermodynamics says hot stuff turns cold, but cold stuff doesn't turn hot. I got to thinking, cold may not turn hot, but it does turn warm or less cold. And if something becomes less cold, doesn't that mean it's getting hotter? Then ask, how hot is hot? Is it a dry hot, like California, or a moist hot, like Florida?

I hope people in one of quantum physics' alternate dimensions don't suffer from the first law of thermodynamics—when excessive energy is displaced by commercial airline jets.

What's funny about Bryce's thermodynamic diatribe is that he writes it as if the first car was built in response to some criteria contained in an 1820s Request for Proposal (RFP).

I understand Bryce is trying to explain the principles behind how something works. And even if the principle wasn't in full view of the original inventors of the automobile or airplane or nuclear turbine, this is like an English teacher telling us what Shakespeare really meant when he wrote *Much Ado About Nothing* hundreds of years before the teacher was born.

Bryce wraps up the whole discussion with this: "Taken together, the first two laws of thermodynamics provide the key to understanding why fossil fuels are so dominant in today's economy."

This statement must be a joke, there's no other explanation for it. But Bryce told the joke the wrong way. It goes much better like this: The reason gasoline is so dominant in today's economy is because of the Golden Rule: He who has the gold makes the rules.

There is one reason and only one reason why gasoline is the dominant engine fuel. In the early 1920s, the world's largest automobile maker, General Motors, invented a formula that added tetra-ethyl lead to gasoline to stop engine knock, thereby emulating one of the best qualities of ethanol. Tetra-ethyl lead, deadly to humans, quieted the knock that prevented the use of gasoline in high compression, high performance spark-induced internal combustion engines. This was the birth of leaded gasoline.

GM filed for and got a patent for this process. GM combined that patent with a similar patent owned by the world's largest oil company, Standard Oil. DuPont, one of the world's largest chemical companies and a large shareholder in GM, joined the group.

John D. Rockefeller and his company, Standard Oil, became fabulously wealthy because America needed a cheap fuel for indoor lighting and heating. The previous, most popular liquid fuel for indoor lighting and heating was alcohol; but because a huge tax was placed on alcohol production during the Civil War, the public had to accept smelly, dirty kerosene being sold for just a fraction of the price of alcohol. Consumer purchasing decisions had nothing to do with any of the three or four laws of thermodynamics.

GM realized they would make billions of dollars during the life of the patents just from their share of leaded gasoline royalties and ceased further efforts to use an ethanol-gasoline blend. They concentrated on building engines optimized for leaded gasoline. Standard Oil also recognized the importance of leaded fuel, that allowed all automakers to design and build high compression, higher performing engines. Having a lock on leaded gasoline, Standard was able to ensure their service stations, and the service stations owned by Rockefeller's other oil companies, were able to remain the dominant filling stations in every geographic market.

Consequently, the world's largest companies in the automobile business and automobile fuel business had the exclusive on the only products of their kind. This happened when alcohol production and consumption were declared illegal in America, so other oil companies couldn't easily develop a gasoline blended fuel to be used in the new high compression engines. And because America happened to have more cars and more paved roads than anywhere else in the world, this made the extremely rich and powerful Standard Oil even richer and more powerful. Being extremely rich and powerful means you can buy more politicians, more voters, more media spokespeople, and more advertising. By the time America shook off the despised effects of the 18th Amendment in 1933, gasoline was enthroned as the king of fuels. The Golden Rule was in effect. Those with the gold, rule.

Here's where Bryce overreached—the two laws of thermodynamics are irrelevant to the dominance of fossil fuels today. Using complicated theories of mechanical action to explain that domination is wishful revisionist history to ennoble what was nothing more than influencing consumer choices brought on by a string of concurrent sociological and political events.

The inventions of the internal combustion engine and the first automobiles came before the discovery of oily black goop (oil) in Pennsylvania that was used to create fuels. The discovery of oil and gasoline didn't beget the internal combustion engine. It would be more correct to say ethanol (alcohol) begat the high-performance engine. Without ethanol or a substance that could emulate its properties, using high-compression engines wouldn't be possible. If the world relied only on pre-leaded gasoline, the automobile industry might not have advanced past the Ford Model T.

On the other hand, there's no reason to believe that if oil had not been discovered, that America and the world would not have gone on their merry way to improve and invent internal combustion machines powered by ethanol. The ethanol of today is the same as ethanol in the 19th century. Ethanol can power the biggest, fastest automobile engines without using poison additives like tetra ethyl lead and other poisons.

Using the laws of thermodynamics or the laws of physics to justify continued reliance on the use of gasoline and oil-based diesel fuel is propaganda and a ruse to fool the public with dubious theories from far too many "energy experts," many of them lured to oil industry defenses by exotic trips to Saudi Arabia and other oil-producing countries that use us.

This inability to understand and recognize simple business precepts is a good example of the problem with people who've had no real business experience. Ironically, John D. Rockefeller, with his grassroots bottom-line business expertise, would probably have understood oil's rise to dominance perfectly. He did, after all, engineer it.

This is also another good example of how the public is being "gulled" by oil industry apologists who want to retard any effort for America to become energy independent.

If you are interested in learning more the history of how gasoline became America's primary engine fuel, please read "The Rise & Fall of General Motors and the Subjugation of the Industrialized World" in this book.

Oil Funds Terrorism—No Matter What Bryce Says

Bryce spends a considerable amount of time arguing that our dependence on foreign oil has little or no real significance to the kind of global terrorism that we have been subjected to over the past 30 or 40 years. First, as I pointed out earlier, he takes the position that since we can't do anything about needing oil, that we will have to suck up the terror and murder of innocent people. His morally bankrupt, horrific ideals—ideals that urge us to lean back enjoy the deaths caused by our gasoline fetish, are repugnant to all human beings who possess a moral compass.

Second, Bryce does what Jerry Taylor of the CATO Institute did to mitigate the issue by saying it doesn't require all that much money to get terrorists to commit terroristic acts. He tries to say that the billion or trillions of dollars that go to oil-fueled regimes that feed and nurture terrorists doesn't mean anything, because the terrorists might get comparatively small amounts of cash to carry out their heinous acts—like blowing up anything from a pizzeria to a passenger train or the World Trade Center in New York City, to hijacking cruise ships and airliners. In other words, Taylor and Bryce argue, if the Saudis had a profitable lemonade stand in the middle of the thirsty desert instead of vast amounts of oil, they would still have funded the attacks on 9/11—so if we love our cars, we must do business with them and their ilk.

That's the definition of insanity. That's what I told Jerry Taylor in 2010, and that's what I'm telling Robert Bryce now. A sane person believes that even a plugged nickel given to terror-sponsoring states is a plugged nickel too much. Any person with a working brain would say, "We must do everything to reduce our dependence on the product that keeps these murderers in business. We must become energy independent."

Even if the Saudi government does nothing more than provide financial assistance (welfare) to unemployed Saudis citizens as social improvement—which they do—and some of those citizens want to improve their lots in this world or the next by committing terrorist acts in the name of Allah, we should not be helping the Saudis. If the Saudi government uses some of the vast amount of money gotten from oil to fund school programs that teach their kids that Muslims must kill the infidels (non-Muslims), it's insane to fund those programs, which is what we do every time we buy a gallon of gasoline. I don't care what someone believes in; I don't care how they worship; and I don't even care that they teach their children that it's okay to kill me and my family, but I don't want to pay for that and neither should any other sane human being.

The Saudis aren't even the worst. Some of the biggest oil producing nations want America destroyed, and make no bones about it. Though they might be living in high style because of our use of their product, it doesn't dampen their criticism of our way of life or their overt support of terrorism aimed at Americans and other western nations.

Bryce states we shouldn't worry about the really evil oil-producing nations since we can refuse to buy their oil. This is absurd because, as Bryce acknowledges elsewhere, the oil industry is a global industry. If we choose to not import oil from Iran, for example, we still need oil; so, it has to come from somewhere else. Then another country, without problems with Iran, fills in their need with Iranian oil. And since it is all priced according to a global, not local, market, the Iranians still get the same price for their oil whether we like it or not. The Iranians don't even blink.

To make matters worse, we are using Iranian oil no matter what our leaders tell us. Some of it will get shipped to a middle-man country before it comes to us, relabeled. Some of this oil comes to us labeled as oil from Canada.

Regardless of whether a particular act of terrorism was funded for $1,000,000 or $1,000, if we rid ourselves of "global oil," we wouldn't have to put on a happy face in light of terrorism or avoid holding the guilty regimes responsible.

Breaking News: No Hostages Taken at Ethanol Plants

Since the October 1973 oil crisis, no hostages were ever abducted by any militant group from any ethanol plant. There was a report in May 2003 that Mary B, a salesclerk at an Iowa-based ethanol plant, was taken hostage; but it turned out to be friends of hers taking her to the local Applebee's for her surprise birthday party.

No lives were lost, no U.S. combat troops were dispatched, no naval assets were deployed, no Tomahawk missiles were fired, no helicopters were shot down, and no drones were needed to get behind enemy lines to protect any ethanol production facilities or the flow of ethanol to American service stations.

America's Love Affair with the Automobile Starts to Sour

Who hasn't heard the old saying a million times, "America's love affair with the automobile." It was so true, and to a large degree it's still true even though young people seem to not have gotten the memo that all Americans love cars and trucks. Interestingly, that bromide isn't just about America, it's also true around the world. The British say, "Britain's love affair with the motorcar." The French are the same, the Italians are the same, the Germans, the Swedes, the Japanese, the Koreans—everyone has a love affair with the automobile. Judging from the way the oil sheiks spent hundreds of millions on their car collections, I'd say the Arab's love for the automobile has spilled into near religious ecstasy.

Bryce writes, "Americans are incredibly mobile. They also love to fly, ride motorcycles, and take boating trips. But they really love cars. ...Americans not only love to drive, but they also love to drive fast. And driving fast requires lots of horsepower." All of this is correct, or at least was correct, until the younger generation decided they were more interested in following their friends on Facebook then in fast cars.

A story published May 1, 2013 by NADA (National Automobile Dealers Association) revealed that fuel economy ranks as the #1 factor in all new car and truck purchases. My business partner, Bob Gordon, remarked, "What a sorry state of affairs! With all of the delicious new car choices American consumers are offered, for them to rank "fuel economy" as their number one purchase criteria is sad beyond measure." (You can read the full story here: *https://www.theautochannel.com/news/2013/05/01/074702-nadaguides-com-survey-ranks-shopping-preferences-new-car-and-truck.html*

To be fair, the statistics pertaining to the younger generation's disinterest in cars and the shift in purchasing criteria weren't available to the same degree when Bryce wrote *Gusher of Lies* as they are today. Statistics in 2008 showed a cooling trend; it's just that no one expected—least of all Robert Bryce—that the trend would become an avalanche.

To be fair on the other side, Bryce never took into account the possibility things could turn sour for the oil industry and fossil fuels. To him, everything related to gasoline was rosy and climbing, and everything related to alternative fuels was crappy and going downhill. It was like the people in The Netherlands five centuries ago who thought tulips would get more and more valuable. All it took was for a few people to notice tulips weren't the only beautiful flowers in the world.

My issues with Bryce's entire position on America's love affair with the automobile is his use of the quote, "There's no such thing as too much sex and too much horsepower," and his follow-up comments about the intoxicating effects of horsepower. I guess none of the people Bryce consulted with admitted or knew that if it's horsepower you want, you need ethanol.

Switzer Performance: A Case in Point

If a person buys a vehicle and they want it to soup it up to go really fast, they can take it to a specialty shop like Switzer Performance in Oberlin, Ohio. Switzer is expert in taking super cars and turning them into super-duper cars.

Years ago, *TheAutoChannel.com* published "Switzer Builds e85-capable 900 hp Nissan GT-R." (*https://www.theautochannel.com/news/2010/06/22/483719.html*) The story was written by Switzer's PR guy, Jo Borras, and tells the story about a client who wasn't satisfied with Switzer's "run of the mill" customization of a Nissan R35 GT-R that put out 800 hp. This client wanted 900 hp.

Switzer re-customized the already customized car by adding a new transmission and making clutch adjustments. But they still had the horsepower wall to climb, and the barrier was gasoline.

Jo reported:

> We didn't want to just deliver a race-gas package that tethered the client to a fuel drum in his garage, but we wanted to be able to turn up the boost, also.
>
> The solution? Readily available E85... In the end we had a fuel that gave us everything we needed to crank up the boost on this car and deliver over 900 hp. The E85 delivered other benefits as well; it is significantly 'greener' than the pump-gas and race-gas versions of the GTR. ...With most high-horsepower cars, you can barely breathe after just a few dyno runs. With E85, we were able to run pull after pull to dial in the tune without feeling any effects of the usual emissions. It's hardly scientific, but there was a noticeable improvement in the air quality in the shop compared to race gas.

Jo Borras did say adjustments were made to make the ethanol-optimize engine, as I noted in many of my articles.

In a phone interview, I asked Jo, who was not aware of The Auto Channel's position on ethanol, why they would use ethanol. He replied, "Why in God's name would you not use ethanol? It's cheaper, cleaner and gives the car more power."

"So why don't you use E100?" I asked.

His answer: "We would if we could, but it's not available to us on a ready basis. It's easy to find E85."

Simply put, if you want the best for those you love, wouldn't you want the best fuel for the vehicles you love to make them drive faster? Gasoline is not the fuel. Serious drivers and race teams love ethanol or at least high ethanol-gasoline blends.

Review and Reply to Robert Bryce's *Gusher of Lies*, A Book Aptly Named for What It Is

Originally published June 10, 2013

Part 4

I Got Your Religion Right Here

Joseph Goebbels, the infamous minister of propaganda for the Nazis, used "doublespeak," which is defined as evasive language that says one thing but usually means the opposite. Given that Standard Oil began their propaganda against ethanol shortly after the passage of the Free Alcohol bill in 1906, and later engaged in un-American activities with the Nazis before and after our entry into World War II, I wondered if Goebbels learned his craft from the oil industry.

In any event, no group or entity has mastered doublespeak like Big Oil and its apologists. They accuse ethanol for causing engine problems caused by gasoline right up to and including Bryce's comment that "Ethanol isn't a motor fuel. It's religion."

Bryce writes, "America is divided into two camps, the believers and the heretics." He adds, "So far it appears that the believers are winning. And their zealotry has created a multi-billion-dollar bushel of subsidies [for ethanol] ..."

Bryce commits character assassination criticisms of Bill Clinton and Al Gore (which I don't generally disagree with), before commenting that "The irrational exuberance over ethanol has led investors to dump bushels of cash into companies that produce ethanol."

While it would be fantastic if ethanol was "winning," how anyone could say that ethanol and ethanol proponents are winning is beyond me. If ethanol was winning, this document would not exist. The effort to get any kind of acceptance of ethanol into the marketplace is always met with fierce attacks by the oil lobby and consumers who have been gulled (as Bryce puts it) into believing that ethanol is the inferior fuel. Even with more than 180 years of experience and proof that ethanol is the superior fuel, Big Oil keeps screaming that more testing is necessary.

If any group is guilty of zealotry, it's Big Oil. People like Robert Bryce just suck in the lines fed to them and spit them back out at American consumers. I seriously doubt that Bryce or Taylor or Searchinger or any oil industry apologists have ever even tried pumping a gallon or two of E85 into their vehicles' fuel tanks just to see what happens. They pretend they know something about the subject, but all their information is second- or third-hand and comes from others who received the information second- or third-hand.

Yes, subsidies have been given to the ethanol industry and investors have invested in ethanol companies, but Bryce is fine with subsidies and investments for electric cars. He and the oil lobby have no problem with the subsidies given to electric car development because they know electric cars are not a threat to the oil industry. "All subsidies are equal, but some are more equal than others. Subsidies for electric is good; subsidies for ethanol is bad." You can almost hear the pig in George Orwell's *Animal Farm* making that statement; it's doublespeak at its finest.

It is absurd to think that an under-funded fuel/energy alternative of any kind could compete with gasoline without significant government assistance. The benefits of developing a domestic alt-fuel industry that causes less pollution and frees us from external forces makes the effort worthwhile, even if it is no more economical than petroleum-based fuels. For private investors to produce ethanol in the quantity needed to retire gasoline, the investors and the alt-fuel industry need the same government assistance and intervention used to make oil/gasoline our primary fuel. He never mentions that the oil industry still receives the highest subsidies of ANY industry in the U.S. even though oil-related companies record astronomically huge profits year after year. The playing field must be leveled.

The staggering amounts of government subsidies for oil are "paid" out in many forms. Government assistance to the oil and gasoline industries continue to this very day. We do not add the costs of waging war and the use our military to defend the oil industry sites around the world as if our military is the oil industry's private security force.

Worst of all, hundreds of thousands of American service men and women have lost their lives in oil-induced wars and military engagements. This is the incalculable cost that I referred to constantly. What value do we put on the life of an American soldier, sailor, airman or marine? Why should we discount their lives so that Arab sheiks and Third World dictators can drive their million-dollar Ferraris in the desert and fund terrorists?

I am not against war; I'm not philosophically opposed to it. I understand the reasons and need to go to war. Sometimes I think we have not gone to war quickly enough. I am, however, against long, drawn-out wars where quick victories are turned into defeat just to secure oil or to protect a bunch of misogynist racists who pretend that they rule by divine right.

We overpay for gasoline and diesel at the pump, we overpay for them via taxes and we overpay with human lives. Bryce talks about morality. He questions the morality of using farmland to grow products that are used to create a fuel to replace gasoline. And to bolster and explain this position, Bryce uses a quote from Fidel Castro—that's right, the same Fidel Castro that made himself a king and then subjected all his fellow countrymen to a life of serfdom. What a role model for morality! "Castro declared that 'converting food into fuel' was a 'sinister idea' and that the expanded use of biofuels would cause hunger among the people in the poorer countries of the world."

So, what's more moral? Protecting the lives of American citizens or protecting an Arab sheiks' right to buy a Maserati to match his Ferrari and use the leftover funds to arm terrorists to take down Americans—as they did on 9/11?

A study released by the Environmental Law Institute in September 2009, covering the years 2002 to 2008 revealed that:

> The federal government provided substantially larger subsidies to fossil fuels than to renewables. Subsidies to fossil fuels—a mature, developed industry that has enjoyed government support for many years—totaled approximately $72 billion over the study period, representing a direct cost to taxpayers.

This figure did not include the cost of our military involvement anywhere in the world, or the funds provided to clean up oil spills, or for water infrastructure improvements in ports, harbors, or waterways that are used to accommodate oil tankers.

"The Military Cost of Our Oil Addiction," from *PUMP* (the movie): *https://www.youtube.com/watch?v=IL3ZdHBrAec*

Subsidies for renewable fuels, a relatively young and developing industry, totaled $29 billion over the same period.

Most of the largest subsidies to fossil fuels were written into the U.S. Tax Code as permanent provisions. By comparison, many subsidies for renewables are time-limited initiatives, implemented through energy bills, with expiration dates that limit their usefulness to the renewables industry.

On July 3, 2010, *The New York Times* published "As Oil Industry Fights a Tax, It Reaps Subsidies" by David Kocieniewski, a Pulitzer Prize-winning writer. Among other findings, Kocieniewski reported:

> When the Deepwater Horizon drilling platform set off the worst oil spill at sea [in the Gulf of Mexico] in American history, it was flying the flag of the Marshall Islands. Registering there allowed the rig's owner to significantly reduce its American taxes.

The owner, Transocean, moved its corporate headquarters from Houston to the Cayman Islands in 1999 and then to Switzerland in 2008, maneuvers that helped it avoid taxes. At the same time, BP was reaping sizable tax benefits from leasing the rig.

According to a letter sent that June to the Senate Finance Committee, the company used a tax break for the oil industry to write off 70 percent of the rent for Deepwater Horizon – a deduction of more than $225,000 a day since the lease began.

When federal officials considered a new tax on petroleum production to pay for the cleanup, the industry fought the measure, warning that it would lead to job losses and higher gasoline prices, as well as increased dependence on foreign oil. But an examination of the American tax code indicated that oil production is among the most heavily subsidized businesses, with tax breaks available at virtually every stage of the exploration and extraction process.

Some of the tax breaks date back nearly a century, when they were intended to encourage exploration in an era of rudimentary technology; when costly investments frequently produced only dry holes. Because of one lingering provision from the Tariff Act of 1913, many small and midsize oil companies based in the United States can claim deductions for the lost value of tapped oil fields far beyond the amount the companies actually paid for the oil rights.

Other tax breaks were born of international politics. To deter Soviet influence in the Middle East in the 1950s, the State Department backed a Saudi Arabian accounting maneuver that reclassified the royalties charged by foreign governments to American oil drillers. Saudi Arabia and others treated some of the royalties as taxes, which entitled the companies to subtract those payments from their American tax bills. Despite repeated attempts to forbid this accounting practice, companies continue to deduct the payments. The Treasury Department estimates that it will cost $8.2 billion over the next decade.

In his encyclopedic book, *Alcohol Can Be A Gas,* David Blume wrote: "Between 1968 and 2000, oil companies received subsidies of $149.6 billion, compared to ethanol's paltry $116.6 million."

On September 23, 2012, Kevin Begos, reporting for Associated Press, wrote a pro-oil story titled, "Decades of Federal Dollars Helped Fuel Gas Boom." The sub-head of the story was, "Pioneers in gas drilling say government research and tax breaks helped launch boom." The story provides valuable insights into why government subsidies were necessary:

"Over three decades, from the shale fields of Texas and Wyoming to the Marcellus in the Northeast, the federal government contributed more than $100 million in research to develop fracking and billions more in tax breaks."

Now, those industry pioneers say their own efforts show that the government should back research into future sources of energy—for decades, if need be—to promote breakthroughs.

"The government has to be involved, to some degree, with new technologies."

The first federal energy subsidies began in 1916; and according to the Congressional Budget Office, until the 1970s, they "focused almost exclusively on increasing the production of domestic oil and natural gas."

So contrary to how Big Oil tries to play down how much government support they get—while yelling their heads off about subsidies given to ethanol—the facts reveal a completely different story.

Bryce wrote: "Tax incentives won't result in energy independence, but they will mean more profits for the world's largest agribusiness outfit, Archer Daniels Midland Co."

Archer Daniels Midland? Bryce is comparing ADM's profits to the profits recorded by the oil and gasoline companies? That is the definition of insanity. Around the time the paperback edition of *Gusher of Lies* was published, ExxonMobil declared a fourth quarter profit of more than $40 billion. That's just ExxonMobil, not the entire gasoline industry; and it was just for one quarter. That came on top of all the subsidies and tax breaks that they received from the United States government. Archer Daniels Midland reported their fiscal fourth quarter profit of $372 million. Although $372 million is nothing to sneeze at, it's less than one percent of ExxonMobil's profit for the same quarter.

Big Oil Gets An $80 Billion 'Bonus' Subsidy

Remember the $80 billion bestowed upon General Motors and Chrysler to bail them out? The taxpayers footed that bill.

But the taxpayers should never have paid for the bailout. The bailout should have come from the automobile industry's number one beneficiary, the oil industry. Without automobiles, the oil industry is done. The oil industry had the most to lose and they had (and have) the resources to save the two Detroit manufacturers. The oil industry was given a gift; a gift that they didn't deserve. A gift that was twenty times more than the Archer Daniels Midland 2008 fourth quarter profit. ExxonMobil could have provided all the bailout funds by themselves from their annual 2008 profits. Isn't that what a company is supposed to do—reinvest their profits to secure future earnings?

By providing the bailout funds, America paid for a second time (and perhaps third time) for the vehicles we bought from GM and Chrysler over the years. If GM and Chrysler went away, only some American pride would have been hurt. Losing the specific vehicles they were producing would have been no shock to the market; they were hardly the best or most popular options for new vehicle buyers. All the other car manufacturers would have gladly picked up the slack. Since we have a very healthy trend towards foreign brands establishing manufacturing plants in the U.S., in all likelihood the building of additional plants would have increased, and they may have acquired some of the old plants for refitting.

Four years later, though the loan was repaid, what do we have to show for it? Two American brand car companies that are more interested in the Chinese market than the American market? Did we put up $80 billion for that? ExxonMobil could have done that on their own.

If ExxonMobil would have bailed out GM and Chrysler, would that have been heroic? I think so; I think it would have been Homerifically heroic. ExxonMobil could have milked that for 50 years: "We Saved the American Auto Industry!" Wow! No one would even think of bringing up the Exxon Valdez oil spill after that. What an incredible marketing and public relations coup it would have been. Instead Bryce and the other oil industry hacks whine about Archer Daniels Midland's comparatively miniscule profits.

Even the CATO Institute agrees that the government bailout wasn't worth the effort. See "The Truth About GM-Chrysler Bailouts." (*https://www.cato.org/blog/truth-about-gm-chrysler-bailouts*)

In the meantime, all of GM's common stock shareholders and Chrysler's bond holders lost everything. How many millions is that? How many billions is that? When do these investors get bailed out?

Robert Bryce's entire recitation on the unfairness of subsidies given to the ethanol industry is exaggerated hypocritical, doublespeak.

Grand National Spitting Contest

Bryce attempts to make hay with a short history of "financial inducements' being given to politicians on behalf of the ethanol industry. Duly noted. Bribes are offered to politicians, and politicians accept them.

Bryce also writes about some wrongdoing that ADM was found guilty of, and had to pay a fine. Big deal.

Other than in the hope that no one will take more than two seconds to consider all of the bribes, scandals, convictions, and fines that the oil industry is responsible for, not to mention all the deaths of those workers who were killed in drilling related accidents and in the tetra-ethyl lead experiments, I don't understand his trying to show us that the oil industry can spit further. We know that. The oil industry can spit further and has been spitting on the public for more than 100 years.

It's more of the Oil industry doublespeak.

Abbreviated list of links related to oil/gasoline subsidies

- *http://www.nytimes.com/2010/07/04/business/04bptax.html*
- *http://www.cato.org/pubs/pas/pa390.pdf*
- *https://www.eli.org/research-report/estimating-us-government-subsidies-energy-sources-2002-2008*
- *http://www.imf.org/external/pubs/ft/spn/2010/spn1005.pdf*

•*https://blumenauer.house.gov/energy-issues#fossil*

•*http://merkley.senate.gov/newsroom/press/release/?id=F68FED0D-08C0-4DC4-9035-211E28DD9B7E*

•*http://blog.ucsusa.org/jim-kliesch/oil-industry-subsidies-are-anything-but-rosy?_ga=1.161769531.686943746.1492723673*

•*https://web.archive.org/web/20110827121616/http://investmentwatchblog.com/ending-oil-subsidies/*

•*https://web.archive.org/web/20110305034655/http://investmentwatchblog.com/house-gop-votes-unanimously-to-protect-big-oil-subsidies*

•*https://www.forbes.com/sites/energysource/2012/04/25/the-surprising-reason-that-oil-subsidies-persist-even-liberals-love-them/#102e12632792*

•*https://thinkprogress.org/big-oils-banner-year-higher-prices-record-profits-less-oil-d337b2dcf58c*

•*http://priceofoil.org/2012/02/07/the-big-the-bad-and-the-subsidized/*

•*http://www.foreffectivegov.org/sites/default/files/budget/Revenue/OilandGas.pdf*

•*http://www.huffingtonpost.com/2011/02/01/obama-moves-to-cut-oil-subsidies_n_816787.html*

•*http://themoderatevoice.com/78678/unemployment-benefits-and-oil-subsidies/*

•*http://www.americanprogress.org/issues/2010/05/oil_company_subsidies.html*

•*http://energydeals.wordpress.com/2011/02/02/obama%E2%80%99s-bid-to-end-oil-subsidies-revives-debate/*

•*http://ilsr.org/wp-content/uploads/2012/05/Oil-Slickers.pdf*

•*https://www.yahoo.com/news/decades-federal-dollars-helped-fuel-141648115.html*

•*http://www.sourcewatch.org/index.php/Estimating_U.S._Government_Subsidies_to_Energy_Sources_2002-2008*

Fool or Fuel?

Bryce's header for his discourse on the issue of using edible corn for ethanol is "Food or Fuel?" But my title is far more appropriate, because anyone who participates in disseminating or believing the food vs. fuel argument is foolish. Bryce's email response to me stated, "I am fundamentally opposed to the use of food to make motor fuel."

The basic premise is that corn is being taken out of the mouths of starving people and turned into ethanol to enrich greedy farmers and distillers. We've all watched those Sally Struthers' "fly-in-the-eye" TV shows where she asks for donations to feed the starving African children. That's the image the oil industry wants the public to visualize, and that's the image far too many fools in the public choose to see.

Unfortunately for the oil industry, it is a totally false image; and when you know the truth, you are astonished they get away with this nonsense.

Let me start with the most palpable image, the starving African children. If corn was no longer used for ethanol, does that mean that we would see images of African children sitting around a campfire eating delicious corn on the cob smothered with butter? No. Why? Because outside of America the world doesn't eat corn the way that Americans do. They don't like it. They eat wheat and rice. Corn is not wheat. Corn meal could be used in some circumstances in place of wheat, but they like wheat and rice.

So why isn't there enough wheat? Partially because Russian and Chinese wheat harvests have been miserable and so wheat is hard to come by. Consequently, there are those who feel we in America have to make up for the farming losses suffered by the two giant communist countries. I read the memo; Russia is Russia now, not the Soviet Union. But their government and economic system stinks, and they still can't get the farming done right. China is still communist China.

There's another problem. If we get the corn to the starving African countries, then what? The local warlords prevent the supplies from getting to the people until and unless they are paid by the starving people—the same starving people who have no jobs and no money.

Assuming that Africans decided that they wanted to make popcorn or corn-on-the-cob as part of their staple diet, and assuming the warlords suddenly were guilt-stricken by how they treat their fellow countrymen, who pays for the corn grown by American farmers? We would have to subsidize the famers to grow the corn so we can give it away as part of foreign aid.

Did I say "subsidize?" Is this another instance where the oil industry is okay with subsidies as long as the subsidies don't go to producing a fuel that competes with gasoline and diesel?

Since the oil industry didn't take advantage of the chance to be heroes by bailing out the auto industry, how about they use all their profits to buy up all the corn before the greedy ethanol distillers can get it, and send the corn to the starving people? It would be terrific PR for them. And the empty tankers that carried Arabian and Nigerian oil to America can carry the corn to the starving countries on the return trip! WOW and double WOW!

Except, this wouldn't work because corn grown for ethanol is not grown for human consumption; it's grown as feed stock for animals. More than 90% of all the corn grown in America is not fit for human consumption. Can it be grown for human consumption? Sure, but what about the animals? Farmers and ranchers need to feed the animals to make them grow so that the animals can get big and fat and provide us with meat. Did we suddenly give up eating meat?

There's yet another problem. Most of our corn is genetically engineered to help corn grow big and abundant. Many countries have regulations against importing food that has been genetically engineered. If we don't use genetic modification, then our abundant crops might not be so abundant. So lower crop yield could result in not enough corn for the starving people who don't eat corn, which results in higher prices or more subsidies for the farmers, which then raises other prices.

Here's the amazing part: much of the corn used to feed animals is not used as you or I eat corn. It's first converted to "distillers grains." The process breaks down the corn so that the starch is removed and only the best parts (in terms of feedstock for cattle) remains. It's rich in nutrients, providing about four times more protein to the animal than straight corn.

As the name "distillers' grains" suggests, the process is conducted by creating a mash and then distilling the mash to get rid of the liquid. This liquid is ethanol. This ethanol is the ethanol that can be used to fuel an internal combustion engine. If the liquid wasn't used, it would be discarded. Why waste something that provides so much nutrient benefits and can also be used in virtually every gasoline-powered engine and to make a profit? This is just part of the story. American farmers grow as much corn as they do because they know how to grow corn, they have the soil and climate conditions that allow for corn to be grown, and they have a marketplace for the corn.

The image the oil industry wants to create is one where the farmer, about to deliver his delicious corn into the mouths of starving people, is ambushed by greedy ethanol producers who swoop in and buy up all the corn. But this isn't the way a manufacturing business is conducted.

Farming is a manufacturing industry, with concerns related to weather conditions in addition to all the other business concerns they have. But it follows the same rudiments of manufacturing used for automobiles, clothing, appliances and a thousand other items. Mature, experienced manufacturers don't make products simply because they have the facilities and workers to do so. They go out and pre-sell the next season's line. They use the pre-sell orders to secure financing that allows the manufacturers to pay the workers, pay for the mechanical aspects of production, pay for marketing, and pay for distribution. Yes, there are instances of people with their family and friends' money who go willy-nilly into production hoping there will be buyers for their finished products. That's what's called risky business.

If there wasn't a known demand for the corn (demand equals money), the farmers would not grow it, or at least not nearly so much of it. They would grow something else, or they just wouldn't grow anything. If the farmers didn't grow anything, they would be unemployed and would need some kind of public assistance, which in itself is a subsidy. If the farmers choose to grow something else, it could cause a glut on the market, causing prices to fall, which would hurt even more farmers. The government may still have to step in with some financial assistance or pay the farmers to not grow anything (which has happened many times in the past). This is a subsidy. If Bryce doesn't like subsidies, why would this subsidy be more acceptable?

The bottom line is that the use of starving African children is a ruse to tug at the heartstrings of people who are or would like to be socially conscientious. The solution for the starving people is to teach them how to grow corn or wheat or cucumbers themselves. And this has been tried, of course. However, a peculiar thing has happened. In the effort to get some of our surplus produce in the hands of people in third world countries, the lower price (and often no-price) of the items winds up under-cutting similar locally grown produce. This makes it unprofitable for the local farmer to continue in business, and it drives him out. See (*https://www.youtube.com/watch?v=-fzj-n8OdGo.*)

This process is really simple to understand and makes all the sense in the world. So why doesn't the oil industry get it? Why do all these high-powered, well-educated people like Robert Bryce and his cohorts not understand this? Why do they insist on painting the food vs. fuel picture? The answer is they either don't know this because they have no real business knowledge and experience, or perhaps, it is because they are paid to not acknowledge it. On the other hand, if I'm the one who is wrong or ignorant, then virtually every manufacturing company in the world is wrong and ignorant.

Food vs. fuel is a foolish non-issue, Mr. Bryce.

It is also true that corn doesn't need to be used to make ethanol. In fact, it is one of the least desirable crops for that because of the relatively low annual yield and the heavy requirement for water. The reason corn is the primary base material is because we have so many farmers who are experienced growing it and because we have so much land. In Brazil they use sugar for ethanol and in Australia they use sweet sorghum and wheat. Some American ethanol farmers are switching to sorghum because of water issues. Sorghum can also be used for distillers' grains too, so we'll still feed animals and produce fuel.

It'll be fun to see what argument the oil industry comes up with then.

One way or the other, I'd rather have my fuel money go to American farmers then to foreign terrorists.

Land Ho … Ha, Ha, Ha

Bryce also says using corn or other produce hogs available land. In doing so, he makes the jump from present ethanol production and use to full-on replacement of all fossil fuels to portray the impossible nature of the effort. He has no room for middle ground. Bryce takes the amount of land currently used to grow corn for ethanol, multiples it by the liquid volume necessary to replace all gasoline, and declares it is impossible.

The amount of land he uses for his statistics was a number given to him by someone with an anti-ethanol agenda. He then uses it against a crop yield number that is out-of-date or was never correct in the first place. Using those figures Bryce then multiples everything with wrong BTU information (see previous essays on BTUs), to arrive at an "energy replacement" number.

Bryce references America's total available cropland as 440 million acres (not just for corn, all crops). He also cites a number used by Dennis Avery of the Hudson Institute as the amount of land that would be needed to replace all imported crude oil (546 million acres). Obviously, there is a significant difference, so Bryce declares it impossible.

However, in addition to the 440 million acres of cropland, America also has 939 million acres of farmland. According to land definitions, farmland is not quite as good as cropland, but suitable for growing things like specialized energy crops.

According to David Blume (*http://www.permaculture.com/node/490*), who's probably the single most experienced and knowledgeable person in the world on ethanol and ethanol production:

> Of its nearly half a billion acres of prime cropland, the U.S. uses only 72.1 million acres for corn in an average year. The land used for corn takes up only 16.6% of our prime cropland, and only 7.45% of our total agricultural land. Even if, for alcohol production, we used only what the USDA considers prime flat cropland, we would still have to produce only 368.5 gallons of alcohol per acre to meet 100% of the demand for transportation fuel at today's (2007) levels. Corn could easily produce this level—and a wide variety of standard crops yield up to triple this.

Compared to Bryce's calculations, this is a vastly different side to the story. And, as Blume points out, if corn is replaced by a higher yielding crop, then calculations become dramatically more favorable.

Furthermore, Blume's figures pertain to 100 percent replacement of gasoline and diesel. As I pointed out regarding Bryce's immediate jump from zero percent to 100 percent, I'd say the same is true for Blume's comments (and I'm sure David would agree); let's first make sure we get E15 into national acceptance—which is just a tiny increase in America's current use of ethanol. Then we can move onto E20, E30, E50, and so on. Even my very aggressive plan for cutting out the production of all new gasoline-powered vehicles takes place in five years and it never fully eliminates the use of gasoline as an engine fuel until the vast majority of gasoline powered vehicles are retired and off the road—that would probably be at least 15-20 years later. Since improvements are regularly being made on crop yields, water requirements, and enzyme technology, the numbers will only get more favorable for ethanol. Remember the teleconferencing video guy who insisted Internet connectivity would never be faster than 28.8k? He must have scrambled eggs all over his face by now.

Review and Reply to Robert Bryce's *Gusher of Lies*, A Book Aptly Named for What It Is

Originally published June 10, 2013

Part 5

Ethanol Production Results in Rising Food Costs—Another Lie

Next to the lie that food is being diverted from the mouths of starving children comes this doozy. Bryce pulls quotations out of the woodwork for it. He's got studies, statistics and equations and wraps it all up in a great big study conducted by the World Bank that said food prices rose because corn prices rose, and the corn prices rose because of ethanol demand. The World Bank retracted that report at least three times since it was written.

Didn't Bryce write that he leans towards libertarianism and that he works for the Manhattan Institute for Policy? What the heck is he doing relying on anything from the World Bank? Bryce did use the Fidel Castro quote to support his position on morality. Now I see the connection, and it goes back to doublespeak.

In the meantime, let me drop this little pearl on you. In 2008, when Bryce wrote *Gusher of Lies*, a $4 retail box of corn flakes cereal contained about 6 cents worth of corn. So, if corn prices rose 50 percent, there would be about 9 cents worth of corn in the box. If the price of corn rose 100 percent, there would be 12 cents worth of corn. Hmm, 6 cents equal $4? Even if the supermarket made 100 percent profit, which they don't, there's a big difference between 6 cents and $4. The price of food did not rise because ethanol pushed the price of corn up too high, food prices rose becausetransportation costs, (paying for gasoline and diesel fuel) and other derivatives from petroleum oil, like printers' ink and marketing costs skyrocketed and so did prices—on everything.

When Bryce wrote *Gusher of Lies,* oil prices were at all-time highs. Gasoline and diesel prices were at their highest levels. Anyone with a brain and a modicum of business sense understood how high transportation costs would affect retail prices. But not Robert Bryce nor all the economic experts that he interviewed for his book. He lived in a bubble world where he must have panicked daily because he believed that ethanol produced from corn had devastating effects on food prices.

Corn prices did rise during that time; and while the World Bank originally said it was because of ethanol production, they admitted they were wrong. In the summer of 2010, a couple of years after they told Bryce and everyone else that corn prices rose because of ethanol production, they issued a retraction.

The rise in corn prices, according to the revised report from the World Bank, was due to investor speculation, not ethanol producers pushing the prices up in order to take the corn-on-the-cob out of the mouths of starving people.

Below are some excerpts from the report:

"Placing The 2006/08 Commodity Price Boom into Perspective"
John Baffes - The World Bank
Tassos Haniotis - European Commission

> The 2006-08 commodity boom was one of the longest and broadest of the post WWII period. The boom, and especially the 2008 rally, when crude oil prices peaked at US$ 133/barrel (up 94 percent from a year earlier) and rice prices doubled within just five months, has renewed interest in the long-term behavior and determinants of commodity prices, and raised questions about whether commodity prices have reversed the downward course that most of them followed during most of the past century. It has also produced numerous calls for coordinated policy actions at the national and international level to address food availability and food security concerns...

> The increasing interaction between the price movements of energy and nonenergy commodities during the boom focused attention on the impact of growing demand for biofuels, including for maize-based ethanol (mainly in the US) and oilseed-based biodiesel production (mainly in Europe). During the boom, maize and crude oil prices moved in tandem, pointing to an emerging new fixed relationship between them. Obviously, maize and its use for ethanol moved into the picture as significant factors affecting price developments. But how much impact was there, and was there a similar one in oilseeds, resulting from their use for biodiesel?
>
> Concluding Remarks: Numerous factors have contributed to the recent commodity boom and have been analyzed extensively in the literature. Yet, their relative weight continues to be an area of contention. In this paper, we examined three key factors whose roles have been somewhat controversial: speculation, the growth of demand for food commodities by emerging economies and the role of biofuels. We conjecture that index fund activity (one type of "speculative" activity among the many that the literature refers to) played a key role during the 2008 price spike. Biofuels played some role too, but much less than initially thought.

Unfortunately, the World Bank retraction did little towards taming the false accusation that ethanol production resulted in higher prices—false accusations that continue three years after the retraction and some five years after the publication of Bryce's book.

I think there are at least three reasons: Like all initial explosive reports in the media, retractions are never covered with the same vigor. Second, the oil industry's public relations propaganda war chest is so large they were able to bury the story. And third, saying that corn and ethanol are causing prices to rise was a great way for snack food manufacturers to explain why they increased their prices and never lowered them.

These three reasons are my educated guesses as an authentic media expert with more than 40 years of extensive hands-on experience in every facet of marketing, advertising, production and sales.

A new report from the World Bank was issued in May 2013 and confirms that the primary driver in rising food prices was the cost of crude oil.

"Long-Term Drivers of Food Prices"
John Baffes
Allen Dennis

The World Bank Development Prospects Group & Poverty Reduction and Economic Management Network Trade Department
May 2013

Conclusions and Further Research

> This paper uses a reduced-form price-determination model on 1960-2012 annual data of five food commodities (maize, wheat, rice, soybeans, and palm oil) to assess the relative contribution of various factors to their respective price changes. The factors include crude oil prices on the supply side, stock-to-use ratios, three macroeconomic indicators (exchange rate movements, interest rates, and inflation), and income on the demand side. The paper concludes that food commodity prices respond strongly to energy prices, stock-to-use ratios, and (in a mixed manner) to exchange rate movements. With a few exceptions, interest rates and income growth do not matter. Yet, crude oil prices mattered the most during the recent boom period because they experienced the largest increase.

However, let me turn the tables on those overly concerned with the use of corn for ethanol, and how corn syrup and corn starch prices might cause rising prices of snack foods, soft drinks and candy. America is in the middle of a national obesity epidemic. Perhaps we should not be eating so much junk food. If Robert Bryce needs a crusade, this would be the one for him because this one is a worthwhile moral cause.

EROEI

EROEI means Energy Returned On Energy Invested. This is sort of an equivalent to the financial community's ROI, Return on Investment. Ultimately, both EROEI and ROI are used to tell you if you're wasting your time.

Here's a simple illustration: Say you want a really good bowl of fruit salad for lunch. You could go to Costco and buy a bunch of fruits that you take home, cut up and eat. Or you can go to Costco and buy one of their ready-made fruit salad bowls. The ready-made costs a little more than the individual fruits, but you save all the cutting time and you only have edible bits, not all the wasted pieces.

You have to make a decision. You ask yourself if it's worth your time to go through the effort of cutting the fruit, particularly the pineapple. You determine what your time is worth, add that pro-rated dollar amount to the cost of the fruit, and then you can compare which gives you better value.

The oil industry used EROEI to make a case for why it's not worth the effort to create an alternative fuel to gasoline. To help do this, the oil industry hired two university professors to create the study. One professor, a professor of entomology (the study of bugs), is David Pimentel from Cornell (a prestigious Ivy League school) and the other is Tad Patzek, who was a professor of geology at University of California at Berkeley (also a prestigious school).

Unfortunately, EROEI is not as straight forward as using ROI. For example, you can say, "I invested $1,000 dollars in a savings account that paid 5 percent interest, and at the end of the year, I had $1,050." So, your ROI was $50.

It's not that straight forward with EROEI. Let's say that in going through the COSTCO salad bowl scenario, that you didn't own any knives. So, you bought the fruit, took it home, realized you don't have any knives, jumped back in the car to drive back to Costco, bought a knife, drove back home and cut your fruit up. Obviously your EROEI would be way off. The cost of the extra time, the fuel, and the cost of the knife all get added in. In this scenario, if the cost of the un-cut fruit and your time originally computed to a lower cost than the ready-made bowl, after all the extra trouble and expense you went through, you would have ended up with a negative EROEI compared to buying the ready-made fruit salad bowl.

Therefore, if you have a situation where someone with an agenda wants to prove that his thing is better than your thing, he could easily do a hatchet job by adding in extra things and extra steps when calculating your costs. The outcome can be complicated by purposely choosing items and materials that are expensive or redundant. Then, if the entire process was conducted under the cloak of secrecy, without you or an objective observer witnessing the process and having some input, you can imagine how problematic the results of the EROEI test could be for you.

Robert Bryce delves into the subject of EROEI with great enthusiasm, perhaps because his buddy Tad is involved. Bryce uses a shell game analogy, like the cardsharp player in Times Square that I have used before.

Bryce titled this section "Corn Ethanol's Enron Accounting." If the first thing that came to your mind is "doublespeak," you're catching on. As you can expect, Bryce uses lots of equations and figures that are irrelevant because, as I showed earlier, it is the process that led to his figures that is incorrect.

Robert Rapier makes another appearance in *Gusher of Lies* when Bryce uses him to prove the EROEI point he is going after. I especially like the part when Bryce calls Rapier a "very clever engineer...who writes one of the best blogs on energy and the energy business." I concur. You may want to go back to watch the video Rapier produced wherein he advocated the production and use of ethanol in order to become energy independent.

The bottom line is that the EROEI studies conducted by Pimentel and Patzek show that the EROEI for ethanol is wildly negative. They say that producing ethanol to replace gasoline is a waste of time. They say it takes more energy to make ethanol than the amount of energy that the ethanol will put out. Moreover, their studies claim that ethanol is far more environmentally hazardous then anyone ever thought, and that it might even cause greater pollution than gasoline. None of that is true.

The Pimentel-Patzek studies were hatchet jobs because they calculated in extraneous materials; they under-estimated crop yields; they over-estimated other aspects of farming (either intentionally or because the information they used was out-of-date); they exaggerated possible worst-case agricultural scenarios; they created deceptive algorithms to calculate and compare statistics; and they purposely ignored certain other aspects of oil refining and gasoline production that tell the truth about gasoline EROEI—that it is horribly negative.

Perhaps the most shocking item Pimentel and Patzek left out of their gasoline EROEI calculations are the hundreds of thousands of American military personnel (past and future) killed or wounded to protect and defend the oil industry's status quo. I'm sure they never calculated in the cost of continually treating those service personnel who sustained long-term and lifelong injuries. Then there are the accidental deaths of oil industry workers (past and future). Then there's the destruction of beaches, the loss of employment for those non-oil industry workers who live near the oil disasters, and then—as in the case of BP's Gulf of Mexico fiasco—the wholesale destruction of masses of birds and fish.

You would have to add in all the energy expended in making every bullet, every gun, every tank and cannon, every jet, ship and pair of shoes worn by our men and women who protect the oil industry. They are all part of the cost of doing business in oil-industry-style.

I'm not playing fair; I'm using tragically true but typically side-bar aspects of oil EROEI. I don't know how you can overlook the deaths of thousands of Americans, but the oil industry does.

I should compare apples to apples, even though Pimentel and Patzek used bad information and irrelevant factors, such as unnecessary farming equipment.

Pimentel and Patzek tried to reinvent the wheel, so to speak, even though the wheel has been with us for thousands of years. They started with a zero-based budget as if America had no farmers. They even calculated the energy required to build some of the equipment that's used in farming into the energy equation. The tractors they selected were unsuitable for the job because they were too large and cumbersome, but they ignored the fact that almost all corn farmers already own the equipment necessary to grow corn or other products (now you can see where I was going with my fruit cutting example). They stacked the deck against ethanol. To be fair, P&P should have calculated every aspect that goes into what it takes to get the gasoline to the pump, and added in all of the costs of war, and every single item right down to the energy it takes to make the zippers on the pants worn by our military service personnel.

Consider the results of some studies that studied the Pimentel-Patzek studies. The first comes from Tad Patzek's own school, UC Berkeley. It was completed in 2006 after the Pimentel-Patzek studies and was conducted by UC Berkeley's Energy and Resources Group. Here are some highlights:

> ETHANOL CAN REPLACE GASOLINE WITH SIGNIFICANT ENERGY SAVINGS AND A COMPARABLE IMPACT ON GREENHOUSE GASES
>
> The analysis, appearing in this week's issue of Science, attempts to settle the ongoing debate over whether ethanol is a good substitute for gasoline and thus can help lessen the country's reliance on foreign oil and support farmers in the bargain.
>
> Dan Kammen and Alex Farrell of the Energy and Resources Group at UC Berkeley, with their students Rich Plevin, Brian Turner and Andy Jones along with Michael O'Hare, a professor in the Goldman School of Public Policy, deconstructed six separate high-profile studies of ethanol. They assessed the studies'

assumptions and then reanalyzed each after correcting errors, inconsistencies and outdated information regarding the amount of energy used to grow corn and make ethanol, and the energy output in the form of fuel and corn byproducts.

Kammen estimates that ethanol could replace 20 to 30 percent of fuel usage in this country with little effort in just a few years. In the long term, the United States may be able to match Sweden, which recently committed to an oil-free future based on ethanol from forests and solar energy. Kammen last year published a paper, also in Science, arguing that even Africa could exploit its biomass to build a biofuel industry that could meet energy needs for the poor and develop a sustainable local fuel supply, a future much better than using fossil fuels.

The goal of the UC Berkeley analysis was to understand how six studies of fuel ethanol could come to such different conclusions about the overall energy balance in its production and use. Farrell, Kammen and their UC Berkeley colleagues dissected each study and recreated its analysis in a spreadsheet where they could be compared side-by-side. The team said it found numerous "errors, inconsistencies and omissions" among the studies, such as not considering the value of co-products of ethanol production—dried distillers' grains, corn gluten feed and corn oil—that boost the net energy gain from ethanol production. Other studies overestimated the energy used by farm machinery.

"The assumptions made by some of the authors were not based on the best data, or were just a little bit too convenient, and had a strong impact on the results," Kammen said.

Included in the UC Berkeley review were the following:

Tad W. Patzek, "Thermodynamics of the Corn-Ethanol Biofuel Cycle," *Critical Reviews in Plant Sciences* 23(6), 519-567 (2004).

David Pimentel and Tad W. Patzek, "Ethanol Production Using Corn, Switchgrass, and Wood; Biodiesel Production Using Soybean and Sunflower," *Natural Resource Research*, 14(1), 65-76 (2005).

Josh Tickell, an award-winning documentary film maker has produced several full-length films that address energy issues. In "Freedom," the film Josh and his wife Rebecca produced in 2011, Josh interviews David Pimentel on the subject of ethanol EROEI. (*https://www.youtube.com/watch?v=pAvUb76eX54*)

Another study, conducted by the United States Department of Agriculture, was presented in 2007 at UC Berkeley by Roger Conway, Office of Energy Policy and New Uses at the United States Department of Agriculture (USDA). The report showed huge discrepancies in the figures that Pimentel and Patzek used to arrive at their conclusions versus the figures used by USDA's efforts to conduct their own studies on ethanol vs. gasoline EROEI. The USDA studies were significantly more favorable towards ethanol production.

A Michigan State University study conducted by Bruce Dale, Professor of Chemical Engineering, found the Pimentel-Patzek methodology was flawed. The measurements of BTUs are irrelevant, and the net energy of ethanol is actually higher than gasoline (in other words, EROEI for ethanol is positive, while the EROEI of gasoline is more negative).

In 2005, Bruce Dale participated in a C-SPAN televised debate with David Pimentel and Tad Patzek. Speaking on Dale's side was John Sheehan, Senior Engineer, National Renewable Energy Laboratory. The debate was supposed to center on the issue of net energy balance. One of the most important points illuminated in the debate was that even if Pimentel-Patzek's finding that corn ethanol production results in a negative net energy equation, gasoline production is far worse and electric energy is horrendously bad compared to both ethanol and gasoline. After watching the video of the debate, it's hard to believe anyone took Pimentel-Patzek seriously, unless all opposing information is kept from the viewer. The entire video can be watched at *http://www.c-span.org/video/?188549-1/ethanol-energy-policy*.

Another report critical of Pimentel-Patzek was published in 2006 by Justus Wesseler, an agricultural economist and professor of Agricultural Economics and Rural Policy at Wageningen University in the Netherlands. He called the Pimentel-Patzek work "flawed" and "misleading." This report can be found at the Elsevier Energy Policy website.

(https://www.researchgate.net/publication/40103027_Opportunities_'costs_matter_A_comment_on_Pimentel_and_Patzek_Ethanol_production_using_corn_switchgrass_and_wood_biodiesel_production_using_soybean_and_sunflower)

Also in 2006, the spring edition of *The New Atlantis* (*Journal of Technology & Society*) had this to say about Pimentel and Patzek's studies:

> Professors Pimentel and Patzek have published several studies on this subject, and these have been thoroughly and repeatedly debunked in the scientific literature, in government reports from the Department of Energy and Department of Agriculture, in congressional testimony, and elsewhere...Reputable scientists have publicly called the work of Pimentel and Patzek 'shoddy,' 'unconvincing,' and lacking in basic scientific transparency. The most recent dissection of their claims, appearing in the journal Science in January 2006, found that their results depended upon "some input data that are old and unrepresentative of current [ethanol-production] processes, or so poorly documented that their quality cannot be evaluated."

The complete editorial from *The New Atlantis* can be found by clicking here:
https://www.thenewatlantis.com/publications/addicted-to-bad-data

In October 2009, David Blume traveled to Cornell University to conduct a workshop on permaculture. While there, he paid a call on David Pimentel and was able to record the conversation. In the 20-minute-plus conversation, Pimentel agrees and acknowledges that many of the conclusions drawn in the studies he conducted are now incorrect, or could be rendered incorrect given advances in farming and ethanol production, advances that already had been proven by 2009, just a year or so after Bryce wrote *Gusher of Lies*. These were also advances that had largely took place since the publication of the Pimentel-Patzek study four years earlier. Bryce's book and his later work, such as the article in *The Daily Beast* towards the end of 2012, made no allowance for the possibility of improvements that could affect the conclusions of the Pimentel-Patzek study. Bryce seems not to have permitted any bothersome details about improvements to interfere with the conclusions his book set in stone. You can watch the entire video by clicking on the window below.

https://www.youtube.com/watch?v=oMGh11bIH3w

David Blume provides additional in-depth discussion of EROEI in his book, "Alcohol Can Be A Gas" and online at *http://www.permaculture.com/node/490.* Since advances in farming and ethanol production have continued in the years since it was written and the videos were made, it would be intriguing to hear what an updated Blume-Pimentel conversation would sound like.

In 2011, Forrest Jehlik, Research Engineer at the Argonne National Laboratory responded to what he felt are the five most prevalent myths about ethanol. He said that ethanol does not take more energy to make than it yields, "Argonne National Laboratory research has shown that corn ethanol delivers a positive energy balance of 8.8 megajoules per liter. The energy balance from second-generation biofuels using cellulosic sources is up to six times better..."

In 2012, the U.S. Department of Energy concluded a report on Lifecycle Energy Balance by stating "...corn-based ethanol shows a clear benefit over gasoline." You can read this report at *http://www.afdc.energy.gov/vehicles/emissions_balance.html.*

In 2016, The U.S. Department of Agriculture released an updated report that verified corn ethanol's positive EROEI and stated that additional gains had been made since 2008. This report can be found at:

https://www.usda.gov/oce/reports/energy/2015EnergyBalanceCornEthanol.pdf

Nonetheless, it's clear that it's not the ethanol industry that's been using Enron Accounting; it's Bryce, Pimentel and Patzek and everyone else fronting for Big Oil.

Robert Bryce, Champion of MTBE

I was not surprised to see Bryce devote a section in his book to defending methyl tertiary-butyl ether (MTBE) and blaming ethanol. Look how many decades it took to get the world to agree that cigarette smoking is harmful. And look how long it took for everyone to acknowledge that tetra-ethyl lead is poison.

When the gasoline industry was finally faced with the mandate to clean up their act and get the lead out, figuratively and literally, they had two choices: The first was to go back to the original formulation, which was to blend in ethanol. The second was to pick another poison, so MTBE was it.

Bryce writes, "...Due to some leaking underground storage tanks, MTBE-blended gasoline was also causing some groundwater problems. And those leaks led to lawsuits over the refiners' use of the additive." Some groundwater problems, as if it was a couple of isolated incidents? Those very few numbers of spills are still with us today. Fresh water lakes and ponds and reservoirs contaminated with MTBEs still have MTBE's in them, years after outlawing its use in gasoline.

Some estimates are that it has taken $30 billion to clean up the more egregious MTBE damage. Guess who gets to pay for that? You and me. Sounds like another subsidy to the oil industry Bryce gets to ignore.

Bryce continues, "Like MTBE, ethanol is a good oxygenate. But it creates big problems for oil refiners because it increases the volatility of gasoline."

Volatility in this instance relates to evaporation. It sounds like Bryce is saying that because ethanol evaporates too quickly, or more quickly than MTBE, the gasoline industry chose MTBE over ethanol.

Interestingly, the Institute for Occupational Safety and Health, located near Bonn, Germany, describes MTBE as "very highly volatile, with a boiling point of 55 degrees centigrade." The same institute labels ethanol as "highly volatile with a boiling point of 78 degrees centigrade." Consequently, it looks like MTBE would evaporate more quickly than ethanol. Therefore, if the problem for refiners was that ethanol was too volatile, then MTBE is also too volatile.

I'm not a chemical engineer; I'm not any kind of engineer; I haven't had a Lionel model train engineer cap and whistle since I was a kid, so I could easily be wrong on my interpretation of the statistics. If so, Bryce has something to bash me with, let him have at it.

But if I'm reading the stats correctly, it sounds like Bryce did the doublespeak thing again: blaming ethanol for gasoline's problem.

Let's not quibble over a few temperature points. Volatility—that is, quick evaporation—and easy solubility with water are two of ethanol's benefits. If there's a large ethanol spill, cleanup is relatively simple. Unlike gasoline, ethanol fumes are not explosive. On the other hand, large gasoline spills often require specialized clean-up assistance and the lingering fumes can explode even after the liquid is removed.

I say again, if my understanding of the chemistry involved is puerile, there's a good reason for it, I don't understand chemistry. But if I'm right, and my childish understanding of the chemistry is even close to being correct, then Bryce and his like-minded colleagues have lots to apologize for.

MTBE use gets even stranger regardless of the volatility issue, because methyl tertiary-butyl ether is manufactured via chemical reaction between methanol and isobutylene. Methanol is an alcohol. Ethanol is an alcohol. The primary difference between the two is the raw materials from which they are derived. In fact, methanol could be a direct replacement for ethanol as a replacement for gasoline.

From my perspective, as I am not an ethanol producer or a farmer, nor am I paid by either the ethanol industry or the farming industry, I would be just as happy if we substituted all gasoline and diesel fuels with methanol fuel. But why do that when there are many more ethanol producers than there are methanol producers? We might as well use ethanol and support our American farmers.

For those of you keeping score, methanol rates at about 57,000 BTUs and isobutylene rates as about 95,000 BTUs. If you combine them and average it out it is 76,000 BTUs. In the earlier discussion on the irrelevancy of BTUs, ethanol rates at 76,000 BTUs. Now I'm not saying that combining of methanol and isobutylene is how you manufacture MTBE, but isn't it strange that the oil industry would go to all that bother to make something that rates no more BTUs than ethanol?

So, if a chemical reaction combining methanol and isobutylene was so attractive to oil refiners as the secret ingredient in gasoline to reduce engine knock and increase octane, why not go straight to methanol, or ethanol?

The answer, I believe, is that the oil industry didn't want to share any of the glory with the ethanol industry—that's glory spelled: P-R-O-F-I-T. That is because in the manufacture of MTBE, the methanol that is used comes from CNG (compressed natural gas), which is an oil refining by-product, instead of coming from biomass, agricultural and timber waste, or solid municipal waste. The people in the oil industry must have figured that giving any kind of "nod" to ethanol was too much, so they figured they could pass off a dressed-up poisonous additive with some chemical mumbo-jumbo.

In the rush to keep all the profits (profits are okay) and to choose anything that would work, except ethanol, the oil refiners chose a carcinogenic substance. Thank you very much, oil industry, for your continued "concern" for the public's health. And thank you, Robert Bryce, for all you've done on the public's behalf to repay the Saudis and Emiratis for their kindness.

Review and Reply to Robert Bryce's *Gusher of Lies*, A Book Aptly Named for What It Is

Originally published June 10, 2013

Part 6

Price Of Ethanol Vs. Gasoline At The Pump

At the time Bryce wrote *Gusher of Lies*, there was a Federal subsidy in place that ostensibly allowed the price of E85 to be priced lower than regular gasoline (usually E10). The subsidy, supposedly showed up as about a 60 cents less per gallon, permitted Bryce and others to use it as an example of how ethanol costs more at the pump (if you added the 50 cents to the price of E85, the price would be just about even with regular gasoline). Then if you calculate the supposed difference in MPG between regular gasoline and E85 according to the BTU nonsense discussed earlier, it shows that E85 (ethanol) cost more at the pump.

However, the reason E85 costs about the same as regular gasoline is because it is formulated and distributed by the oil industry, and they refuse to sell something that could hurt their normal pricing and profit structure for oil products. Of course, different states have different regulations and taxes, so the 50 cents I'm referring to may not be the same everywhere. I'm using California as my reference.

A good example of this problem is what is happened in Iowa in 2018. Here's the story from *Ethanol Producer Magazine* by Holly Jesson:

> Linn Co-op Oil Co. and Iowa's other E15 retailers were recently forced to stop selling E15 because oil refiners won't supply the gasoline blendstock

required for the summer months. "Consumers who want a higher-grade ethanol blend, in this case E15, are being denied that choice," said U.S. Rep. Bruce Braley, D-Iowa. Braley was part of a June 3 conference call hosted by the Iowa Renewable Fuels Association (IRFA) that was held to discuss the petroleum distribution monopoly, one of the tools Big Oil is using to block the sale of higher ethanol blends, said Monte Shaw, executive director of IRFA.

In the summer months the U.S. EPA allows E10 to exceed the 9.0 pounds per-square-inch Reid Vapor Pressure (RVP) requirement by 1 pound from June 1 to Sept. 15 but did not give E15 the same waiver. In order to offer E15 to vehicles model year 2001 and newer, the renewable fuel must be blended with low-volatility gasoline, which is used in markets that require reformulated gasoline. The catch is, although that fuel travels through Iowa via pipeline, E15 retailers are being denied access and, therefore, must only sell E15 to flex-fuel vehicle drivers or shut down their E15 pumps, as Linn Co-op did. "One of the key things about this is we're not asking for something that doesn't exist," Shaw said. "The gasoline blendstock we need to make E15 in the summer is available; it flows through the very pipeline system that services Iowa, but they will not let us take it out of the pipeline here."

"Earlier in the year, a group of Iowa's E15 retailers sent a letter asking oil refiners supplying Iowa to provide the summer gas blendstock required for E15, IRFA said. However, as of June 1, 2013, no oil refiner had done so. In the meantime, the necessary blendstock travels through Iowa, on its way to Kansas City and Chicago, where reformulated gas is required due to high smog levels. Retailers in the vicinity of a market that requires reformulated gas can continue to offer E15 by trucking in the blendstock, but transportation costs to Iowa make that cost prohibitive," Shaw said.

> "Oil refiners are using RVP seasonality to block retailers from selling E15 because they don't want to give up the five percent market share," Shaw added. In the meantime, the petroleum industry is attacking the renewable fuels standard (RFS), claiming it can't scale the 10 percent blend wall because retailers don't want to sell E15 and consumers don't want to buy E15. "And that's simply not true," he said, adding that recent events actually demonstrate the need for the RFS. "The high profit scenario for Big Oil is almost always not the low-cost fuel scenario for consumers," he said.
>
> Jim Becthold, service manager for Linn Co-op, the first retailer to sell E15 in Iowa, said customers started asking why the station couldn't sell E15 right after the pumps were bagged. The fuel is more economical, cleaner and higher performing. "Consumers are starting to really demand that; I'm seeing it in my sales," he said, adding that he hopes the situation will be resolved so the station can again sell E15 in the summer months, at least by next summer.

At the beginning of 2012, the subsidy ended, and anti-ethanol people predicted that E85's price would rise to be at or higher than regular gasoline. The prices didn't really change. Even without the subsidy, I'm still buying E85 at 50 to 60 cents cheaper than regular (E10) and 75 to 90 cents cheaper than premium. When you factor in the real difference in MPG between using gasoline in a gasoline optimized engine and using ethanol in a gasoline-optimized engine, not the theoretical 33% difference, the difference is usually enough to provide a net savings when using the E85.

In other situations, such as those I described earlier in this review, independent service stations who wish to bring in ethanol-gasoline blends other than E10 and E85, like E15, are often blocked by their gasoline vendors from doing it, although the vendor doesn't offer E15 themselves. The reason is that the vendor doesn't want to lose sales of their products to another vendor. You can understand this happening in a corporate franchised station, but not at stations that are supposed to be independently owned and operated.

As it turns out, the removal of the subsidy didn't affect retail prices of ethanol-gasoline blends because the subsidy went to the oil industry (the blenders) to make up for the lost revenue from adding ethanol to the blend in place of gasoline—so consumers never received a discount. The oil industry lobbied hard to get rid of the subsidy in an effort to drive ethanol-gasoline blend prices higher than E0 (gasoline without ethanol). It didn't work, it backfired. They cost themselves about 50 cents per gallon.

Back in the early 1900s after the passage of the Free Alcohol Bill, which removed the onerous $2-plus per gallon tax on the manufacture of alcohol, ethanol sold at about the same price as gasoline, sometimes less. The ethanol could be sold at a competitive price because it is inexpensive to produce. It is still "inexpensive" to produce, although times have obviously changed. If ethanol could be sold directly to the public, there's no reason why it wouldn't be priced very competitively to gasoline, even if the price per barrel of oil plunged—as long as the ethanol could be freed from the tether of the oil refiners and gasoline distributors.

Bob Gordon, my business partner at The Auto Channel, and I have long imagined a scenario where local ethanol producers are able to dispense pure ethanol (E100) directly to the consumer at their facilities or at non-traditional filling stations in conjunction with some big box retailers like Home Depot and Loews Hardware (harkening back to the very early days, before dedicated service stations existed and automobile owners would buy fuel at hardware stores). Similarly, at-home or office deliveries like water and coffee services could open a whole new category of entrepreneurial opportunities. There are already some fuel services like this used to deliver fuel to race teams at smaller tracks. And, of course, roadside assistance services usually carry fuel for those motorists stranded because they ran out of fuel. This business model is incorporated into the business plan of smaller ethanol producers and manufacturers of "community-sized" distillation plants. David Blume's Blume Distillation is one example.

Although any unconventional ethanol distribution would most likely only offer straight ethanol, not ethanol-gasoline blends, Bob and I can imagine interested consumers would do what I do—that is, pump the desired amount of E85 into the tank of my non-flex fuel Ford Taurus and then pull up to the E10 pump, or go across the street to a cheaper E10 station—and fill the tank the rest of the way. Sure, it requires more time, but to me it's worth it to save some money and do my part to help make us less dependent on international regimes that support terrorists. There's no question that not everyone would do this. But we believe there could be a few dozen million patriotic vehicle owners across America who would love to do the same, if they knew they could do it. I think this could be a really good grass roots movement that would have wide-ranging beneficial effects on their wallets and the environment (not beneficial to the oil companies, and that's one of the best benefits).

Water Usage

One of the last topics in Bryce's book that I'll address is the issue of water. To be honest with you, this one made me laugh. Can you imagine the oil industry and its apologists pretending they are concerned about water? It is the ultimate in hypocrisy. It's not just doublespeak, it's doublespeak to the 10th power.

The people who brought us global fresh and saltwater disasters, who have killed untold millions of wildlife, who have destroyed untold miles of beautiful beaches—they're raising a red flag warning about the water requirements to produce ethanol. It's a classic definition of *chutzpah*.

I'm glad Bryce did raise it because it may be the single best example of the advances made in ethanol production.

The premise is this: Ethanol requires water, not just in the actual distillation process, but in the growing of the crops used to make the mash that gets fermented and then distilled. It's true; water is required for these tasks. Bryce cites various studies that reveal the enormous amount of water required. He then uses various voodoo equations to paint a picture that would show the Earth turning into the desert planet Arrakis from Frank Herbert's science fiction classic, *Dune*.

As expected, Bryce foresees absolutely no circumstances in which the *Dune* scenario wouldn't happen. He doesn't consider that desalination plants could mitigate the problem; he doesn't consider any wild-eyed ideas like towing giant icebergs from the Arctic or Antarctic regions and using them as water sources; and since he doesn't believe in man-made global warming, he would never see the iceberg solution as being something that could mitigate the rising seas issue. I don't believe in man-made global warming either, but at least I have vision—I was able to visualize Internet connectivity faster than 28.8k, even though some "expert" in the industry told me it was impossible at the time.

POET Ethanol, the largest producer of ethanol in America has the response to Bryce:

> The first plant that POET purchased took 17 gallons of water per gallon of ethanol in 1987. Today, POET averages less than 3 gallons of water, an 80 percent reduction from where it started. But we're not done yet.
>
> POET believes that water is a precious resource and is committed to using as little of it as possible in our ethanol production process. So, we have set a goal, to be attained by 2014, of decreasing the overall annual water intake at our plants 22% by targeting an average usage rate of 2.33 gallons of water taken in per gallon of ethanol produced. This will reduce our total annual water use by one billion gallons based on POET's 2009 production capacity of 1.5 billion gallons of ethanol.
>
> POET engineers have invented a new way for its facilities to use water which results in a decreased intake with a reasonable economic payback. Named

Total Water Recovery, it ensures that our facilities don't have to choose between profit and planet. As of December 2011, eighteen POET plants had installed Total Water Recovery.

Most of the lifecycle water used in ethanol production is for growing the feedstock. According to the National Renewable Energy Laboratory, almost all of that is from rainfall as less than five percent of corn used for ethanol production is irrigated. Still, water requirements of the feedstock represent the vast majority of the total lifecycle water use for ethanol production. POET is working with our feedstock producers to understand the amount of irrigated corn coming into our facilities in order to identify opportunities to collaborate for additional water conservation.

With many parts of the world facing water shortages, water conservation has become a global priority. That's why POET chose reducing our water usage at the first Ingreenuity™ goal. As of December 2011, we had cut water use by approximately 770 million gallons.

To support POET's comments, consider these comments made in 2011 by Forrest Jehlik, Research Engineer at Argonne National Laboratory:

> The amount of water used to make ethanol has declined dramatically. Today, producing one gallon of ethanol requires about 3.5 gallons of water. That's a little more than it takes to process a gallon of gasoline. Much of the criticism about ethanol's water requirements stem from the need to irrigate feedstock crops in drier climates. But most ethanol is produced from rain-fed crops grown in the Midwest.

In addition, ethanol is not carcinogenic and doesn't poison

groundwater or the ocean. Ethanol rapidly biodegrades. Concerns over ethanol spills are muted by ethanol's low toxicity. In fact, you'll find ethanol in beer, bourbon and other happy-hour beverages you've probably consumed.

Water advancements are possible; they have been made, and they will undoubtedly continue. The key to it in Mr. Jehlik's comments is where he wrote, "Today, producing one gallon of ethanol requires about 3.5 gallons of water. That's a little more than it takes to process a gallon of gasoline." Yes, it also requires lots of water to make gasoline, but Bryce would never think of telling that to his readers.

Compressed Natural Gas

I like CNG; Bob Gordon likes CNG. We've written about it and talked it up for several years. We've noted in several stories how even oil rich Iran is relying on CNG for their domestic energy needs. What's more, a couple of years before I purchased the non-flex fuel Ford Taurus that I use with various ethanol-gasoline blends, I bought a CNG-powered 2001 Dodge Ram 1500 van. I wanted a Honda Civic CNG car but couldn't get one, so I went online and found the van.

I wanted a CNG vehicle for the same reason I bought the Taurus, I wanted to do my own testing. I wanted to see if the good things I heard about CNG were correct, and I wanted to see if the negative things I heard about CNG were correct.

Many of the negative things I heard about CNG paralleled the negative comments made about ethanol: Less energy content (the BTU nonsense), too many government subsidies, filling station availability is very limited, it causes more pollution than gasoline, and it damages engines. Guess who spread these rumors? Yes, the oil industry. They concocted a few others, such as that it takes too long to fill the vehicle (we've had people write to us saying that they heard it could take 30-minutes or longer to fill a CNG car at a commercial station).

Going back before the oil industry realized there was big money to be made by pushing CNG, they were frightened by it in the same way they are about ethanol. To stave off government pressure, the oil industry invented the Natural Gas Vehicle Association (NGVA) to pretend they were exploring alternative fuels. At the same time, the oil industry PR people started distributing negative comments like those above about CNG to play both ends against the middle. In time, NGVA moved out of the API office building to give it a little more distance, and Richard Kolodziej, the association's president, was given enough room to operate like he meant it—and I think he does a pretty good job, although I wish he was fully independent. Unfortunately, the oil lobby didn't call off their PR dogs until recently, so the lies about CNG continued and festered.

From personal experience, I can say that all the negative comments are lies. The issue over energy content is wrong (BTUs is just as irrelevant with CNG as with ethanol). An engine optimized to run on CNG will get the same mileage as if it was gasoline-optimized and using gasoline. My van gets virtually the same MPG if I had purchased the gasoline-powered version. The price of a "gallon" of CNG can be considerably less than gasoline; in some facilities the CNG costs a bit more than half the price of gasoline. The CNG keeps the engine cleaner and the emissions are cleaner than those from gasoline.

CNG does not damage engines any more than ethanol does, and both of them are better for an internal combustion engine than gasoline. CNG, like ethanol, does not cause engine knock.

There is however one great big glaring problem with CNG (there are a few "small" ones too, which I'll address shortly). The big problem is that CNG is controlled by the oil industry. This means that no matter what, if we rely on CNG, we will be screwed blue and tattooed by the exact same people that are keeping our economy in chains. It is this connection that has soured me on CNG. I still love and use my van, and I always enjoy saving money every time I fill it up, but CNG is not the solution we need.

When Robert Bryce wrote *Gusher of Lies*, he obviously hadn't yet received the memo from the oil gods that CNG was good, after all. Although it has been known for many years that the U.S. and Canada have tremendous reserves of natural gas, Bryce writes about dwindling Canadian and America resources and foresees America becoming a major importer of natural gas. He uses the expectation of our becoming a CNG importer to argue that it's not worth struggling for energy independence; that we should just give in to the inevitable.

Somewhere around 2010 Bryce must have checked his email and found the "CNG is OK" memo because that's when he started to write about America's huge natural gas resources. This is around the time the oil industry started looking for large scale government aid to help them develop the "suddenly" discovered domestic reserves. The government aid that the oil industry is seeking are subsidies in various forms. So once again, because the subsidies would be for the oil industry, the subsidies are okay. In his most recent writing about natural gas and CNG I wasn't able to find any criticism related to any type of subsidy for natural gas and CNG.

Remember how Bryce decries the "heavy" water usage needed for ethanol, and the part about how much water is needed to make gasoline? Well, fracking also requires huge amounts of water, and some of the fracking is taking place in areas where water is particularly scarce. Bryce never mentions this; as with all the other hypocritical positions Bryce takes, it's suddenly okay to "waste" water, because it's for the benefit of the oil industry.

In 2013, Bryce and the oil industry and its mouthpieces are hyping the heck out of CNG. They're even talking about America becoming an exporter of natural gas and CNG. Armed with this enthusiasm, the oil industry wants to frack here, frack there, frack everywhere. Simply put, they are fracking crazy.

There's a lot of bad publicity about fracking. Some say it's very harmful to ground water, lakes, ponds, and reservoirs; and some, like Boone Pickens, says that there has never been a confirmed fracking disaster. I've never done any fracking, so I have no personal experience.

But there are the problems with CNG above and beyond the fact that it's controlled by the oil industry. First, there is only one automobile maker building a CNG-powered car for the American market. That company is Honda. The CNG car they make is the Civic GX and for several years their entire annual production was about 1,000 to 1,500 cars (the GX designation has since been dropped and it's just Civic CNG). This small number of cars is more like the output we would expect for a hand-built non-assembly line car. It's almost not worth calling it a "production run." This very small availability was why I couldn't buy a Honda Civic. I had the Honda PR people looking for a car for me, but none were available. And I was buying the car, not looking to trade one for a favorable review.

Then all Civic GX production stopped while Honda was building a new factory in Indiana. Honda lost about a year in building new CNG Civics. In 2012, Honda upped Civic CNG production to around 3,000 cars, but then stopped production in 2013 because they still have too many 2012 models available.

The Civic is a great car. It's generally inexpensive, it drives very nicely, it looks good, and it is a Honda, so you know you can rely on it. However, the Civic is small-ish, which means it doesn't have much room for the over-sized CNG tanks (they're located in the trunk). This is one of the more unlikely vehicles that Honda should have chosen from their lineup for CNG production. What they ended up with was only one passenger vehicle made for the American market, and its availability is close to none, when it's being built. On top of that, because of the CNG tanks, it has very limited cargo room. (Note: all Honda Civic CNG production ended with the 2015 model year.)

Dodge announced production of a hybrid CNG-gasoline Ram pickup truck and a very small run of CNG vans that GM built under special order last year. That's it in the way of CNG vehicles. There was a time 10-plus years ago when Ford, Chrysler and GM were building several CNG models, but that was for show to get carbon credits and other brown-nosing points from Congress.

Consequently, if the oil industry went ahead with fracking and the U.S. produced all this CNG, what would we do with it? CNG and natural gas is already used in millions and millions of American homes for heating, and there's no gas shortage to overcome. The new home market has been dead for a few years, and it's just barely breathing now.

There's no talk of any great spurt in CNG filling stations, so even if there were more CNG cars produced, there would be a fueling station availability problem.

We do have a couple of hundred million passenger vehicles on the road that could be converted, but aftermarket engine conversion is terribly expensive (roughly $10,000) and limited to just a few models. The cost and the restrictions are because of EPA regulations. The funny part is that these regulations exist due to oil lobbying to make it difficult for vehicle owners to convert from gasoline. So, while the oil industry is up in arms against the EPA for the E15 waiver, they were thrilled with the EPA action on CNG conversions. To make matters worse, if you live in California and you have the money and the right model vehicle, California Air Resources Board (CARB) won't allow you to make the conversion. If you buy a CNG-converted vehicle outside of the state and bring it into California, CARB won't allow you to register it.

So again, what would we do with all the natural gas and CNG from all the new fracking?

Ah, the oil industry lackeys say, "We'll export it and make a profit." Okay, but who are they going to export it to? Canada doesn't need it. Mexico doesn't need it. Venezuela and its friends down south don't need it. Brazil doesn't need it. China and Russia have their own vast supplies of natural gas, and Russia has oil, too. So where? Western Europe? Why buy the U.S. gas? Israel (and Lebanon and Cyprus by proximity) discovered huge natural gas reserves and has already started bringing the gas up. It's cheaper for Western Europe to import the gas from Israel, Lebanon or Cyprus.

In short, there's no market for an increased supply of American natural gas and CNG. All that will happen is that the government will give billions to the oil industry. It will be another diversion like electric cars, and we will have wasted valuable time and money that should be spent on getting rid of our dependence on fossil fuels.

Let me put it as succinctly as I can: No fracking way!

Wrap Up to My Review Of *Gusher of Lies*

The 19th century was a milestone in human existence because it was the century of freedom: social, national and economic. All the social awakenings triggered by the European renaissance a couple of centuries earlier came to life with American independence, followed by the French struggles to find their own brand of independence from autocratic rule, then the unification of disparate Italian and German states into cohesive nations, and the abolition of slavery in the western world. These factors combined with the burgeoning age of machines to allow common people to go forth on their own. Millions left the reasonable comfort of familiar surroundings to journey thousands of miles away in search of a better life.

The invention of the automobile was the crowning achievement in the age of machinery because individuals could range further and longer at a moment's notice. It personalized the spirit of freedom and independence.

The automobile wasn't invented because of any particular liquid or solid fuel; it was invented by the fuel of hope and imagination. Automobiles allowed people to move out of crowded festering cities. It allowed people to travel every day to work in locations that just a short time before were considered a week's journey away. It allowed every strata of the public to shop, to play, to visit, to spend money, which allowed others to earn money to spend and buy, repeating the cycle over and over. The automobile engendered powered-human flight and personal power boating; it was the great catalyst. The automobile, not its fuel source, was the ultimate expression of freedom.

Then we lost our way. People don't aspire to use a prettier fuel, or a faster fuel, or a more luxurious fuel. The machine is the key; the fuel is a minor element. Sure, without fuel, the automobile won't run; but an automobile also won't run without a spark plug. However, we don't fight wars over spark plugs. Spark plug manufacturers aren't paying terrorists to hijack airplanes.

Shockingly, more people today are basing their automobile purchasing decisions on fuel economy rather than on the criteria that are really important: Hard top or convertible, 2-doors or 4-doors or 5-doors, with or without GPS navigation, business or pleasure, how many seats, what color; these are the decisions that are important. It's a world gone mad; a world turned upside down; our freedom to come and go as we please has been greatly compromised. We would all laugh if someone said he purchased one vehicle over another because spark plug A had a longer life than spark plug B. This is essentially what has happened with fuel, more correctly and precisely with gasoline fuel, but it's not a laughing matter.

There are people who are not going on job interviews, because they can't afford the gasoline. There are people who are turning down educational opportunities, because they can't afford the gasoline to travel to school. The freedom the automobile gave us is limited by fuel controlled by foreign, terrorism-supporting regimes, who don't respect their own fellow countrymen and coreligionists; and certainly do not respect us.

The great truth that the oil industry and people like Robert Bryce have been desperately trying to squelch is information about the fuel, and the fuel itself that can do the job--a fuel that can cost less, is more environmentally friendly, and can be produced within our borders by our own citizens. It is available right now, today. That fuel is ethanol.

Moreover, the great preponderance of the passenger vehicles on our roads can use high quantities of ethanol immediately without any conversions or adjustments to their engines. This is a remarkable technological feat. It's akin to saying that you can take a music DVD and insert it into a cassette tape player and hear the music. You can't put diesel in a gasoline engine and expect it to work. You can't put CNG in a gasoline engine and expect it to work. You can't put milk in a gasoline engine and expect it to work. But you can put ethanol in your gasoline engine, and it will work; and it will perform almost as if it was gasoline. The only real difference is that the ethanol will cost less and be better for the environment—and we won't have to fight quite so many wars.

Even if ethanol was not as economical, and even if ethanol in any blend level did create engine damage, the answer is not to abandon ethanol; the answer is to immediately mandate that all applicable new passenger vehicles and light trucks to be sold in America be built with engines that are optimized to run on ethanol, not gasoline.

Ethanol is at least as economical as gasoline and it doesn't cause the mechanical problems alleged by the oil industry and its apologists, so we should make it our primary engine fuel. When and if electric passenger vehicles are finally ready to play their part as a viable alternative to internal combustion powered vehicles, we (or our grandchildren) should welcome them. But until then, we must free ourselves and our country from the yoke of foreign fossil fuels—which means all fossil fuels.

The myths disseminated by the oil industry with the help of Robert Bryce, David Pimentel, Tad Patzek, Jerry Taylor, Tim Searchinger, David Fridley, Richard Rahn, Mark Mills and others like them are tantamount to slave chains. America must break free from those bonds for our economic health and social wellbeing. This is also true for all other nations; not because it's "the American thing to do," but because they need their own economic security. Ethanol can be made anywhere by anybody from a very wide variety of raw materials. It is the freedom fuel.

Let freedom ring throughout America, again.

Conclusion

November 23, 2019

There are lots of words in this book, lots of pages, lots of videos and external links. There's a lot of repetition of facts and my replies to false and misleading information. This is intentional. The oil industry and its supporters don't make a false claim and then retire from the claim when it is challenged and set aright. They simply find another chump to pick it up and spread the same lie in exchange for goods, services and money.

My feeling is that every time a lie or gross exaggeration is told in favor of petroleum oil fuels, it must be countered with the truth. This is what I have attempted to do, sticking to the same basic facts that leading scientists, inventors, and engineers from previous centuries knew: that ethanol is the best internal combustion engine fuel. As the years pass, and I learn of new information from additional research or from the colleagues I acknowledge throughout this book, and discover better ways to articulate a point, I add them to my arsenal. In perusing or skimming through this book, I hope you will note that I didn't change my tone and invent new ways to defeat the new telling of the old lies. I stayed rock solid with the truth.

If there is one point in this book worth repeating and remembering, it's that for several decades, while the petroleum oil industry lied to Americans about the use of ethanol, the oil industry's biggest and richest companies actively marketed ethanol-gasoline blends as being safer, cleaner, more powerful, healthier, and more economical than ethanol-free gasoline. It's all in the chapter "The Hypocrisy of Big Oil and API."

I'm not the first to reveal this hypocrisy. Bill Kovarik and his co-authors told this story in 1982 when they published *The Forbidden Fuel*. Dave Blume has written and talked about this, as have Bob Falco and Michael Carolan, to name a few advocates of veracity. My contribution is that I have repeated it anew and married the narrative to additional elements used by the oil industry to aggressively promote ethanol during all those years.

The bottom line, is that we have been lied to, deceived, dishonored, and poisoned by the oil industry and their cronies. America has been betrayed by the so-called American oil companies. Britain was betrayed by their oil companies. The Netherlands was betrayed by their oil companies. We fought long wars for them; millions of heroic brothers, sisters, sons, daughters, mothers and fathers were sacrificed so oil companies could get richer and poison us more. It's gone on long enough.

It's time to stop the lies against ethanol. The information contained in this book obliterates every single anti-ethanol claim and exaggeration. It's time to hold the petroleum oil industry accountable for the non-stop death and destruction they have wrought for over a hundred years.

It's time to take the shackles off ethanol fuel availability in America. After mandating E25 for a few years in Brazil, they upped the mandate to E27 as their basic ethanol-gasoline blend, and it created no problems. Their next step is to move to E40 as their basic fuel. We should be at least on par with Brazil, although we really should be at the stage that I advocated in 2008: no new gasoline-powered engines should be manufactured for use in America. All new internal combustion engines should be running on E85 to E95. Petroleum oil should be relegated to use only in non-fuel products, and never again as combustible fuels.

Ethanol is already the number one renewable energy resource in the world. It can be the number one internal combustion engine fuel in the world, as it should have been and as it was predicted to be by the leading scientists and automobile people in the world.

Ethanol is cleaner than gasoline, it is safer than gasoline, it is healthier than gasoline, it provides more horse-power than gasoline, it can provide better miles per gallon than gasoline, it is less expensive than gasoline, it is better for engines, it will free America and all countries from dependence on foreign fuels, and it can be completely domestically produced. Electric vehicles are not the answer and won't be for a long time. Fracking is not the answer. Tar sands are not the answer.

ETHANOL IS THE ANSWER. This is the story on ethanol, the whole story.

Major Sources

The full spectrum of journalistic and historical resources was employed for *The Ethanol Papers*, including interviews, field reporting, official contacts, newspapers and other media, historical investigation, and the uncovering of obscure and unpublished materials. A list of selected sources and resources follows.

Books

Bernton, Hal, William Kovarik, and Scott Sklar. *The Forbidden Fuel: A History of Power Alcohol.* Omaha: Bison Books, 2010.

Black, Edwin. *British Petroleum and the Redline Agreement: the West's Secret Pact to Get Mideast Oil*. Washington DC: Dialog Press, 2011.

Black, Edwin. *Internal Combustion: How Corporations and Governments Addicted the World to Oil and Derailed the Alternatives*. New York: St. Martin's, 2004; Washington DC: Dialog Press, 2017.

Black, Edwin. *The Plan: How to Save America When the Oil Stops – or the Day Before*. Washington, DC: Dialog Press, 2008.

Blackmore, D.R. and Alun Thomas (ed). *Fuel Economy of the Gasoline Engine: Fuel, Lubricant and Other Effects*. Palgrave Macmillan, 2014.

Blume, David and Michael Winks. *Alcohol Can Be a Gas: Fueling an Ethanol Revolution for the 21st Century*. Santa Cruz: International Institute for Ecological Agriculture, 2007.

Bradley Jr., Robert and Richard W Fulmer. *Energy, The Master Resource: an Introduction to the History, Technology, Economics, and Public Policy of Energy*. Dubuque, IA: Kendall/Hunt Pub. Co., 2004.

Branham, Russ. *The Ford Century: Ford Motor Company and the Innovations that Shaped the World.* Artisan, 2002.

Bryce, Robert. *Gusher of Lies: the Dangerous Delusions of "Energy Independence."* New York: Public Affairs, 2009.

Clean Fuels Development Coalition. *Ethanol Fact Book: A Compilation of Information about Fuel Ethanol.* Washington, DC: CFDC, 2010. http://cleanfuelsdc.org/wp-content/uploads/2018/05/CFDC-2010-Ethanol-Fact-Book.pdf (12/12/2019).

Dewsbury, Suzanne and Ian M. Dewsbury. *At Issue: Ethanol.* Detroit: Greenhaven Press, 2009.

Donovan, John. *Sir Henri Deterding and the Nazi History of Royal Dutch Shell.* 2016.

Evans, Klassy and Adam Khan. *Fill Your Tank with Freedom: How Fuel Competition in America Could Change the World and How You Can Help Make It Happen.* Amazon Digital Services LLC-KDP Print US, 2012.

Flink, James J. *America Adopts the Automobile, 1895–1910.* Cambridge, MA: MIT Press, 1970.

Goettemoeller. Jeffrey and Adrian Goettemoeller. *Sustainable Ethanol: Biofuels, Biorefineries, Cellulosic Biomass, Flex-fuel Vehicles, and Sustainable Farming for Energy Independence.* Maryville, MO: Prairie Oak Publishing, 2007.

Hale, William J. *Prosperity Beckons: Dawn of the Alcohol Era.* Minneapolis: Rutan Pub., 1979.

Holm, Dale, Herb Seto, and Cathy Travaglini. *The Chemistry of Corn into Alcohol.* Cornville, AZ: Desert Publications, 1980.

Jacobs, P. Burke and Harry P Newton. *Motor Fuels from Farm Products.* Washington DC: USDA, 1938. https://archive.org/details/motorfuelsfromfa327jaco (12/12/2019).

Luft, Gal and Anne Korin. *Petropoly: The Collapse of America's Energy Security Paradigm.* Charleston: CreateSpace, 2012.

Luft, Gal and Anne Korin. *Turning Oil into Salt: Energy Independence through Fuel Choice.* Charleston: CreateSpace, 2009.

McCarthy, Tom. *Auto Mania: Cars, Consumers, and the Environment.* New Haven: Yale University Press, 2010.

Mitre Corporation, et al. *Comparative Economic Assessment of Ethanol from Biomass.* Washington DC: US Department of Energy, 1978.

O'Sullivan, Meghan L. Windfall: *How the New Energy Abundance Upends Global Politics and Strengthens America's Power*. New York: Simon & Schuster Paperbacks, 2018.

Pagé, Victor Wilfred. *The Model T Ford Car, Its Construction, Operation and Repair*. New York: Norman W. Henley, 1917. https://books.google.com/books/about/The_Model_T_Ford_Car_Its_Construction_Op.html?id=70mJeb_q19EC (12/13/2019).

Sampson, Anthony. *The Seven Sisters: The Great Oil Companies and the World They Made*. London: Coronet Books, 1993.

Schmer, Robert. *The Motor Gasoline Industry: Past, Present, and Future.* Washington, DC: US Energy Information Administration, 1991.

Shapouri, Hosein, James A. Duffield, and Michael Wang. *The Energy Balance of Corn Ethanol: An Update.* Washington, DC: US. Department of Agriculture, 2002.

Strong, Robert M. *Commercial Deductions from Comparisons of Gasoline and Alcohol Tests on Internal-Combustion Engines.* Washington, DC: US Geological Survey, 1909. https://pubs.usgs.gov/bul/0392/report.pdf (12/13/2019).

Weisberg, Barry. *Beyond Repair: The Ecology of Capitalism.* Boston: Beacon Press, 1972.

White, Kathleen Hartnett. *Fossil Fuels: The Moral Case.* Austin: Texas Public Policy Foundation, 2014. https://files.texaspolicy.com/uploads/2018/08/16101046/2014-06-RR03-FossilFuelsTheMoralCase-KathleenWhite-ACEE.pdf (12/13/2019).

Yergin, Daniel. *The Prize: The Epic Quest for Oil, Money, and Power.* London; New York: Simon & Schuster, 2012.

Yergin, Daniel. *The Quest: Energy, Security, and the Remaking of the Modern World.* New York: Penguin Random House, 2010.

Forthcoming Books

Epstein, Alex. *The Moral Case for Fossil Fuels*. New York: Portfolio Penguin, 2020.

Theses and Dissertations

Kovarik, William J. "The Ethyl Controversy: How the News Media Set the Agenda for a Public Health Controversy over Leaded Gasoline, 1924–1926." PhD diss., University of Maryland, 1993. https://drum.lib.umd.edu/handle/1903/16750 (12/13/2019).

Journal Articles

Carolan, Michael S. "Ethanol versus Gasoline: The Contestation and Closure of a Socio-technical System in the USA." *Social Studies of Science* 39 (2009), 421–448. https://journals.sagepub.com/doiabs/10.1177/0306312708101049 (12/12/2019).

Kovarik, Bill. "Henry Ford, Charles Kettering, and the Fuel of the Future." *Automotive History Review* 32 (1998) 7–27. http://environmentalhistory.org/people/henry-ford-charles-kettering-and-the-fuel-of-the-future/ (12/12/2019).

Films and Videos

The Big Fix. Directed by Josh Tickell. Santa Monica: Green Planet Productions, 2012.

Freedom. Directed by Josh Tickell and Rebecca Harrell Tickell. Irvine: Green Planet Productions, 2011.

David Blume. "Ford's 1st Flex Fuel Car: the Model 'T.'" Jul 19, 2009. https://www.youtube.com/watch?v=5qDYoEupI28 (12/10/2019).

Gashole. Directed by Scott D. Roberts and Jeremy Wegener. Burbank: Cinema Libre Studio, 2011.

"Make Your Own Ethanol At Home with the E-Fuel 100 MicroFueler." May 29, 2008. https://www.youtube.com/watch?v=rmFwSogNAEc (12/12/2019).

PUMP: The Movie. Directed by Josh Tickell and Rebecca Harrell Tickell. Irvine, CA: Fuel Freedom Foundation/Submarine Deluxe.

ADDITIONAL RESOURCES (LINKS ACTIVE AS OF PRESS TIME)

AgWired: Energy, http://energy.agwired.com
Al-Corn Clean Fuel, http://al-corn.com
American Coalition for Ethanol, https://ethanol.org
American Energy Independence, http://americanenergyindependence.com
Bioenergy Australia, https://www.bioenergyaustralia.org.au
Blume Distillation, https://www.blumedistillation.com
Bobby Likis Car Clinic, https://www.carclinicnetwork.com/_ethanolupdate-original.htm
DynoTune Speed & Performance, http://dynotuneusa.com
E-Fuel Personal Energy System, http://www.microfueler.com
Fuel Freedom Foundation, https://www.fuelfreedom.org
Fuels Institute, http://www.fuelsinstitute.org
Glacial Lakes Energy, http://glaciallakesenergy.com
Growth Energy, https://growthenergy.org
ICM Energy, https://www.icminc.com
Ignite Racing Fuel, http://www.igniteracingfuel.com
The Institute for Energy Resourcefulness, http://www.energyresourcefulness.org
Minnesota Bio-Fuels Association, https://mnbiofuels.org
Minnesota Corn Growers Association, https://www.mncorn.org
National Farmers Union, https://nfu.org
Nebraska Ethanol Board, http://ethanol.nebraska.gov
Pearson Fuels, https://pearsonfuels.com
POET Biofuels, https://www.poet.com
Propel Fuels, https://propelfuels.com
Renewable Fuel Association, https://ethanolrfa.org
Show Me Ethanol, LLC, https://www.showmeethanolllc.com
South Dakota Farmers Union, https://www.sdfu.org
Queensland Renewable Fuels Association, https://www.qrfa.com.au
Urban Air Initiative, https://fixourfuel.com
US Enviro Fuels, http://usenvirofuels.com
Western Plains Energy, http://wpellc.com

Made in the USA
Monee, IL
31 October 2025

33413790R00343